Myths and Realities

OF CRIME AND JUSTICE

What Every American Should Know

STEVEN E. BARKAN, PHD

Professor of Sociology
The University of Maine
Orono, Maine

GEORGE J. BRYJAK, PHD

Former Professor of Sociology
University of San Diego
San Diego, California

JONES AND BARTLETT PUBLISHERS

Sudbury, Massachusetts

BOSTON   TORONTO   LONDON   SINGAPORE

World Headquarters

Jones and Bartlett Publishers
40 Tall Pine Drive
Sudbury, MA 01776
978-443-5000
info@jbpub.com
www.jbpub.com

Jones and Bartlett Publishers
Canada
6339 Ormindale Way
Mississauga, Ontario L5V 1J2
Canada

Jones and Bartlett Publishers
International
Barb House, Barb Mews
London W6 7PA
United Kingdom

Jones and Bartlett's books and products are available through most bookstores and online booksellers. To contact Jones and Bartlett Publishers directly, call 800-832-0034, fax 978-443-8000, or visit our website www.jbpub.com.

Substantial discounts on bulk quantities of Jones and Bartlett's publications are available to corporations, professional associations, and other qualified organizations. For details and specific discount information, contact the special sales department at Jones and Bartlett via the above contact information or send an email to specialsales@jbpub.com.

Library of Congress Cataloging-in-Publication Data
Barkan, Steven E., 1951–
  Myths and realities of crime and justice : what every American should know / by Steven E. Barkan and George J. Bryjak.
      p. cm.
  Includes bibliographical references and index.
  ISBN 978-0-7637-5574-4 (pbk.)
  1. Criminal justice, Administration of—United States. 2. Crime—United States. 3. Criminal law—United States. I. Bryjak, George J. II. Title.
  HV9950.B355 2009
  364.973—dc22
                                        2008013289
6048

Production Credits
Acquisitions Editor: Jeremy Spiegel
Editorial Assistant: Maro Asadoorian
Production Director: Amy Rose
Senior Production Editor: Renée Sekerak
Associate Marketing Manager: Lisa Gordon
Manufacturing and Inventory Control Supervisor: Amy Bacus
Assistant Print Buyer: Jessica DeMarco
Cover Design: Brian Moore
Photo Research Manager and Photographer: Kimberly Potvin
Cover Image: © JustASC/ShutterStock, Inc.
Compositor: Northeast Compositors, Inc.
Printing and Binding: Malloy Incorporated
Cover Printing: Malloy Incorporated

Printed in the United States of America
12 11 10 09 08   10 9 8 7 6 5 4 3 2 1

Preface

This short book provides an understanding of crime and justice in the United States that will surprise many Americans. Most people obtain information about crime and justice from the evening news, television shows, newspaper stories, movies, and detective novels. Unfortunately, these sources often provide a greatly distorted picture that perpetuates certain myths about crime and justice while ignoring the realities of these issues. Drawing on our many years of research and writing in criminology and criminal justice, we offer an easy-to-read primer on what every American should know about crime, criminals, the police, the courts, and the prisons.

To keep the book short and to avoid superficial coverage, we do not cover every possible issue of crime and justice today—for example, we do not discuss the complexities and ramifications of organized crime. We hope we succeed in telling you something you did not know, and in whetting your appetite for learning more via the lists of recommended readings found at the end of most chapters.

Acknowledgments

This book would not have been possible without the help of numerous individuals. For their assistance in bringing our work to print and enabling it to reach the public, we extend our heartfelt gratitude to Maro Asadoorian, Jeremy Spiegel, Renée Sekerak, Wendy Thayer, and their colleagues at Jones and Bartlett Publishers. We are pleased that Jones and Bartlett shared our vision for this book and agreed that the issues it explores deserve a wider audience.

As always, our deepest thanks go to our significant others, Diane M. Bryjak and Barbara Tennent. We are grateful for their understanding, support, and love as well as for enduring yet another writing project.

Finally, we dedicate this book to past, present, and future victims of crime, and to the police, judges, prosecutors, public defenders, correction officers, and other criminal justice professionals who deal with the reality of the crime problem in this country. We hope readers of this book will agree that a greater public understanding of the myths and realities of crime and criminal justice will result in lower crime rates, fewer victims of crime, and a safer working environment for police and other members of the law enforcement community.

Contents

"Speed Bump" © Dave Coverly/Dist. By Creators Syndicate, Inc.

chapter one

What No One Is Telling You About Crime and Justice

This book is written for every American who is concerned about or interested in crime and criminal justice. This short, readily accessible book discusses what crime is all about and how the criminal justice system works, as opposed to how it is supposed to work. Most Americans obtain their information about crime and justice from television shows, movies, novels, true crime books, and the news media. Unfortunately, public beliefs about crime and justice in America are often inaccurate, because the sources of these beliefs often distort the true nature of criminal behavior and provide a misleading picture of the police, the courts, and the penal system.

Some quick examples should illustrate why this book discusses, as its subtitle indicates, what every American should know. Think about the last few times you watched police shows on television. It is virtually certain that the crimes depicted in these shows were violent offenses: usually a murder, often an especially vicious rape or sexual assault, sometimes a strong-arm robbery. In actuality, only about 10 percent of all street crimes (which includes violent and property combined) are serious violent crimes, i.e., murder, robbery, rape, or aggravated assault. By the end of the shows you watched, the person who committed the crimes usually was identified and arrested. In actuality, only about 10 percent of all street crimes result in an arrest, and only about 40 percent of all serious violent crimes result in an arrest.

The misinformation we gain from television and other sources undermines a reliable understanding about the reality of crime and justice in the United States. Perhaps because violent offenses are featured so heavily in the news media and television police dramas, Americans overestimate the amount of violent crime that actually occurs in this country. In one 1994 study, several hundred college students in an introductory criminal justice class were asked to estimate the annual number of criminal homicides across the country. Almost half the students responded that at least 250,000 individuals are murdered even though fewer than 24,000 homicides occurred in the year the study was done. In other words, these students were wrong in their perception of criminal homicide deaths by a factor of ten.

There is also widespread misunderstanding about how the criminal justice system works. Consider the criminal trial, one of the most compelling events in fiction or in real life. In the world of television, attorneys in old-time programs such as *Perry Mason* and newer ones such as *Boston Legal*, the original *Law and Order*, and *The Practice* match wits and legal skills in an escalating courtroom drama wherein the defendant is convicted or the real culprit is uncovered and justice triumphs. But what percentage of criminal cases are actually decided in a court of law? Contrary to what TV shows portray, the criminal trial is actually a rare legal event; nationwide, more than 90 percent of criminal cases are resolved by means of a plea bargain. Although viewers of *Law and Order* and other shows may be somewhat familiar with plea bargaining, most probably are unaware of how deals between prosecutors and defense attorneys are struck and why bargaining is the pivotal feature of the American criminal court system.

CRIME, CRIMINALS, AND WHY WE ALL NEED TO KNOW ABOUT THEM

Why does crime occur? Why do some people commit crime while others abstain from engaging in criminal activity? Are parents at fault? Is the criminal justice system too lenient? Do criminals have defective genes? Is lack of education the culprit? Are poverty and urban overcrowding to blame? In order to know how to reduce crime, we need to know what causes it. If we believe the main problem is that the criminal justice system is too lenient, then as a society we must crack down on crime and criminals, with substantially more arrests, more convictions, and longer prison terms. If we think that parents are at fault, then mothers and fathers must change their child-drearing strategies. If we believe that poverty and urban overcrowding are at fault, then governments at the local, state, and/or federal level must implement social programs that will mitigate these problems. If several factors are to blame for crime, then multiple interventions will be necessary.

These possibilities underscore the need to determine the causes of crime so that we can effectively reduce criminal behavior. An analogy comes from the field of public health and the treatment of cancer. Although it is important to provide cancer patients with the best medical care possible, it is equally critical to determine what causes cancer in the first place. Possible causes include defective genes, improper diet, obesity, smoking, and environmental pollution. Researchers recognize the necessity of treating people who already have cancer, but they also recognize that no matter how many cancer patients might be treated successfully, identifying and addressing the causes of cancer will reduce its occurrence in the first place. Thus, public health researchers urge people to stop smoking and change their diets as well as make other lifestyle alterations to reduce the odds of developing cancer.

The analogy to crime is apparent: an important strategy for reducing the occurrence of crime is to identify and address its causes with appropriate and effective policies. No matter how strongly we feel about crime and the individuals who may have victimized us or our loved ones, it is important to step back and examine the evidence gathered by criminologists and other social scientists. Sound social policy that will reduce crime demands an informed citizenry as well as an informed government.

GETTING TOUGH? THE U.S. CRIMINAL JUSTICE SYSTEM IN CRISIS

The federal and state governments are certainly concerned about crime and have tried to reduce it since the 1970s with a get-tough approach involving a crackdown on minor offenses, longer prison sentences, and the expenditure of tens of billions of dollars on the police, prisons, and a variety of law enforcement strategies. The criminal justice system now costs about $200 billion annually, compared to roughly $36 billion in the early 1980s. More than 2.2 million individuals are in jail or prison on any given day, yielding the highest incarceration rate in the Western world. A number of states (especially California, with the nation's largest prison population) have had to reduce higher education budgets or to undertake other drastic actions in order to finance the costs of incarceration and prison construction. The American criminal justice system is in crisis, and governments at the local, state, and federal level are struggling to finance the ever-increasing cost of this largely dysfunctional system.

High rates of incarceration and a $200 billion yearly expenditure might be worth it if the get-tough approach helped keep Americans much safer from street crime, corporate crime, identity theft, and other crimes. However, this is hardly the case. As we discuss in a later chapter, the get-tough approach has resulted in only a modest dent in crime rates and has cost far too much for the modicum of crime reduction it has achieved. Most criminologists favor a public health strategy that attacks the causes of crime, arguing that such strategies will be more cost-effective than the get-tough perspective and do a much better job of making the nation safer.

WHO WE ARE AND WHY WE WROTE THIS BOOK

We wrote this book for two reasons. First, as American citizens, we are naturally concerned about the crime problem in the United States and about the condition and effectiveness of our criminal justice system. Second, as social scientists, we are also concerned that most Americans know too little about crime and justice through no fault of their own, and we believe that an effective crime strategy demands accurate information about crime and justice in our nation.

Who are we and why do we believe we are well-suited to write this book? We are both sociologists who have written textbooks and editorial commentaries about crime and criminal justice for several newspapers. George Bryjak was on the sociology faculty at the University of San Diego from 1979 to 2003 and is currently a Research Associate at USD. He is coauthor with Steve Barkan of *Fundamentals of Criminal Justice* (forthcoming from Jones and Bartlett Publishers) and coauthor of two other textbooks: *Social Problems: A World at Risk* (Allyn & Bacon), and *Sociology: Changing Societies in a Diverse World* (Allyn & Bacon). George has published numerous articles in both scholarly journals and other academic outlets as well as more than 80 oped pieces in newspapers including *USA Today, The San Diego Union-Tribune, The Los Angles Daily News, The Baltimore Sun, The Seattle Post-Intelligencer, The Orange Country Register,* and *Houston Chronicle.*

Steven Barkan has taught sociology at the University of Maine since 1979. He is coauthor with George Bryjak of *Fundamentals of Criminal Justice* and author of *Criminology: A Sociological Understanding* (Prentice Hall). He also authored *Law and Society: An Introduction* (Prentice Hall) and *Protestors on Trial: Criminal Justice in the Southern Civil Rights and Vietnam Antiwar Movements* (Rutgers University Press), along with several journal articles on various crime and justice topics. He is the 2008–2009 President of the Society for the Study of Social Problems and was chair of the American Sociological Association's Task Force on Sociology and Criminology Programs.

We share a love of our country, but also a perception that it can do much better in numerous areas, not the least of which is crime and criminal justice. We hope this book will help you understand the nature and causes of crime in the United States; how our police, courts, and prisons work; what the criminal justice system can and cannot do to reduce crime; and promising strategies for slashing crime rates so that we can finally have a safer society. We hope you will find our discussion enjoyable and informative and that it stimulates your thinking about these important issues.

"Speed Bump" © Dave Coverly/Dist. By Creators Syndicate, Inc.

chapter two

The Crime Problem

Americans are very concerned about crime, and rightly so. Crime is all around us: we see it on our television sets and in our movie theaters, and many of us have been victims of crime or know someone who has been a victim. Americans suffer more than 20 million violent and property crimes every year. In our lifetimes, according to a government report, almost one-third of us will be a robbery victim, and about three-fourths will have our residences burglarized. National survey evidence finds that one-fourth of women have been victims of domestic violence or rape and sexual assault. The United States has the highest homicide rate in the Western world and very high rates of other serious street crimes. Reflecting this high volume, our nation spends about $200 billion every year to fight crime, including about $90 billion on policing, $43 billion on the courts, and more than $62 billion on jails and prisons.

Because crime is a very serious problem, many Americans have strong opinions and even fears about it. What are their views? How might mass media coverage about crime shape these views? How effective has the government's "war on crime" been since the 1970s? We discuss all these topics in the pages ahead before ending with a look at the wedding-cake, funnel, and other models of the criminal justice system.

FEAR AND LOATHING: DON'T BELIEVE EVERYTHING YOU SEE IN THE NEWSPAPERS OR ON TELEVISION

Americans may have many beliefs about crime and criminal justice, but it turns out that much of what we think we know about crime may not be true or may at least overlook its complexity. For example, Americans underestimate the extent to which convicted offenders go to prison and the length of their prison terms. In addition, although the majority of Americans in any one year think that crime has been rising, in fact the U.S. crime rate declined steadily from 1993 through 2000 and was then fairly stable through 2004 before rising slightly since then. And although Americans often have a sense that crime is a more serious problem than ever before, it turns out that Americans have *always* considered crime a serious problem. As a presidential commission reported in 1967, "There has always been too much crime. Virtually every generation since the founding of the Nation and before has felt itself threatened by the spectre of rising crime and violence." Mob violence was a common sight in major U.S. cities in the decades before the Civil War, for example, and teenage gangs were a common threat. Concern over crime guided criminal justice policy after the Civil War and in the early decades of the twentieth century. In short, the United States has never been free from crime, and Americans have always been concerned about it.

It is also true that our current concern about crime may both exaggerate the extent of violent crime and minimize the harm caused by other types of crime. Consider homicide, for example, a crime featured in countless news stories and popular entertainment. According to the FBI, just over 17,000 homicides occurred in 2006. This is not a small number, but does not even place homicide among the top ten causes of death (which are led by heart disease, and include cancer, motor vehicle accidents, and suicide). Homicide thus receives much more media attention than its actual occurrence might merit.

Homicide and other street crimes probably receive so much attention because they threaten us personally and violate our sense of security. Yet this fact leads us to neglect the gravity of white-collar crime, which in many ways is more harmful than street crime. Examples of white-collar crime include medical fraud, false advertising and price-fixing, financial fraud, and corporate practices that threaten the health and safety of their workers or of the public. Estimates of the costs of white-collar crime dwarf those of street crime. For example, although the Federal Bureau of Investigation (FBI) estimates that property crime costs the public about $18 billion in direct costs, the monetary cost of white-collar crime, depending on how it is defined, may exceed $800 billion. And although about 17,000 people were murdered in

2006, the annual death toll from white-collar crime (from such sources as pollution and unsafe workplaces and products) may exceed 100,000. Although white-collar crime is undoubtedly more costly and deadly than street crime, it receives far less media and public attention than street crime and much less attention from the criminal justice system.

What the News Media Tell Us

What accounts for this false or oversimplified understanding of crime? Many observers blame the news media. Because most people learn about crime from the media, the media should, of course, provide an accurate picture of the amount of crime, the nature of crime, trends in crime rates, and the operation of the criminal justice system. Yet many studies confirm that the media do not provide an accurate picture.

A major problem is that the media overdramatize crime in at least two ways. First, they report many crime stories in order to capture viewer or reader attention. Local television newscasts often report more stories about crime than about any topic other than sports. If a particularly violent crime or spurt of crimes occurs, the media give these heavy attention and thus contribute to a false belief that crime is becoming more frequent. The media create the perception of crime waves by devoting so much attention to one crime or a small number of crimes that the public becomes more alarmed about the menace of crime than is warranted by the actual crime rate.

Second, the media overdramatize crime by giving disproportionate attention to violent crime: If it bleeds, it leads. Many studies find that most crime stories on television news or in the newspapers feature violent crime, even though violent crime is much less common than other types of crime. In particular, studies find that homicides are more than 25 percent of the crime stories on the evening news even though homicides make up much less than 1 percent of all crimes. Research also finds that media coverage may distort actual trends in violent crime rates. For example, the number of homicide stories on the national television network news shows increased 473 percent from 1990 to 1998 even though homicides actually decreased by 33 percent during those years.

Such coverage yields the false impression that most crime is violent, and such coverage can affect public perceptions of the amount of violent crime and concern over it. As recounted in the previous chapter, half the students in one study greatly overestimated the number of homicides per year. Other studies of Baltimore and Philadelphia residents find that those who watch television news shows often are more likely to worry about crime than those who watch such shows less often.

Another problem is that media coverage often highlights crime committed by minorities and teenagers. In television and newspaper stories about crime, a greater proportion of the offenders are African American and Hispanic than is true in actual crime statistics, and a greater proportion of the victims are white than is true in actual statistics. For example, although 80 percent of Los Angeles homicide victims are African American or Hispanic, a study found that murders of white victims were much more likely to be covered by that city's largest newspaper than murders of African American or Hispanic victims. Newspaper stories about white victims of homicide are longer than those about African American victims, and these stories also disproportionately depict crimes in which African Americans are the offenders and whites are the victims, even though most crime involves offenders and victims of the same race.

In all these ways, media coverage exaggerates both the degree to which African Americans and Hispanics commit crime and the degree to which whites are victims of crime. Perhaps for this reason, one poll found that whites think they are more likely to be victimized by minorities than by whites, even though about 75 percent of all crimes against whites are committed by other whites. Another provocative study focused on subjects who watched news stories that did not depict the offender. Sixty percent of the subjects falsely remembered seeing an offender, and 70 percent of these subjects said the offender was African American.

A similar bias exists in the coverage of youth crime. Media crime coverage features more teenagers than is true in actual crime statistics. For example, one study in California found that almost 70 percent of television news stories on violence featured teenagers, even though only 14 percent of all violent crime arrests in that state are of teenagers. In a related problem, newspaper and television news stories about teenagers tend to show them committing violence rather than prosocial acts. Such coverage exaggerates the degree to which teenagers commit violence. Not surprisingly, therefore, 62 percent of respondents in a 1998 national poll said youth violence was increasing even though youth homicides had actually been declining significantly. Respondents in other polls say that teenagers commit most violent crime, even though they actually only commit about 14–16 percent of it.

Crime Myths and the Drama of Violence

These problems in media coverage help contribute to crime myths, or false beliefs about crime and criminal justice. Like the fable about the emperor's new clothes, people believe certain things are true when, in fact, they are not. Crime myths include the following: (1) the public believes crime is rising when, in fact, it is not; (2) people think that most crime is violent when, in fact, it is not; (3) people overestimate the involvement of people of color and youths in crime and underestimate their victimization by crime; and (4) people worry more about street crime than perhaps the facts warrant.

Statements by public officials also contribute to **crime myths**. Politicians regularly have used strong words about crime and criminals, some of it racially coded, to win public support for harsher criminal sanctions and, not incidentally, for the politicians' campaign efforts. A memorable example involved a 1988 television commercial aired on behalf of the presidential campaign of then-Vice President George Bush. The commercial featured convicts passing through a revolving door and showed a picture of an African American Massachusetts prisoner, Willie Horton, who had committed a vicious murder while on a prison furlough. The commercial was widely credited with hurting the campaign of Bush's Democratic Party opponent, Massachusetts Governor Michael Dukakis, and was heavily criticized for its racial overtones. After Bush became president, he gave a speech about drugs on Labor Day 1989 in which he held up a bag of crack cocaine that, he said, had been bought by undercover agents in Lafayette Park across the street from the White House. It was later revealed that the White House had asked federal agents to buy the crack in the park in order to provide a dramatic prop for the President's speech. When the agents were unable to find anyone selling crack in the park—because, they were told, it was too near all the police at the White House—they had to trick a drug dealer into going to the park so that the agents could buy it there. Although both these examples involved a Republican president, Bush's successor, Democrat Bill Clinton, also engaged in harsh crime rhetoric when he sought Congressional support for a major crime bill during his first term as president.

Whether promoted by the news media or by elected officials, crime myths help distort public understanding about crime. Media coverage fuels public concern about crime and takes attention away from white-collar crime. It also contributes to negative stereotypes about African Americans and Hispanics. Crime itself is a real problem, but beliefs about crime and criminal justice gained from the media and from official statements may not be accurate.

Views About Punishment: A Punitive and Prejudiced Public?

If media coverage of crime contributes to negative stereotyping of racial minorities, it also makes the public more punitive when it comes to the treatment of criminals. Americans are indeed fairly punitive in this regard, and more so than citizens of other Western nations. In 2006, about 64 percent of the public thought the courts in their area did not deal harshly enough with criminals, and 69 percent favored the death penalty for persons convicted of murder, although the latter figure represented a drop from the high of 80 percent who supported the death penalty in 1994.

Although a clear majority of Americans want harsher courts and favor the death penalty, it should be kept in mind that many Americans also want to rehabilitate criminals and to address the underlying causes of crime. In a 2003 Pew Research Center survey, 73 percent completely or mostly agreed that "the criminal justice system should try to rehabilitate criminals, not just punish them." A 2006 Gallup poll asked respondents which approach to lowering the crime rate came closer to their own view: "attacking the social and economic problems that lead to crime through better education and job training," or "deterring crime by improving law enforcement with more prisons, police, and judges." Americans showed a clear preference in answering this question: 65 percent chose the social problems response, while only 31 percent chose the law enforcement response. Thus, while many Americans do hold punitive views, they also recognize the need to rehabilitate offenders and to address the underlying causes of crime.

When we consider race and views about the criminal justice system, a troubling picture emerges. First, there are strong racial differences in views about some aspects of criminal justice. For example, although 69 percent of whites favor the death penalty for persons convicted of murder, only 35 percent of African Americans hold this view. African Americans are thus much less likely than whites to support capital punishment. They are also much more likely than whites to perceive that racial bias and other problems of unfairness exist in the criminal justice system.

Second, and more troubling, a growing body of research finds that racial prejudice affects whites' views about several aspects of crime and criminal justice. To be more precise, racial prejudice has been found to be a significant factor that raises the support of whites for the death penalty, for harsher courts, for police brutality, and even for the spending of more money to fight crime. To put this another way, racial prejudice is among the factors influencing whites' support for all these measures.

This finding is troubling because of its implications for criminal justice policymaking in a democracy. In a democracy, public opinion should affect policymaking, including policymaking about criminal justice. But it violates democratic ideals if racial prejudice influences policymaking, as appears to be the case for criminal justice. There are many reasons why people might favor the death penalty, harsher courts, police brutality, and spending more money to fight crime. We might agree or disagree with these various reasons, but racial prejudice is a reason that is antidemocratic. As a 2005 study of this problem by one of this book's authors concluded, "policymakers must be careful not to be unduly swayed by public opinion on crime, as such opinion rests in part upon the racial prejudices of white Americans."

Research on views about the police also reveals a racial divide. African Americans and Hispanics are much more likely than whites to hold negative views about the police. They have less confidence in the police and hold a dimmer view of the honesty and ethics of the police. African Americans are also almost three times more likely than whites to think that police brutality occurs in their neighborhoods. Scholars attribute these differences in perception about the police in part to the fact that so many people of color are stopped by the police, questioned, and sometimes searched. Because many of the subjects of these police stops interpret them as racial harassment, it is no surprise that they hold such negative views about police.

IF YOU BUILD IT, THEY WILL COME (AND THEN LEAVE): THE WAR ON CRIME AND THE GROWTH OF PRISONS

How should we deal with crime? In the 2006 Gallup Poll discussed earlier, Americans favored "attacking the social and economic problems that lead to crime" over "improving law enforcement with more prisons, police, and judges" by 65 percent to 31 percent. Since the 1970s, however, the United States has emphasized the latter strategy. In this respect, U.S. policy differs from that of other Western nations, which give much more emphasis to rehabilitation and the social factors underlying crime. The approach in other nations is more akin to the public health model discussed in the previous chapter, which stresses the need to identify and address the causes of disease in order to reduce its occurrence. The public health model thus focuses on the prevention of disease, not just on its treatment.

The get-tough approach instead treats crime after it occurs by focusing on the arrest and imprisonment of criminals above all else. Criminals must be apprehended, of course, just as cancer patients must be treated, but the get-tough approach neglects underlying causes that the public health model emphasizes. The get-tough approach also has brought with it many economic and social costs.

The Growth and Cost of Criminal Justice

What are some of these costs? The United States now incarcerates more than 2.2 million people in prisons and jails at any one time. More than 7 million citizens are under correctional supervision of some form—incarceration, probation, or parole—amounting to more than 3 percent of all adults. The 2.2 million inmates are about twice the number behind bars in 1990 and almost five times the number behind bars in 1980. The U.S. incarceration rate—about 750 jail or prison inmates per 100,000 Americans—is more than five times greater than that of any other Western nation.

As the surge in imprisonment should suggest, the get-tough approach has cost the United States tens of billions of tax dollars for more police, prisons, and jails. As noted earlier, the nation now spends about $200 billion annually on the criminal justice system. In 1990, we spent $79 billion, and in 1980 about $35 billion. Thus, criminal justice expenditures have risen by about 150 percent since 1990 and 470 percent since 1980. Even adjusted for inflation, this is a huge increase, and several states have had to divert money from higher education and social programs to pay for new prisons and other criminal justice costs.

Truth and Consequences: Has This War Been Worth It?

In a word, no. By neglecting the underlying causes of criminal behavior, the get-tough approach makes it likely that crime will continue no matter how many offenders are arrested and imprisoned. This is a harsh charge, to be sure, and one with which get-tough advocates would doubtless disagree. In their view, the get-tough approach has reduced crime by imprisoning so many offenders. This view probably has some truth, and we examine it more fully in Chapter 11. Yet the bulk of the evidence indicates that the get-tough approach has achieved only modest reductions in crime and that these reductions have come at a very heavy social and financial cost. Dollar for dollar, much greater crime reductions can be achieved if the nation's tax dollars were spent on programs that address the underlying causes of crime and the rehabilitation of offenders than on more police, and especially, on more prisons and jails.

In addition, the get-tough approach has fallen disproportionately on racial and ethnic minorities. About 12 percent of African American males and 4 percent of Hispanic males in their late twenties are in prison or jail at any one time, compared to just under 2 percent of white males. Although the government estimates that more than 5 percent of all Americans will be put in prison at least once during their lifetime, this figure rises to 16 percent for Hispanic men and almost 30 percent for African-American men. About one-third of young African American men are currently under correctional supervision. These racial disparities might make sense if they reflected actual racial differences in offending, but, as we shall see in later chapters, the differences in offending are not nearly large enough to explain the differences in imprisonment.

In additional problems, the get-tough approach has destabilized urban neighborhoods by putting so many of their young males into prison. It has also created a force of hundreds of thousands of inmates who are released every year from prison back into their communities. Many of these young men have the same, or even worse, personal problems as those that helped put them into prison, and most have bleak chances of stable employment and good social relationships. Observers fear continued crime from these ex-inmates and call for more social programs to ease prisoner reentry.

We discuss these issues further in later chapters, but spend the rest of this chapter sketching the operation of the criminal justice system. The controversy over the get-tough approach notwithstanding, the criminal justice system plays a fundamental role in the United States, and it is impossible to imagine our nation without it. It employs hundreds of thousands of people, costs billions of dollars, and processes millions of offenders annually. In one way or another, it touches all of our lives directly or indirectly. For all these reasons, it is important to understand the criminal justice system's operation, its strengths, its weaknesses, and its impact on the crime rate and other aspects of American society.

DIRTY SECRETS OF CRIMINAL JUSTICE

Television cop and lawyer shows would have you believe that the police usually capture criminals and that defense attorneys and prosecutors go at it tooth and nail before a jury. However, the way the criminal justice system actually works differs greatly from what we see on television.

The Funnel Effect

One way of understanding the criminal justice system is to think of it as a
funnel, with many cases entering the top of the system, and only a very few
trickling out into prison at the bottom. Of the serious crimes or felonies (violent and property offenses for which the possible punishment is at least one
year in prison) that occur, fewer than two-thirds are investigated by the
police, because many victims do not report their crimes to the police. Of the
crimes the police do investigate, only about one-fifth end in an arrest. Fewer
still are fully prosecuted because of evidence problems, and fewer of those
that are prosecuted end in convictions. For every 1000 felonies that occur,
only about 11 to 20 (or 1.1 to 2.0 percent) result in someone going to prison
or jail. Some of the remaining cases are dismissed by prosecutors or judges,
and others result in guilty pleas to misdemeanor charges that avoid incarceration. Most convictions for serious crimes do result in heavy punishment for
the defendant. But the funnel model reminds us that most serious offenses
do not lead to imprisonment, because so many crimes are not reported to the
police and because so many of those that are reported do not end in arrest.
This has important implications for the ability of the corrections system to
make a large dent in the crime rate through increased incarceration rates.

The Wedding-Cake Model

The typical wedding cake has a small layer on the top, a larger layer just
beneath it, and larger layers below that. The criminal justice system may also
be understood through a **wedding-cake model**. A very small number of celebrated cases lies at the top of the model. These are cases that, because of the
enormity of the crime or the fame of the offender and/or victim, receive
heavy media attention and the utmost attention from criminal justice officials. The most celebrated case since 1990 was probably the 1994 arrest and
subsequent trial of football star and television and movie celebrity O.J. Simpson for allegedly murdering his ex-wife and a friend of hers. A close second,
perhaps, was the arrest and trial of Timothy McVeigh for bombing the Oklahoma City federal building in 1995. Celebrated cases usually involve actual
trials instead of just guilty pleas and can often be very dramatic. Because
they receive so much publicity, these cases unduly influence public perceptions of criminal justice, but most criminal cases do not necessarily proceed
the way the celebrated ones do.

The second layer of the wedding cake involves the most serious felonies. These cases are distinguished by the seriousness of the offense and the extent of the injuries involved, the extent of the defendant's prior record, and whether the victim knew the offender. Many homicides and robberies and some rapes fall into this category, although offenses in which the offender and victim were acquainted are apt to fall into the next layer. Although serious felonies represent only a small percentage of all crimes, they demand a disproportionate amount of the time, money, and energy of the criminal justice process. Because prosecutors believe that defendants in these cases should receive harsh punishment, prosecutors are reluctant to plea bargain, and a greater proportion of these cases than those in the remaining layers are likely to go to trial. In this second layer of the wedding cake, the criminal justice system imposes long prison terms because these cases involve serious crimes against strangers by offenders with long prior records.

The third layer of the wedding cake involves less serious felonies. These are cases involving less serious charges, such as property crimes and violent crimes where little or no injury occurred and where the victim knew the offender. These are also cases where the defendant has little or no prior record. Compared to cases in the top two layers, cases in this third layer are more likely to be dismissed or to end in guilty pleas as a result of plea bargaining. Defendants convicted in these cases are more likely than those in the upper layers to avoid incarceration.

The fourth layer of the wedding cake involves misdemeanors, which make up the bulk of all criminal acts. The major violent and property crimes—homicide, aggravated assault, rape, robbery, burglary, larceny, motor vehicle theft, and arson—total only about 16–17 percent of all arrests in any given year. About half of all arrests involve minor offenses such as disorderly conduct, public drunkenness, prostitution, simple assault, and petty theft. Because so many cases are in the fourth layer, they proceed very quickly. Public defenders or assigned counsel spend only a few minutes with their clients, if that, and the cases are resolved through plea bargaining with defendants appearing before a judge one after another like a factory line. Very few defendants in this layer are incarcerated.

The wedding-cake model helps us understand that different kinds of cases in the criminal justice system are treated very differently. The heavy attention given to the celebrated cases at the top of the cake may lead to misperceptions of how the bulk of criminal cases in lower layers are handled.

Decisions, Decisions: The Discretionary Model

Yet another way to understand the criminal justice is with the **discretionary model**, as the criminal justice system can be understood as a series of decisions at every stage of the process. Crime victims and witnesses must decide whether to report the crime; police must decide how much to investigate a crime and whether to arrest a suspect; prosecutors must decide whether to prosecute a case and which charges to file; judges must decide whether to dismiss a case and whether to let defendants out on bail or their own recognizance; defendants must decide whether to plead guilty, and prosecutors and defendants must decide on a plea bargain; if a trial occurs, juries or judges must decide whether to convict a defendant; if a conviction occurs, judges must decide on the appropriate sentence; and, finally, various corrections officials must decide whether to release inmates early from prison or, for those in the community under supervision, whether to send them to prison for violating the conditions of their supervision.

The criminal justice system could not operate without all these kinds of discretion. No two cases are alike, and no two defendants are alike. Criminal justice officials recognize this, and they also recognize the need to keep the whole process working as smoothly and efficiently as possible. Thus, discretion in the criminal justice system is necessary. Yet with so much discretion at every stage of the process, the opportunity for conscious or unconscious abuse of discretion arises: criminal justice officials may make biased decisions, either consciously or unconsciously, based on such factors as a defendant's or victim's race, ethnicity, gender, or social class. Officials also may allow their views of the nature of certain criminal behaviors to affect their decision-making. For example, some officials in the past, and perhaps still in the present, did not take crimes like rape and domestic violence seriously and failed to arrest or prosecute to the fullest extent of their law. More generally, discretion also may mean that similar cases receive very different outcomes in different jurisdictions of the nation, raising important questions about the basic fairness of these outcomes. Discretion is thus a double-edged sword with profound implications for the operation of the criminal justice system.

The Consensual Model

We often hear that the United States has an adversary system of justice. Under this **adversarial model**, court proceedings are viewed as a contest between the prosecution and the defense in which both parties do their best to win, often with fiery rhetoric. Many television shows about lawyers feature this model with lawyers sharply questioning witnesses and defense attorneys doing everything possible to get their client off the hook.

The adversarial model is certainly highlighted in many law school courses and generally characterizes the celebrated cases at the top of the wedding cake and some of the serious felony cases below them. But, for most criminal cases, the adversary model does not apply. Instead, a **consensual model** prevails, as prosecutors, defense attorneys, and judges generally cooperate to push cases through as quickly and efficiently as possible. They share ideas of what normal crimes are and what appropriate punishment should be, and these shared ideas permit most cases to be resolved through plea bargaining. Without such efficient case processing, the criminal court system would quickly break down. Ironically, then, U.S. criminal courts would be severely hampered if they actually did follow the adversary model. (We discuss plea bargaining further in Chapter 10.)

Crime Control and Due Process Models

Two final models of the criminal justice system in a sense compete with each other, both in the legal system itself and in the court of public opinion. They remind us of the tensions that exist as the legal system strives to keep the public safe while preserving the rights of due process that are embodied in our Constitution and Bill of Rights.

As its name implies, the key objective of the **crime control model** is to prevent and punish crime, and, by so doing, to keep society safe. Because this model assumes that most criminal suspects are guilty, it emphasizes the need to process cases as quickly and efficiently as possible. The image of an assembly-line conveyor belt best captures the operation of the criminal justice system under the crime control model. As with products on a conveyer belt, cases are passed as quickly as possible from one stage to another. The task of everyone on the criminal justice conveyor belt is clear and simple: to make sure that offenders are punished as quickly and as easily as possible.

The **due process model** stands in sharp contrast. Its key goal is to prevent government abuse of the legal system against guilty and innocent people alike. This model assumes that some suspects are indeed innocent of the crimes with which they are charged. It also assumes that even guilty suspects deserve fair treatment in a democracy. Given these assumptions, the due process model emphasizes the need for the government to follow specific procedures to arrest, prosecute, and convict suspects.

These two models have been in tension at least since crime began to rise during the 1960s. The get-tough trend of the last few decades indicates that the crime control model has won out over the due process model. The reverse was true during the 1960s, when the Supreme Court under the leadership of Chief Justice Earl Warren expanded the legal rights of suspects, defendants, and prisoners. Due process rights remain but have since been somewhat curtailed, and the crime control model is now more popular.

The clash between the two models became especially evident in the aftermath of the September 11 terrorist attacks. In the wake of that day, Congress passed and President George W. Bush signed the USA Patriot Act, which authorized the detention and deportation of immigrants, the seizure of financial and medical records, and the designation of various domestic groups as terrorist organizations. The government quickly detained hundreds of Middle Eastern immigrants for intense secret questioning and denied many of them legal counsel or monitored attorney-client conversations of those who were allowed counsel. President Bush also announced that immigrants accused of terrorism could be tried by secret military tribunals where they would lack many due process rights, including the right to a jury trial and to the presumption of innocence. In a policy called extraordinary rendition, suspected terrorists were kidnapped from Canada, the United States, and other nations and taken to secret detention camps in the Middle East and Eastern Europe, where they were reportedly tortured. Critics charged that all these measures violated several amendments to the U.S. Constitution, while supporters claimed they were all necessary to ensure public safety against the threat of new terrorist acts.

KEY TERMS

adversarial model: A model of the criminal justice system that emphasizes fierce competition between prosecutors and defense attorneys

consensual model: A model of the criminal justice system that emphasizes cooperation among prosecutors, defense attorneys, and judges to expedite case processing

crime control model: A model of the criminal justice system that emphasizes the need to prevent and punish crime and, by so doing, to keep society safe

crime myth: A false belief about crime and criminal justice

discretionary model: A model of the criminal justice system that involves a series of decisions at each stage of the process

due process model: A model of the criminal justice system that emphasizes the need to prevent government abuse of the legal system

funnel: A model of the criminal justice system in which many cases enter the top of the system and only a very few trickle out into prison at the bottom

wedding-cake model: A model of the criminal justice system in which there are a very small number of celebrated cases at the top of the model, a greater number of serious felonies on the next tier and many less serious felonies and misdemeanors below

SUGGESTED READINGS

Bohm, Robert M., and Jeffery T. Walker. 2006. *Demystifying Crime and Criminal Justice.* Los Angeles: Roxbury Publishing Company.

Kappeler, Victor E., Mark Blumberg, and Gary W. Potter. 2005. *The Mythology of Crime and Criminal Justice.* Prospect Heights, IL: Waveland Press.

Muraskin, Roslyn, and Shelly Feuer Domash. 2007. *Crime and the Media: Headlines vs. Reality.* Upper Saddle River, NJ: Prentice Hall.

Reiman, Jeffrey. 2007. *The Rich Get Richer and the Poor Get Prison: Ideology, Class, and Criminal Justice.* Boston: Allyn & Bacon.

Sacco, Vincent F. 2005. *When Crime Waves.* Thousand Oaks, CA: Sage Publications.

Surette, Ray. 2007. *Media, Crime, and Criminal Justice: Images, Realities, and Policies.* Belmont, CA: Wadsworth Publishing Co.

"Speed Bump" © Dave Coverly/Dist. By Creators Syndicate, Inc.

chapter three

How Much Crime Is There and Who Commits It?

Have you ever had anything stolen from you? Has anyone ever assaulted you? If either of these things happened to you, did you tell the police about it? If someone else you know was ever victimized by crime, did that person tell the police about it?

If the police never heard about a crime committed against you or someone you know, that crime never became part of the nation's official crime count. The police hear about only 40 percent of all serious street crimes, meaning that 60 percent of these crimes never come to their attention and thus do not get counted as crimes. Even when the police do hear about a crime, that does not guarantee that it eventually gets counted as an official crime.

However, it is important to get as accurate account as possible. If we do not have good information about the number of crimes, then we cannot know for sure whether crime is increasing, decreasing, or staying about the same, and we cannot know for sure whether crime is higher in one location or in another location. And if we do not know these things, then it becomes more difficult to know whether various efforts to reduce crime are working or failing.

HOW CRIMES ARE COUNTED

Every fall the newspapers dutifully report the FBI's latest information on crime statistics. In September 2007, for example, the Washington Post reported the FBI's finding that violent crime rose 1.9 percent in 2006 and that this increase came on the heels of a 2.3 percent rise in violent crime in 2005. The FBI reported that more than 1.4 million violent crimes (homicide, aggravated assault, robbery, rape) and almost 9.9 million property crimes (burglary, larceny, motor vehicle theft) occurred in 2006, for a total of about 11.4 million serious crimes.

This is a large number, to be sure, but another agency of the federal government, the Bureau of Justice Statistics (BJS) in the Department of Justice, reported in December 2007 that a much larger number of crimes occurred in 2006. Based on its interviews of a random sample of Americans nationwide, the BJS estimated that about 2.3 million violent crimes (excluding minor assaults) and 19 million property crimes occurred in 2006, for a total of more than 21 million serious crimes. This latter figure is almost twice as high as the FBI's count, yet the BJS statistics usually receive much less news media coverage than the FBI's statistics. If the BJS statistics are more accurate than the FBI's, however, then the number of serious street crimes in the United States is almost twice as high as the public would believe from what it hears from the news media.

The Official Version: U.S. Government Certified

The FBI and BJS crime statistics are the U.S. government's official estimates of the number of crimes in the nation, and it is important to understand their methodologies so that we can in turn understand what each estimate can and cannot tell us.

The FBI publishes its crime statistics every fall in a document called *Crime in the United States*. The data in this document come from the FBI's crime-reporting system, called the **Uniform Crime Reports** (UCR). The UCR compiles several kinds of crime data, including the numbers and rates (number of crimes per 100,000 residents) of various types of crimes, the numbers of people arrested for these crimes, the percentage of all crimes that are cleared by arrest, the rates of crime by geographical region, and the age, gender, and race of the persons arrested. When you read in the newspapers or on the Web or see on the television news that crime is going up (or down), you will usually be hearing about UCR data.

UCR data are based on police reports and, thus, on **crimes known to the police**, to use a common phrase. When police hear about a possible crime, they investigate it and fill out appropriate paperwork if they determine that a crime has indeed occurred. They then file regular reports with the FBI about the number of the different crimes that have become known to the police, about whether someone was arrested for a given crime, and about the age, gender, and race and ethnicity of those arrested.

However, as noted just above, many people fail to tell the police about crimes they have experienced. Because these crimes—about 60 percent of all serious offenses—do not come to the attention of the police, they remain hidden from the FBI's crime statistics and are called the dark figure of crime by criminologists.

This dark figure exists because police discover only about 3 percent of crimes on their own and must rely on citizens to tell them about crimes they experience as victims or notice as witnesses. But as we noted earlier, overall only about 40 percent of the victims of serious crime report these crimes to the police. The reporting rate varies by type of crime, and the reporting rate is higher, say, for motor vehicle theft than for rape. The reasons for non-reporting also vary by type of crime, but include fear of reprisal, the belief that the police could not capture the offender, and the belief that the alleged offense was a private or personal matter. Because about 60 percent of crime victims fail to file a police report, the crimes known to the police that become part of the FBI's official crime estimate are less than half of all crimes that actually occur.

A better estimate of the actual number of crimes comes from the Bureau of Justice Statistics methodology, which relies on an annual survey of tends of thousands of Americans regarding criminal victimization. This survey is called the **National Crime Victimization Survey** (NCVS). Interviewers draw from a random selection of households nationwide and ask a series of questions to determine whether that person or someone in that household has been the victim of crime during the prior six months. To avoid influencing responses, interviewers do not use crime terms (like burglary) in their questions and instead read the respondents descriptions of the crimes. Respondents who report having experienced a crime are then asked another series of questions about when the crime occurred, whether a weapon was involved, and so forth. Because the NCVS is a random sample, NCVS researchers use the number of victimizations reported by their respondents to estimate the number that actually occur in the entire nation. Although the UCR and NCVS do not measure crimes in exactly the same way (for example, the NCVS does not include homicide, because homicide victims obviously cannot report their victimization), the NCVS estimate of the number of crimes is considered by criminologists to be more accurate than the FBI's UCR estimate and is usually at least twice as high as the FBI's estimate.

Mistaken Crime Statistics

Although the UCR underestimates the actual number of crimes, that is not its only problem. Other aspects of the UCR lead to inevitable slippage in the number of crimes that become known to the police, and sometimes police themselves make it appear that fewer crimes are occurring than is really true. They do this in several ways.

Police Underrecording of Crime   When police hear about a crime, they must determine whether a crime indeed occurred. This is called the founding stage. Sometimes citizens make up the details of a victimization they report to the police, in order to get someone in trouble or to grab attention for themselves. Sometimes citizens sincerely feel they were victimized by a crime, but the circumstances do not lead the police to conclude that an actual criminal law was violated. These two types of false or mistaken reports are thought to account for about 2 to 4 percent of all reports to the police, and these reports do not get counted in the UCR as crimes that actually occurred.

However, even if police do believe that a crime has occurred, that does not necessarily mean they always record it as a crime. If the police are very busy and think the crime was minor, they simply might not record it in order to save the time and energy of dong the paperwork. Although the extent of police non-recording of crimes they hear about and believe to have occurred is unknown, some evidence indicates that the police record only about two-thirds of all citizen complaints about crime. If so, the UCR fails to report all the crimes "known to the police" because the police did not bother to do the necessary paperwork.

Another problem at this stage involves the fact that one criminal incident often includes many crimes. For example, suppose a burglar breaks into a house with the intent of stealing cash or jewelry. The homeowner returns and surprises the burglar, who hits the homeowner with a club and escapes. At least two felonies have occurred here, burglary and aggravated assault, but, following UCR procedures, the police record only the more serious of the two crimes, the assault. Thus although a burglary also occurred, only the assault gets counted, and the burglary disappears from the official crime count.

A worse problem involves deliberate efforts by police to make it appear that fewer crimes are occurring; this manipulation of crime statistics makes it appear that the police are doing a good job of controlling crime. For example, police may deliberately fail to record crimes they hear about or downgrade serious crimes into misdemeanors: a burglary might be recorded as a disturbing the peace offense, or a rape might be recorded as a simple assault. Either procedure makes it appear that fewer serious crimes are occurring.

This manipulation of police reports has led to serious crime-reporting scandals since 1990 in cities such as Atlanta, Baltimore, Boca Raton, New York City, and Philadelphia. One officer in Boca Raton downgraded almost 400 felonies to misdemeanors and thus artificially reduced the city's crime rate by more than 10 percent in 1997. In Philadelphia, the scandal was especially serious. Police there regularly downgraded about 10 percent of all serious crimes into minor ones. They also counted crimes only when the crimes were finally recorded, not when they actually took place. Thus many crimes were counted as occurring in 1997 when they actually occurred in 1996, artificially lowering Philadelphia's 1996 crime rate. As a result the FBI disregarded the city's 1996 and 1997 crime statistics.

About two years after Philadelphia's scandal broke, a related one emerged in the city. Investigators found that since the early 1980s the city's police sex-crimes unit had either failed to record or else downgraded thousands of reports of rape and sexual assault. In order to keep the reported rape rate low, the unit coded many of these reports—about half of them in some years—as "unfounded" and others as "investigation of person" or similar category. The victims were unaware of this and the police never followed up on victims' complaints. Some of the perpetrators later raped and sexually assaulted other women.

Manipulation of crime data may also occur at college and university campuses, as critics say that college administrators and campus security try to keep rapes and other crimes from coming to the attention of the public. The campuses handle them internally in student discipline hearings rather than tell the police about them. Although the federal government requires colleges and universities to provide accurate data of the crimes occurring on and near their campuses, at some schools many crimes, especially those occurring just off campus, do not get reported to the government.

Other Problems in Police Recording   Several other problems affect the accuracy of UCR crime data. First, despite UCR instructions, police agencies around the country differ to some extent in their understandings of the definitions of crime and thus in their likelihood of recording an incident as a particular type of crime. For example, consider the difference between an aggravated assault and a simple assault. An aggravated assault must involve a serious injury and/or the use of a weapon, while a simple assault involves no weapon and only a minor injury. Often, however, it is not clear whether an injury is serious or minor. Some police agencies appear more likely than others to classify assaults with such unclear injuries as aggravated rather than simple.

Police also respond differently to victims' complaints about crimes, based on several characteristics of the victims. In a well-known study from the 1960s, researchers rode around in police cars and observed police-citizen encounters. The police came into contact with many citizens who alleged that a crime had been committed against them. All other things being equal, the police were more likely to record the citizens' claims as a crime when: (a) the citizen complainant (victim) was polite rather than impolite; (b) the complainant and suspect were strangers or only acquaintances rather than relatives or close friends; and (c) the complainant preferred the police to do something about the crime rather than do nothing. Once again we see that citizen reports of crime are not automatically translated into official crimes.

In a final problem, the police sometimes stage crackdowns in which they flood a high-crime neighborhood and make many arrests for such things as prostitution and sale of illegal drugs. The number of these sorts of crimes thus soars in police records and in the UCR, even though the actual number of these crimes had not changed at all.

Problems in Interpreting Crime Trends   For all the reasons discussed, the UCR does not provide an accurate count of the number of crimes in the United States. This makes it even more difficult to assess or identify trends in the crime rate. If the police do not record all crimes, or if they otherwise manipulate reporting, the crime rate will appear to fall. If citizens become more or less likely to report a crime, the crime rate will rise or fall respectively. For example, the UCR rates of rape rose steadily in the 1970s and 1980s, but this increase probably did not mean that the actual rate of rape was increasing. Instead, it probably reflected both greater reporting by rape victims and the greater likelihood of police taking their reports seriously. Similarly, the introduction of the 911 emergency number enabled more crime victims and witnesses to report crimes to the police. Crimes that thus used to remain hidden now became known to the police, causing the official UCR count of crimes to rise when in fact no actual increase had occurred. Because UCR data are flawed in these ways, criminologists often rely on homicide data to understand crime trends. Most homicides do become known to the police, because corpses are hard to hide. Thus, if a change in the UCR homicide rate occurs, we can be fairly certain that the number of actual homicides did indeed change.

The Absence of Corporate Crime in the UCR   A final problem with the UCR is that it does not count crime by corporations, including financial corruption, dangerous workplaces and products, and illegal pollution of our air, ground, and water. This omission sends the message that corporate crime is neither as important nor as serious as the street crime that the UCR covers.

The NCVS Revisited   Although, as noted earlier, the NCVS provides a more accurate estimate than the UCR of the actual number of crimes in the United

States, the NCVS, too, is inaccurate in several ways. First, it does not include two very important types of crime, commercial crime (e.g., break-ins into stores) and white-collar crime, and it does not cover crimes whose victims are under age 12. Because of these omissions, the NCVS underestimates the actual amount of crime in the United States even if it provides more accurate estimates than the UCR does.

Second, NCVS respondents may forget about at least some of the victimizations they experienced during the previous six months, or they may simply choose not to tell NCVS interviewers about a victimization. This latter problem may be especially true for crimes like rape, sexual assault, and domestic violence. Whether respondents forget about their victimizations or just decide to remain silent about them, the NCVS fails to uncover these crimes and again underestimates the actual number of crimes.

If these problems mean the NCVS might underestimate the actual number of crimes, other problems might lead it to provide an overestimate. Some respondents might mistakenly interpret some things they experienced as crimes, when in fact the circumstances would not fit the definition of a crime. Other respondents might mistakenly telescope crimes by telling NCVS interviewers about victimizations that happened before the survey's six-month reporting period. Despite these possibilities, most researchers think that underestimation is a more likely problem in the NCVS than overestimation.

WHO COMMITS CRIME?

Although crime is ultimately an act committed by individuals, it is also true that some kinds of individuals are more likely than other kinds to commit crimes. By "kinds" of individuals we mean individuals with certain social backgrounds or characteristics that affect the likelihood they will commit crime. Criminal justice researchers have examined several such characteristics: gender, age, social class, race and ethnicity, and geographical location. We discuss all these characteristics below. The information provided comes from UCR and NCVS data and also from **self-report surveys**, in which samples of youths report anonymously on various crimes they have committed. Although self-report surveys are, like the UCR and NCVS, not a perfect vehicle for assessing the amount of crime and who commits it, they are a very common source of data for criminological research and have provided some very important information.

In discussing how social characteristics relate to crime rates, note that we are not saying that everyone with a certain characteristic will commit crime, or that anyone without the characteristic will not commit crime. We are simply saying that our social characteristics affect our chances of committing crime. Thus when we say just below that gender affects crime rates, and in particular that men have a much higher crime rate than women, we do not mean to imply that every man commits crime, or even that most men commit crime, and that no woman ever commits a crime. We are simply saying that a significantly greater percentage of men than women commit crime.

Geography: Location, Location, Location

People are more or less likely to commit crime depending on where they live. In the United States, the South and West have higher violent crime rates than the Northeast, and the Midwest has the lowest rate; the West also has the highest property crime rate. The reasons for these regional differences are not clear. Most scholarship on this issue tries to explain why the South's violent crime rate is so high. Some researchers say the South has a regional **subculture of violence** in which slights and insults are taken very seriously and can escalate easily into violence. Other researchers dispute this point, and say that the high rate of southern violence instead stems from its high rate of poverty.

Crime rates are also higher in urban areas than in rural areas. The violent crime rate is about three times greater in our largest cities than in rural areas. Why does this difference exist? Because crime rates control for the number of people, it is not just that urban areas have more people than rural areas. Rather, there is probably something about urban living conditions that leads to higher crime rates. Although poverty is commonly cited as a reason for high urban crime rates, many rural areas are also poor and yet have lower crime rates than their urban counterparts. Although poverty may be a contributing factor, other features of urban life also contribute to higher rates of crime. According to some research, these features include population density (many people on a street living closely together), household overcrowding (crowded conditions within a household), and the presence of bars and other businesses where people congregate and where violence and other crime can thus occur.

Of these features, population density is certainly very important. A hundred thousand people living within a few square miles in a city are obviously living in very different circumstances than a hundred thousand people living spread out over hundreds of square miles in a more rural area. They come into contact with one another much more often than their more rural counterparts. Because, practically speaking, much crime cannot occur unless people come into contact, the more frequent social interaction among people crowded into a city helps to account for the fact that crime rates are much higher in cities than in rural areas. In a city, it is very easy for young people to congregate and to get into trouble. In a rural area where the houses might be thousands of feet from each other, it is less easy to do so.

Climate may also affect crime rates. Generally, assaults, rapes, and burglary and larceny tend to be higher in the summer and lower in colder weather. Warm weather causes us to be more irritable and to lose our tempers, and it also prompts us to spend much more time outdoors, where we interact with more people. The more interaction, the greater opportunities for arguments, and the greater the opportunities for arguments, the more violence will occur. Then too, in the summer people spend much more time away from their homes, leaving their households vulnerable to burglars. If warm weather does seem to promote several types of crime, perhaps that helps explain why the South and West have such high crime rates: these regions are generally warmer than the Midwest and Northeast.

Age: What's the Matter with Kids Today?

A well-known song from the classic 1960 musical "Bye Bye Birdie" lamented, "Why can't they be like we were, perfect in every way?/What's the matter with kids today?" In the world of crime, that is still a relevant question. Simply put, young people commit a disproportionately large amount of street crime, which in general peaks between ages 17 and 19. Rates decline somewhat but remain high as individuals move into their early 20s, and decline further as individuals move into their late 20s and beyond.

Why do young people have a higher crime rate? Researchers cite several reasons, all of them having to do with the nature of adolescence and young adulthood in our culture. First, adolescence and young adulthood are times of adventure, excitement, rebellion, and, sometimes, a lack of maturity. What seems to be fun and daring to youths is often regarded much more negatively by older generations. For better or worse, breaking the law is something many youths find fun and exciting. In their need for excitement, they sometimes turn to crime.

Second, adolescence and young adulthood are also times when peer influences are especially important. For better or worse, peer influences also play a strong role in decisions to commit crime: if your friends like getting into trouble, you are that much more likely to do so yourself. If peer influences play such a role, it is no surprise that adolescents have high rates of offending.

Next, adolescence and, to a lesser extent, young adulthood, are times when we want to own a lot of things but often do not have enough money to buy them. To some extent, the property crimes that young people commit stem from their desire for money and possessions that they otherwise would have trouble obtaining.

Finally, adolescence is a time when youths have few stakes in conformity. Most are not married and are not parents, and most do not have full-time jobs. They have not developed the sense of obligation that arises from all these responsibilities, and they are still free of the time and energy constraints that go along with these responsibilities. They thus feel freer, and in fact are freer, to hang out with friends and to break the law.

For all these reasons, adolescents and young adults have higher rates of offending than older people. Most youthful offenders age out of crime as they age chronologically. The number of people in the 15-30 age group especially prone to crime can affect the nation's crime rate. One possible reason that crime rose in the 1960s was that the baby boom born after World War II was entering its crime-prone years. By the same token, crime declined during the 1990s, and one possible reason for this decline was the decline in the number of young people during that decade owing to declining birth rates 15–30 years earlier.

Gender: It's a Man's World

Crime scholars disagree on many things, but one area on which they very much agree is that men have much higher rates of serious crime than women. Evidence for this gender difference comes from both the UCR and NCVS. The UCR reports the percentage of all arrestees who are male and the percentage who are female; the NCVS asks its respondents who were victims of personal crime to report the offender's gender, age, and race. While sometimes victims might report the offender's age or race inaccurately, they are less likely to be mistaken about the perpetrator's sex.

Both the UCR and the NCVS report almost identical results for the percentage of male offenders, lending confidence to their findings. For robbery, both data sources find that about 90 percent of all arrestees are male, and for aggravated assault both sources find that about 80 percent of offenders are male. Because males are just under one-half the population, their representation among offenders is much higher than we would expect by chance. Self-report data involving adolescents confirm that males are involved more heavily in serious offenses, but also find that there is a much smaller or even no gender difference for minor offenses such as drinking and vandalism.

Why do males have much higher rates of serious crime? Scholars have tried to answer this question for at least a century, and their early answers now sound antiquated. A noted 19th-century Italian physician, Cesare Lombroso, said that women committed less crime because they were naturally passive; his evidence for this was that sperm move around a lot more than does an ovum. Sigmund Freud and his early followers thought that women were also naturally more passive, but when they did commit crime they did so out of penis envy: because they were frustrated that they did not have a penis, they committed crime to be more like men. In 1950 one sociologist even wrote that women committed more crimes than people thought, but were especially good at hiding evidence of their crimes and thus more likely to evade arrest. His "evidence" for this supposed skill at being deceitful was that girls and women learn at an early age to hide evidence of their menstrual flow and to pretend that they are having orgasms during sexual intercourse.

Contemporary explanations of the gender difference in crime rates sound much more plausible. Girls and boys and women and men differ in many ways besides crime rates, and researchers cite many of these differences as reasons for the gender difference in criminality. Citing biological differences between the sexes, some researchers say that males are naturally more aggressive than females, and also have much higher levels of testosterone that boost their aggressiveness. We discuss this and other biological views of crime in Chapter 8, but for now will simply note that many crime scholars say biology has little, if anything, to do with the gender difference in crime rates. Instead they cite two other factors: socialization and opportunity.

Boys and girls are still socialized differently. At the risk of oversimplifying a large body of evidence, boys are raised to be more assertive and aggressive, while girls are raised to be more gentle and nurturing. Girls also become more attached to their parents and to value their schooling more, and both these factors help keep them from offending. During adolescence, boys are given more freedom than girls to be on their own, and hence have greater opportunity than girls to get into trouble. Many researchers say that boys' socialization thus helps them to learn values, attitudes, and behaviors that contribute to criminal behavior, while girls' socialization helps them to learn values, attitudes, and behaviors that contribute to a lack of criminal behavior. They emphasize that our crime problem is largely a male crime problem: if our crime rates were no higher than women's crime rates, crime would be considered much less serious.

For more than a decade, female arrests have been rising at a greater rate than male arrests, and this trend has prompted some observers to claim that girls and women are becoming more violent as gender roles have changed. However, many criminologists instead attribute the rise in female arrests to changing police practices and also to the legal war on drugs, which has led to the arrest of many poor women for possession and use.

Income: Blue Collar or White Collar?

If criminal justice researchers mostly agree on gender and age differences in criminal offending, they very much disagree on social class differences. Their disagreement reflects the complexity of the evidence on social class and crime rates. For example, although the UCR and NCVS do not measure the social class of offenders, it is true that most criminals in prison and jail are poor and lack even a high school degree. But does this fact reflect true social class differences in offending, or does it instead reflect bias by the criminal justice system against the poor?

Many researchers argue that social class differences in serious offending do exist and that the poverty of jail and prison inmates reflects these differences, regardless of criminal justice system biases. Other researchers, however, argue that the supposed relationship between social class and criminal behavior is merely a myth that stems from large biases in the criminal justice system. As further evidence, they point to several self-report studies that find similar offending rates among poor and middle-class youths.

Critics of the myth view challenge it on two counts. First, the self-report studies that find no relationship between crime rates and social class included only minor offenses and thus were finding only that no social class differences exist for these types of offenses. When serious offenses are examined, as they are in more recent self-report surveys, social class differences do emerge: poor youths have higher rates than non-poor youths of violence and other serious crime. Second, it is important to measure social class adequately by looking at youths from the most deprived backgrounds. Although this measurement issue is complex, when very poor youths in self-report studies are compared to other youths, social class differences in serious offending are more apt to emerge than when such youths are not isolated for comparison. After discussing the views of scholars who dismiss the relationship as a mere myth, one researcher wrote pointedly, "Yet, through it all, social scientists somehow still knew better than to stroll the streets at night in certain parts of town or even to park there . . . [and they] knew that the parts of town that scared them were not upper-income neighborhoods."

Why do the poor commit more street crime? Some of the explanations of crime discussed in the next chapter try to answer this question. For now we can make a few points. Children raised in poor families tend to grow up in higher-crime neighborhoods, where they are influenced by delinquent peers into committing crime themselves. The conditions in these neighborhoods also cause frustration and despair that sometimes drive people into

criminality. These and other problems help to produce higher rates of delinquency and crime among the poor.

Resolution of the debate on the social class-crime relationship is critical
for an understanding of crime and for criminal justice policy. If social class
does make a difference for serious street crime, this difference reflects the
fault lines of a society with large pockets of poverty, unemployment, and
other structural conditions that many researchers think contribute to crime.
To the extent this is true, measures to improve these conditions should help
reduce street crime significantly. On the other hand, strong evidence for the
lack of a relationship between class and crime would mean that these conditions do not matter for crime, and the measures to improve them would do
relatively little to reduce the crime rate.

One thing does seem clear: if we include white-collar crime, especially that
committed by corporate executives, no social class-crime relationship exists.
The people involved in such crime do not come from the ranks of the poor. On
the contrary, the executives who commit corporate crime rank among the
wealthiest individuals in the nation. Although the poor commit more than their
fair share of street crime, the rich commit more than their fair share of white-
collar crime. The poor are certainly not the only ones who break the law.

Race and Ethnicity: It's Not All Black and White

When we ask who commits crime, perhaps the most controversial discussion
centers on the issue of race and ethnicity. African Americans represent about 13
percent of the U.S. population but account for almost half of state prisoners.
About one-third of young African American men are under correctional supervision (in jail or prison or on probation or parole) at any one time. Clearly
African Americans are overrepresented as offenders in the criminal justice system. Hispanics are also overrepresented. The fundamental question is this: does
this overrepresentation reflect actual racial and ethnic differences in offending,
or does it reflect racial and ethnic bias in the criminal justice system?

Focusing mostly on African Americans, criminal justice researchers
again disagree on their answers to this question. Some argue that the African
American overrepresentation mostly represents bias by police and other
criminal justice professionals. To these researchers, the United States has a
system of "unequal justice" and is on a "search and destroy" mission to put
African Americans, especially young males, behind bars. Other researchers
concede that some bias exists but say that the overrepresentation of African
Americans reflects actual racial differences in offending much more than it

reflects any bias. Even if no bias existed, they add, African Americans would still be overrepresented as offenders.

Where does the truth lie? The evidence on offending and on bias is again complex. We address the evidence on bias in later chapters, but for now state our agreement with the researchers who concede some bias but still believe that African Americans have higher rates of offending. Evidence for this view again comes from UCR arrest data and NCVS data victims' perceptions of offenders. African Americans are overrepresented both in arrest data and in offenders reported in the NCVS. Even allowing for the possibility that some NCVS victims misperceive the race of their offenders, African Americans do seem to have higher rates of offending for serious street crime.

What do self-report data indicate? The answer depends on which types of offenses are studied. Self-report studies that examine only minor offenses find only small racial differences in offending, but studies that examine serious violent offenses usually find substantial racial differences. Most crime researchers would probably agree that African Americans do, in fact, have higher rates of serious street crime, notwithstanding any criminal justice system bias that might exist, and would draw the same conclusion regarding Hispanics. If such a racial difference does exist, the important task then is to explain it.

Researchers began to advance such explanations almost a century ago, and these early explanations, now considered racist, blamed African Americans' high crime rates on biological inferiority. Contemporary explanations center on other factors: a subculture of violence in black communities; the absence of fathers in black families; and structural problems facing African Americans, including extreme poverty and segregation.

The term **subculture of violence** refers to a set of beliefs and values that lead people to be willing to use violence to resolve interpersonal disputes. Evidence on whether African Americans are more likely than whites to hold such values is mixed. Several surveys find that African Americans and whites give the same kinds of answers to questions asking about their approval of violence in various situations. The researchers who analyze these survey data conclude that a subculture of violence does not exist among African Americans, and, further, that their higher rates of offending thus cannot derive from a subculture of violence. Yet other researchers do say that urban African Americans have values such as the need for respect that promote the use of violence. However, these researchers are careful to add that if these values do exist, they arise from the structural deprivation in which many urban African Americans live.

The evidence on the African American family structure is also mixed. Although many more black children than white children grow up without fathers in the home, evidence of the effects of single-parent households on juvenile offending is unclear. Some studies find that single-parent households greatly affect the likelihood that adolescents will break the law, but other studies find that such households have only little, if any, effect.

The evidence for the importance of structural problems for African American crime rates is more consistent. African Americans are much poorer than whites, much more likely to live in segregated communities with high crime rates, and are worse off in other structural ways. These structural problems all lead to higher rates of offending. Some scholars say that whites' crime rates would be just as high as blacks' crime rates if whites lived amid the same criminogenic (crime-producing) structural conditions. Sociologist Rodney Stark says the higher crime rate of African Americans also stems from where they live: outside the South, African Americans are much more likely than whites to live in inner-city neighborhoods "where the probabilities of *anyone* committing a crime are high." This sort of structural explanation allows us to explain and understand higher African American crime rates without resorting to a racist explanation that says African Americans are biologically or culturally inferior.

If African Americans and Hispanics do have higher rates of street crime, they certainly have lower rates of white-collar crime. Most white-collar crime, and especially corporate crime, is committed by whites, because whites monopolize the ranks of corporate managers and executives. If whites have higher rates of such crime because of their greater opportunities to be in white-collar positions in the first place, then perhaps African Americans have higher rates of street crime because of their lack of opportunity to escape the kinds of structural conditions that lead to such crime.

Before leaving the subject of race and ethnicity, it is important to comment on the issue of immigration and crime. Many Americans think that the increasing immigration the nation has experienced during the past two decades is leading to higher crime rates. In a 2000 national survey, more than 70 percent of the respondents said that a higher crime rate would be a "very likely" or "somewhat likely" result of "more immigrants coming to this country." However, the evidence on this issue paints a very different picture. First-generation Hispanics—those born in other nations and then immigrating to the United States—actually have lower crime rates than people, including non-Hispanic whites, who were born and raised in the United States. Some scholars think that the increasing numbers of new immigrants in many of our large cities during the 1990s helped account for the decline in crime in those cities during that decade. Ironically, second-generation immigrants—the children of new immigrants—have higher crime rates than their parents, and third-generation immigrants have higher crime rates still. To put this more simply, living in the United States seems to raise immigrants' crime rates.

Two Points to Remember

We have discussed how and why different kinds of people have different crime rates and now want to end with two very important points. First, and to reiterate something we noted earlier, even if a certain age group, gender, social class, or race has a higher crime rate, it remains true that most people in that grouping do not commit crime. For example, although men are arrested at a much higher rate than women for violent crime, only about three out of every thousand men (or 0.3 percent) are arrested in any given year for violent crime. This means that 99.7 percent of men are *not* arrested. Similarly, although African Americans are arrested at a much higher rate than whites for violent crime, only about five African Americans of every thousand African Americans (or 0.5 percent) are arrested in any given year. This means that 99.5 percent of African Americans are *not* arrested for violent crime. Just as we should not stereotype all men for the crimes of a statistically few men, neither should we stereotype African Americans, Hispanics, the poor, or youths for the crimes of a statistically few people in each group.

Second, our discussion implies that your chances of committing crime depend to a large extent on your gender, social class, and race/ethnicity. These characteristics heavily influence how and/or where you are raised and socialized, and how and where you are raised and socialized in turn heavily influence your chances of committing a crime. If we can imagine cloning a baby and the baby's parents and then having one baby clone raised by its cloned parents in a rural area and the other baby clone raised by its cloned parents in an urban area, the urban baby will more likely than the rural baby to commit crime by age 20 because of the urban environment in which it lives. African Americans and Hispanics might have higher crime rates, but these higher rates are largely a function of the urban location and socioeconomic circumstances in which people of color reside. By stressing how social backgrounds and environments influence criminality, we do not mean to excuse the crimes that result. But we do mean to stress the importance of social background and environment, because crime reduction efforts need to address the criminogenic aspects of social background and environment to achieve real reductions in crime.

KEY TERMS

crimes known to the police: The UCR's term for the official number of crimes that the police report to the UCR and that the UCR in turn reports to the public

National Crime Victimization Survey (NCVS): An annual survey taken by the government in thousands of randomly selected households nationwide that asks residents of these households questions about crimes that they, other household members, and the household itself have experienced during the past six-month period

self-report surveys: Surveys in which respondents are asked to provide information on various offenses they may have committed

subculture of violence: Beliefs and values that make it more likely that people will use violence to resolve interpersonal disputes

uniform crime reports (UCR): The U.S. government's official source of data on crime that relies on police reports of crimes reported by citizens

SUGGESTED READINGS

Belknap, Joanne. 2007. *The Invisible Woman: Gender, Crime, and Justice.* Belmont, CA: Wadsworth Publishing Company.

Catalano, Shannan M. 2006. *The Measurement of Crime: Victim Reporting and Police Recording.* New York: LFB Scholarly Publishing.

Gabbidon, Shaun L., and Helen Taylor Greene. 2005. *Race and Crime.* Thousand Oaks, CA: Sage Publications.

Mauer, Marc. 2006. *Race to Incarcerate.* New York: New Press.

Reiman, Jeffrey. 2007. *The Rich Get Richer and the Poor Get Prison: Ideology, Class, and Criminal Justice.* Boston: Allyn & Bacon.

Walker, Samuel, Cassia Spohn, and Miriam DeLone. 2007. *The Color of Justice: Race, Ethnicity, and Crime in America.* Belmont, CA: Wadsworth Publishing Company.

"Speed Bump" © Dave Coverly/Dist. By Creators Syndicate, Inc.

chapter four

Robbers, Rapists, and Serial Killers: Violent Crime in America

The 19th-century French sociologist Emile Durkheim stated that crime is normal, because it is impossible to imagine (or find) a society that does not have criminal behavior. Durkheim did not mean that crime is normal in the sense that it is good. Rather, he used the term in the sense of statistical normalcy; crime is everywhere. Certainly, that is true in a modern industrial society such as the United States, where millions of people commit crimes each year, and millions more are victims of crime.

Because of the overall impact crime has on society, most people know something about it. Unfortunately, many beliefs about crime are false. For example, most homicide victims are not killed by assailants who methodically plan their crimes down to the last detail. In this chapter we examine violent crimes and ask a series of questions about such crimes: Is the violent crime rate increasing or decreasing? What is the profile of the typical violent offender? How are these crimes committed? What motivates violent offenders?

HOMICIDE

Homicide is the killing of one human being by another and is not necessarily a crime. The state- sanctioned death penalty and police officers taking a life in the line of duty are examples of noncriminal homicide. Most states make distinctions among five types of *unlawful* killings: first degree murder; second degree murder; voluntary manslaughter; involuntary manslaughter; and felony murder.

First degree murder is a crime made famous by countless mystery novels and movies—the classic "whodunit?" theme. This unlawful killing is willful, deliberate, and premeditated. In other words, the offender makes a conscious decision to kill his/her victim, prior to committing the crime. This decision to take another life could be made months prior to the act or no more than a split second before the crime takes place. In states with capital punishment, first degree murder carries the death penalty. In second degree murder, while the offender may or may not have intended to take the life of the victim, the act occurs without deliberation and premeditation. In Boston, British-born au pair Louise Woodward was convicted in 1997 of second degree murder in the death of an eight-month-old baby for whom she was caring. The prosecution argued that although the 19-year-old probably had not intended to kill the infant when she hit and shook him, she was responsible for his death.

Manslaughter refers to those killings that, although unlawful, are committed without malice aforethought. These homicides are considered less severe. Voluntary manslaughter is an intentional killing under circumstances that make it less blameworthy, and, therefore, reduce the offender's culpability. The classical example of this crime as legal scholar Steven H. Gigis notes is when the accused killed in a state of intense emotion "caused by the deceased's provocation. The defendant must have been in the heat of passion, such as rage, fright, terror, or wild desperation when he killed the victim." Involuntary manslaughter is the death of an individual that results from the negligent or reckless behavior of the offender. The most common version of this crime in modern industrial societies is death resulting from reckless and/or drunk driving.

Felony murder is an unlawful homicide that occurs during the commission (or attempted commission) of a felony, most notably a robbery, rape, or arson. Two important aspects of this crime must be noted. First, in many states, the sentence for felony murder is the same as that for first degree murder, which means that felony murder is a capital offense in death penalty states. Second, this crime does not require malice aforethought. If, during the commission of a convenience store robbery, an offender panics and shoots and kills the clerk, the defendant can be charged with felony murder even though he had no intention of hurting, much less killing, anyone. As Steven Gifis notes: "The evil mind or malice that is necessary to find some-one guilty of murder is implied or imputed from the actor's intent to commit a felony." Keep in mind that murder and manslaughter categories are not always clear and may require interpretation on the part of prosecutors regarding which charges to bring against a defendant. (Prosecutorial discretion will be discussed in a later chapter.)

Homicide rates (the number of homicides per 100,000 population) increased from 4.1 in 1955 to 10.2 in 1980. Murder rates remained high through the early 1990s, when they began a gradual decline to 5.6 in 2005. In 2006 there was a slight increase to 5.7. Some criminologists think that an increase in the number of young people in the crime-prone years (15 to 30 years of age) over the next decade, the so called echo-boom generation, will result in escalating homicide rates.

An examination of the Uniform Crime Reports data over the past 40 years reveals the following characteristics about murder in the United States.

1. Historically, in this country, the killer and the victim know each other (family, friends, or acquaintances). However, in 1991, for the first time, more than half of the nation's murders were committed by strangers or under circumstances in which the relationship between the victim and offender was unknown. In 1965, one of every three murders was family-related, a figure that has dropped to one in eight.

2. This changing nature of the offender-victim relationship has had a significant impact on the number of murders solved by police. In 1965, police could rightly assume that the killer knew or was related to the victim, and the police arrested someone in connection with the crime in 91 percent of the cases. By the mid-1990s, police made an arrest in only 65 percent of all murder cases (60.7 percent in 2006), a number that dropped to 33 percent in some urban neighborhoods.

3. Most victims and perpetrators of homicide in this country are male with both male and female offenders most likely to kill males. Consider the following offender-victim gender homicide relationship in a recent year when the gender of the two parties was known to police.

    Male offender/Male victim accounted for 65.3 percent of homicides

    Male offender/Female victim accounted for 22.7 percent of homicides

    Female offender/Male victim accounted for 9.6 percent of homicides

    Female offender/Female victim accounted for 2.4 percent of homicides

4. Although slightly less true today than in previous years, murder is overwhelmingly an intraracial crime. From 1976 to 2005, 86 percent of white victims were killed by whites while 94 percent of black victims were killed by blacks. In 2006, 91.8 percent of black victims were slain by black offenders and 81.5 percent of white victims were killed by white assailants. Murders by strangers are more likely to cross racial lines than killings involving friends and acquaintances.

5. Although African Americans made up only 13 percent of the population in 2005, almost one out of two murder victims was black. In 2005, homicide victimization rates for African Americans were six times as that of white Americans. Blacks are less often the victims of sex-related homicides, workplace homicides, and killings by poison, while they are over represented in drug related killings. The lifetime risk of being an unlawful homicide victim is 1 out of 26 for black males, 1 out of 125 for black females, 1 out of 170 for white males, and 1 out of 503 for white females. If social class and residence were taken into consideration, the lifetime risk of being a homicide victim for poor, inner city, young black males could be as high as 1 out of 20. Using the 1 out of 26 figure, black males in this country are 19.3 times more likely to be murdered than white females.

6. The average age of people arrested for murder has declined from 32.5 years in 1965 to 27 in the mid 1990s with many of these killings gang and drug related. Young adults between 18 and 24 years of age historically have the highest murder rate and these rates doubled between 1983 and 1993. Since 1993, offending rates for this age group have declined but remain at slightly higher levels than prior to the mid 1980s. Juvenile homicides are overwhelmingly an inner-city phenomenon.

7. Nearly three of every four children murdered in the 26 industrialized countries of the world are slain in the United States. Another way of looking at this relationship is that a child in this country is five times more likely to be murdered than a child in the rest of the industrialized world.

8. The geography of criminal homicide reveals this behavior is disproportionately a big-city crime with homicide rates in the nation's largest urban areas about twice as high as rates in suburban areas and rural districts. From 1976 to 2005, over half of all homicides occurred in cities of 100,000 or more while almost one-quarter of the nation's murders happened in cities with populations over one million. While homicide rates are lowest in the Midwest, traditionally they have been the highest in the south with the western United States a close second.

9. Peak periods for homicide are July, August, December, and the weekends when "routine activities" bring people together. As interaction increases, so, too, does the possibility that conflict will occur, which in some cases will result in murder.

10. Firearms are the weapons of choice of murderers. Approximately two of three murders in 2006 involved a firearm, and in 82 percent of these cases the weapon was a handgun. The proportion of homicides involving guns differs by circumstance. In recent years almost all gang related killing are committed with guns while nearly four out of five felony related murders (a killing in the commission of a robbery, for example) are gun related. Guns are used in approximately 60 of killings resulting from arguments.

As noted above, homicide rates traditionally have been the highest in cities and the Southern states. Sociologists David Luckenbill and Daniel Doyle posed the question: "What is there about residing in an urban or southern area that generates a high rate of violence?" They noted that a significant number of people in these areas (young, male, lower-income) have a lifestyle characterized by disputatiousness; that is, these individuals share a culturally transmitted willingness to settle disputes that are perceived as a threat to their masculinity or status by using physical force. In an earlier and related work, Luckenbill stated that homicide is the product of a character contest.

During the course of an argument, insults and threats are traded until escalating tension brings the matter to a point of no return, with participants and bystanders agreeing that the contest can be resolved only by violence. Obviously, the more often people settle disputes by physical force, the greater the likelihood that someone will be killed, even if the intent to take a life is absent. A significant number of homicides can be characterized as overly successfully aggravated assaults. That is, in the course of a fight, A wants to stab B to win the fight, but the difference between stabbing someone in the shoulder and a fatal thrust to the chest is only a matter of inches.

SERIAL KILLERS

While most Americans could name half a dozen serial killers, or are familiar with their media tags ("Son of Sam," the "Green River Killer," "Night Stalker," the "BTK Strangler"), few individuals have more than superficial knowledge about the fundamental issues surrounding these criminals. How many serial killers are there in the United States? How do these predators choose their victims? And, perhaps the most intriguing question, why do they kill?

The FBI's definition of serial killers is utilized by the law enforcement community and criminologists: individuals who take the lives of three or more people in separate events. The emotional cooling-off period between murders can last for weeks, months, or years. In the vast majority of cases the killers and their victims are strangers. According to one estimate, serial killers claim between 3500 and 5000 victims annually in this country. These figures are based on the fact that each year some 25 to 28 percent of murder circumstances are classified as unknown. That is, these killings cannot be attributed to factors such as arguments over property or money, a romantic triangle, or murders committed during the commission of a felony (for example, a drug deal or robbery). Twenty-five percent of the approximately 17,034 criminal homicides known to police in 2006 yields 4,258 unexplained murders—an overwhelming majority of which are believed to be serial killings.

The many critics of this perspective contend that because law enforcement cannot provide a specific explanation for a murder, it does not follow that the crime in question was motiveless, and, therefore, likely the work of a serial killer. Also, murders solved months or in some cases years after they occurred may not be removed from the circumstances-unknown category by overworked urban homicide detectives.

The FBI estimates that there are approximately 35 but possibly as many as 100 serial killers active in the United States at any one time. (This does not mean that every year there are 35 to 100 new killers.) Collectively, these individuals are thought to murder about 200 to 300 people a year and account for between 1 and 2 percent of known murders annually. Almost all homicide experts accept this significantly lower estimate of serial killings.

Serial Killers—International Laws

Although reliable international statistics of serial killing do not exist, this form of homicide has been reported in countries throughout the world with some of the most egregious examples to be found in Latin America and Russia. Mexican brothel owners Maria and Delfina de Jesus Gonzalez killed 80 prostitutes they hired after the women no longer proved to be an economic asset. These murdering siblings also took the lives of at least 11of their wealthier clients before they were convicted of their crimes in 1964.

In October 2007, a Moscow jury convicted Alexander Pichushkin for the murder of 48 people over a 15-year period. Most of the 33-year-old Russian's victims were dispatched in a stretch of woods near his home. "One's first murder, like first love" he told the court at his trial, "is impossible to forget. I was always killing for one reason—because I liked life so much. As soon as you kill, you want to live more."

Columbian born (1949) Pedro Alonso Lopez, The Monster of the Andes, may be the most prolific serial killer in modern history. In 1980, Lopez confessed to Ecuadorian officials that he had murdered over 300 teenager girls in Ecuador, Peru, and Columbia over the past few years, killing as many as three children a week. Authorities refused to accept the high body count until a flash flood unearthed a mass grave containing the remains of over 50 young girls. After serving almost 20 years in prison, mostly in solitary confinement, Lopez was taken to the Columbian border in 1998 and released.

A Typology of Serial Murder   In their book *Overkill*, criminologists James Fox and Jack Levin offer a typology of serial killers based on motivation.

Thrill-oriented killers murder for the perverse fun of it, relishing the complete control they exercise over their victims. Fox and Levin note these individuals rarely use a gun, because a firearm results in a relatively rapid, painless death, a demise that would deprive the killers of their greatest pleasure—watching the victim die slowly and with as much suffering as possible.

In a 2005 prison interview, Dennis Rader, the BTK (Bind-Torture-Kill) Strangler who killed at least 10 women in Wichita, Kansas, between 1974 and 1991, spoke of the pleasure he experienced planning his crimes and killing victims as well as reliving the murderous incidents in his mind. "My sexual fantasy is . . . if I'm going to kill a victim or do something to the victim, is having them bound and tied. In my dreams, I had what they called torture chambers . . . I had more satisfaction building up to it and afterwards than I did the actual killing of the person."

Serial killer Douglas Clark's ultimate fantasy was to murder a woman he was having intercourse with, to feel her death spasms during intercourse. Some thrill killers record their torture sessions on videotape so they can be viewed repeatedly. Many take a souvenir of the crime, such as jewelry or a piece of intimate clothing, something that will connect them to the torture and kill for future enjoyment. The overwhelming majority of serial killers are thrill-oriented.

Mission-oriented killers are fanatics engaged in a campaign to rid the world of evil and filth. As opposed to experiencing a measure of psychological or physical ecstasy via their crimes, these individuals are attempting to further some social, political, and/or religious agenda. Mission killers usually target a particular group or category of individuals they deem reprehensible—often prostitutes, homosexuals, homeless people, or drug users. One mission-oriented killer specifically targeted young black males, all of whom had white girlfriends. A killer of prostitutes in Louisville, Kentucky, boasted to police that he was performing a valuable service for the community. These individuals see themselves as killing in the defense of some noble idea or eternal truth rather than as murderers. In their minds they are heroes, not monsters.

The profit-driven serial killer takes the lives of his victims in connection with another crime, typically a robbery. In a 1992 midwest crime spree, thieves held up a number of small convenience stores, killing the proprietors and witnesses along the way. Unlike thrill-oriented killers, profit-motivated murderers favor a gun and kill their victims quickly. In 1989 and 1990, 35-year-old Aileen Wuornos murdered seven motorists along Florida's highways. The hitch-hiking prostitute would have sex with the men who picked her up, shoot them several times, take their money, then dump the bodies. Her motive, as Fox and Levin note, was greed. Profit-driven killers dispatch their victims because they believe it is necessary in order to avoid identification and apprehension. For them, murder is not a form of pleasure but a requisite for survival.

The 2002 Washington, D.C. area sniper killers do not easily fit into this typology. Because the victims were murdered from a distance and the shooter and his accomplice had no intimate contact with them before or after the attacks, these killings cannot be considered thrill-oriented murders. A possible profit motive was noted in a demand for $10 million after the 13th shooting, but experts are not convinced that money was the prime motivation for the murderous rampage. The victims ranged in age from 14 to 72 and were members of different racial and ethnic groups; the victims did not appear to be the target of a mission-oriented killer.

However, on occasion John Allen Muhammad spoke of his hatred of the United States and support for the terrorists who carried out the September 11th, 2001 attacks. The random shooting of 13 people from a distance with a high-powered rifle in three weeks is more akin to the philosophy of terrorists who embark on a deadly plan to psychologically and socially paralyze a population. The murderous spree of Muhammad and John Lee Malvo may represent a new form of serial killing/terrorism.

Cruel or Crazy? While serial killers have come from all racial and ethnic backgrounds and may be young or old, the typical offender is a white male (about 80 percent) between 20 and 39 years of age. No more than 10 to 20 percent of known serial killers in this country have been females. Fox and Levin argue that, with the exception of serial killers who murder because they hear voices and/or actually see their victims as devils, most of these individuals are more cruel than crazy, with a disorder of character rather than of the mind. They know right from wrong and can control their murderous desires. This latter point is of particular importance. If serial killers (especially thrill-motivated murderers) could not restrain their urges, they would make foolish mistakes and be apprehended quickly.

Often possessing above-average intelligence, serial killers have masqueraded as stranded motorists and police officers or posed as utility company employees or interested buyers answering classified newspaper ads. Serial killers can be charming, persuasive, and manipulative. In a prison interview, Ted Bundy, who murdered at least 30 young women, stated there was a rational, methodical element to his killings. He noted that serial killers learn to stalk, wait, and kill, and with each episode grow increasingly proficient.

BTK killer Dennis Rader thought of his victims as "a project." Shortly after his incarceration he told an NBC Dateline reporter that "the stalking stage is when you start learning more about your victim or potential victims. I went to the library and looked up their names, address, cross-reference and called them a couple of times, drove by there whenever I could." Because so many of these criminals (especially thrill-motivated serial killers) are rational and cunning, when they are caught it is often a matter of luck rather than as a result of some mistake on their part.

No Simple Answers To date, attempts at constructing profiles of serial killers have not proven useful. For example, a number of psychiatrists believe that severe injury to the brain's limbic system predisposes an individual to violent behavior. A study of 15 inmates on Florida's death row revealed that all of these men exhibited signs of neurological damage. What is missing from this research (by way of comparison) is the prevalence of individuals in the non-criminally violent population who have also suffered severe head injuries. However, Fox and Levin argue that "if head trauma were as strong a contributor to serial murder as some would suggest, we would have many times more serial killers than we actually do."

A number of violent offenders have an extra Y chromosome, producing what some investigators believed was the key to understanding and predicting masculine hostile behavior, the XYY "super male." However, further research discovered that a significant number of men who possess an additional male chromosome do not engage in criminal violence, and most violent offenders lack a second Y chromosome. Alcohol and pornography have been linked to violence against women, but as serial killer expert Eric Hickey notes, there are many individuals "who drink heavily and indulge in pornography even violent pornography—and never become serial killers."

Childhood physical and sexual abuse are considered by some to be crucial factors in producing thrill-oriented serial killers. However, hundreds of thousands of Americans have been abused as children and the overwhelming number of these individuals do not grow up to be murderers. Fox and Levin state that while head trauma and abuse may be important "risk factors . . . they are neither necessary nor sufficient to make someone a serial killer." Rather, they are part of a long list of components that may predispose an individual, "but not predestine, him or her toward extreme violence." Similarly, psychiatrist Dorothy Lewis argues that "neuropsychiatric problems alone don't make you violent. Probably the environmental factors in and of themselves don't make you a violent person. But when you put them together, you create a very dangerous character."

Understanding why some people choose to murder in a systematic and often horrific manner, and ridding society of the factors that bring about this deadly motivation are not the same thing. Even if social science and medical science knowledge advance to a level where an accurate profile of thrill-oriented or mission-oriented serial killers can be constructed, preventive detention of everyone who conforms to that psychological portrait is unacceptable. Addressing some risk factors may be possible, but the chances of significantly reducing, much less eliminating, serial killing are almost zero. In a society of more than 300 million people it is inevitable that a very small number of individuals will be aberrant to the extent that they find killing pleasurable, profitable, or justified in the pursuit of some sexual perversion, cause or twisted notion of religious purity. We can only be thankful that whatever ultimately causes this murderous behavior is present in so few people.

MASS MURDER

Mass murder is the killing of four or more people by one or more assailants within a single event lasting from a few minutes to several hours. Unlike serial killings, there is little controversy regarding how many mass murders occur annually or the demographic profile of the assailants. With four or more victims per incident, these crimes hardly go unnoticed and usually garner immediate national media coverage. Using FBI data, James Fox and Jack Levin examined 636 massacres between 1976 and 2002 involving 861 offenders and 2869 victims. On average there are two mass murders a month in this country, accounting for approximately 100 deaths a year.

A Typology of Mass Murder   Like killers of single victims, mass murders are young, with 87 percent of these assailants under 40 years of age. Three of five are white, and almost all—93.8 percent—are male. Guns are the weapons of choice for mass murderers in 75 percent of the slaughters. Almost 60 percent of victims are family members or acquaintances of the killers.

To better understand mass murder, numerous observers have created typologies that encompass the various manifestations of this crime as well as the motivation of the killers. The following is a listing of mass murder types by Ronald M. Holmes and Stephen T. Holmes, who acknowledge the earlier work of Park Dietz.

The disciple mass killer is a member of a group and commits crimes because of his or her relationship with the group leader. In non-traditional religious organizations (usually referred to as cults), the group may regard the leader as a prophet or as someone who receives communications from a deity; the leader may instruct members to carry out a particular plan. It is important to note that the motivation to kill is extrinsic to the killers. Rather, group members commit mass murder to please the leader and gain a closer relationship with him or her. In some cases the killings may have no utilitarian necessity from the point of view of the leader, but serve as a loyalty test for followers. The Manson murders of 1969 are an example of this type of killing.

In 1969, four members of the Charles Manson "family," broke into the southern California home of actress Sharon Tate and brutally murdered the pregnant actress and three of her friends. Two days later Charles Manson and several family members entered the home of grocery chain owner Leno LaBianaca and killed him and his wife, Rosemary.

The family annihilator selects members of his or her own family for execution. Over a four day period beginning Christmas Day, 1987, Ronald Gene Simmons, former Air Force Master Sergeant and recipient of the Bronze Star, killed 14 family members, including his wife, children, and grandchildren. On June, 20, 2001, Andrea Yates methodically drowned her five children ages 6 months to 7 years in the bathtub of their home in Houston, Texas while her husband was at work. She called the police, then placed the bodies of her dead children on a bed covering them with a sheet. James Fox and Jack Levin argue that it is unlikely that family annihilators suddenly go berserk as their crimes " were highly selective, well planned out, and purposeful. Like most other mass murderers, he didn't 'just snap.'" Fox and Levin suggest these killings may be "perverted acts of love." In line with psychiatrist Shervert Frazier's notion of suicide by proxy, these murderers see family members as extensions of themselves. For Fox and Levin, depressed and despondent, but not necessarily deranged, family annihilators feel personal responsibility for the well-being of their loved ones. They kill family members to spare them lives of frustration, anguish, and torment in an indifferent world and regard death as the only salvation.

Disgruntled employees are out for revenge, seeing themselves as victims of a work-related injustice. In January, 1993, Paul Calden killed three people and wounded two others in 30 seconds. Before firing the first shot he announced: "This is what you get for firing me." Calden drove to a local golf course walked to the 13th hole and shot himself in the head. Joseph Wesbecker returned to his former place of employment (he was fired after seven months leave for psychiatric problems) in Louisville, Kentucky killed 13 people and wounded 7 more before taking his own life.

Fox and Levin argue that despite changing gender roles, men define themselves largely in terms of their workplace identities more than women do. While only a minuscule number of men who are terminated engage in workplace violence, the loss of a job is especially troublesome for loners, single males, or men who are separated or divorced. "Socially isolated, they regard work as the only meaningful part of their lives. When they lose their jobs, they lose everything." Many also have mental problems.

The ideological mass killer convinces followers to commit suicide. Jim Jones, in Jonestown, Guyana, and Marshall Applewhite, in southern California, convinced their followers that suicide would result in salvation of some kind. Convinced that government authorities were closing in on him and that his South American retreat would come to an end, Jones decided the only way out was mass suicide. His followers were commanded to drink a concoction of cyanide and Kool Aid; those who resisted were shot or stabbed by a loyal cadre of disciples.

Under the guidance of Marshall Applewhite, 39 members of the Heaven's Gate group committed suicide in Southern California in 1997. With shaved heads and baggy pants, the unisex (castrated) men and women believed that in death they would be united with space aliens approaching earth.

Jones and Applewhite preached the same message: If you desire everlasting salvation, follow me. Holmes and Holmes argue that indoctrinated members of these groups, convinced of the righteousness of their leaders' beliefs, could not refuse their death commands lest they risk "everlasting damnation." In this sense both men were mass murderers. While most Americans are not at risk from ideological mass killers, "There will be other Jonestowns and Heaven's Gates. That is a certainty . . ."

The set-and-run mass killer does not want to die, and victims are not known to the killer. Rather, the victims are at the wrong place at the wrong time, although the site of the killings is chosen with care. The venue may be chosen for its symbolic value, and the motivation may be political. The motivation is political as opposed to financial gain or getting even with specific individuals. As opposed to Jim Jones and Marshall Applewhite who were concerned specifically with their followers, by way of his murders, Timothy McVeigh was interested in bringing about political change at the societal level.

The disgruntled citizen mass killer is seething with rage and pent-up hostility; these murderers are lashing out at the world. Victims are selected randomly and they have no previous relation with their killer. On July 31, 1966, Charles Whitman killed his mother, returned home and fatally stabbed his sleeping wife four times, allegedly one cutting blow for each year they were married. At noon the next day Whitman made is his way up a tower at the University of Texas at Austin, lugging an arsenal and 700 rounds of ammunition with him. Over the next four hours the former Marine shot 44 people, killing 13 individuals before he was gunned down by three police officers and a civilian.

In October, 1991, George Hennard drove his pickup truck through the front window of a crowded Kileen, Texas, restaurant. When some of the patrons rushed to help what they believed was an accident victim, Hennard opened fire. Well known for his hatred of women, especially feminists, Hennard shouted "Wait till those fuckin' women in Belton, Texas see this! I wonder if they'll think it was worth it!"

Fox and Levin note that while a murder spree was likely motivated by a particular attitude or hatred, a "'spillover effect' took hold: His anger and resentment became generalized to include just about everyone in Belton, County." This line of reasoning is probably correct for all disgruntled citizen mass murderers. Group-specific hatreds are unleashed on the entire community.

The psychotic mass killer may hear voices or believe that other people are after him or her. In 1980, Priscilla Ford claimed to have experienced a vision of Jesus Christ who told her to drive her automobile on the sidewalk in downtown Reno, Nevada, and kill anyone blocking her path. Acting on this godly revelation, the 50-year-old woman killed 7 people and injured 21 more. James Curse believed his neighbors were continually talking about him, spreading gossip the retired Florida librarian was a homosexual. When a shopper allegedly stuck his tongue out at him, Curse flew into a murderous rage killing 6 people including two police officers. Holmes and Holmes note that mass murder commonly occurs after a "psychotic episode," convincing the killer "that a spiritual apparition or ecclesiastical voice has commanded the person to kill." As is the case with the set-and-run killer and disgruntled citizen murderers, victims of these psychotic individuals are in the wrong place at the wrong time.

School shootings with youthful killers over the past 10 years have occurred multiple times. The most infamous of these events occurred at Columbine High School in Littleton, Colorado in 1999. After months of planning, 18-year-old Eric Harris and 17-year-old Dylan Klebold killed 12 fellow students and one teacher. The boys, who died by their own hands, had plans to blow up the school then fly a hijacked plane into the Manhattan skyline. School shootings typically occur in rural or suburban communities and are committed by white youths between 11 and 18 years of age. Holmes and Holmes note that these individuals are not from deprived backgrounds and are likely to have secure economic futures. They are often fascinated by guns and bombs. The assailants are almost always students at the school where the crime takes place and are familiar, at least as passing acquaintances, with their victims. Youthful school killers are often described after the murderous incident by classmates as "weird." They are loners or members of an outcaste school group.

Fox and Levin believe that youthful killers who work together may receive from each other two of most valuable commodities in the youth culture, commodities they did not receive from their classmates: respect and admiration. School settings "can sometimes breed feelings of inadequacy, anxiety, fear, hostility, rejection, and boredom." They are also venues "both logistically and symbolically, for getting even or settling a score."

A Final Note   Typologies are valuable in organizing data of a complex and multifaceted phenomenon such as mass murder, but they do have drawbacks. Some mass killings to not fit neatly into any of the categories. The killing of 32 Virginia Tech students in April, 2007 by Cho Seung-Hui is a prime example. Was Cho a disgruntled citizen? A psychotic mass murderer?

One interpretation of Cho's murderous rampage and the crimes of other mass murderers is that these individuals are attempting to exercise power over a world wherein they feel powerless, a world that has done them a great injustice. Forensic psychologist Louis Schlesinger states, "They're angry and they want to take it out on the world. They develop the idea that murder will be the solution to whatever their problem is and they fixate on it."

In a self-made video Cho alludes to "martyrs like Eric and Dylan," presumably Columbine High School killers Eric Harris and Dylan Klebold. To what extent did Cho pattern or copy his killings from the tragedy in Colorado? While the events at Columbine had little if anything to do with the South Korean killer's deep seated hatred and rage, they may well have impacted how and where he committed his murderous assaults. This line of thought inevitably leads to the role and responsibility of the media in covering mass murderers. While the public has every right to know what is happening, is the saturation coverage these events bring fuel for the next killing rampage?

Clinical psychologist Jana Martin notes that some mass murderers are searching for one more chance to give their life meaning. "They may think, 'I'm never going to amount to much, but I'm going to die amounting to something. This is my final mark on the world, my final statement.'" The paraphrased suicide note of Robert Hawkins, the troubled 19-year old who killed eight people then himself at a Nebraska shopping mall stated that "he was a piece of shit all his life" and now he'd be famous. A high body count killing is a guarantee that media coverage will be thorough and long lasting, that one's name will never be forgotten.

RAPE: A CRIME OF POWER, NOT SEX

The National Crime Victimization Survey (NCVS) defines rape as "forced sexual intercourse including both psychological coercion as well as physical force." This definition includes attempted rapes and male as well as female victims both heterosexual and homosexual. Attempted rape also includes verbal threats of rape. According to the NCVS in 2006 there were 272,350 victims of rape, attempted rape, or sexual assaults. Because of the methodology used by researchers, these figures do not include victims 12 years-of-age or younger.

Almost three of four sexual assaults are perpetrated by a nonstranger, that is, an intimate (spouse or boyfriend, for example), relative, friend, or acquaintance of the victim. Approximately 44 percent of rape victims are under age 18, and 80 percent have not reached their 30th birthday. Young females are four times more likely then members of any other age/gender group to be sexual assault victims. Although they comprise an estimated 10 percent of all victims, males are least likely to report a sexual assault. Since 1993, rape/sexual assault has declined by more than 60 percent in the United States.

Serial Rapists

To learn more about stranger rape, FBI Special Agents interviewed 41 men responsible for raping 837 victims. These "serial rapists" typically used one of three premeditated strategies to approach their targets, the con approach, the blitz approach, or the surprise approach.

Offenders using the con approach have confidence in their ability to interact with their victims. They may present themselves as police officers or other trustworthy people to get close to the victim.

The blitz approach features a sudden, violent attack, possibly when the victim is less able to respond or get away. Used less frequently then the con approach, blitz rapes usually result in more extensive injuries to victims. Violent attacks are often fantasy components of the crime and sexually stimulate rapists.

Offenders using the surprise approach may scrutinize a victim's comings and goings, entering the victim's apartment and await their return, or approach his victims while they sleep. The surprise attack was the most frequently used strategy by offenders who lacked the confidence or ability to subdue their victims by other means.

FBI researchers wanted to know how much pleasure rapists experienced during the commission of three specific sexual attacks and asked the assailants the following: "Assuming that '0' equals your worst sexual experience and '10' your absolutely best sexual experience, rate the amount of pleasure you experienced." Agents were surprised to learn the majority of offenders reported low levels of pleasure (3.7), and slightly over one-third of the offenders admitted to a sexual dysfunction during the attack. These findings lend credibility to the well-accepted position that rape is a crime of violence, not sex, and that it has nothing to do with the sexual appetites of offenders. Rather, rape is a mechanism for exercising power over victims, to control, dominate, and humiliate the subjects of their attacks. Although aggression and sexuality are components of all rapes, the sexual assault is the mechanism by which deep seated feelings of hostility and rage toward women are expressed. By way of his assault the rapist is telling the victim: "I can violate your body, mind, and spirit and you are powerless to stop me."

Rape Myths

The rape as sex crime misconception is the bedrock of at least two firmly entrenched myths about this crime. Rapists have been portrayed as hypersexed individuals with no access to female partners. By way of examining more than 500 rapists over a 10-year period, psychologist A. Nicholas Groth found that approximately one-third of these men were married and sexually active with their wives. The majority of single male subjects were involved with sexually consenting partners at the time they were committing sex offenses.

The most pernicious rape myth is the notion that the rape victim was somehow responsible for the rapist's behavior. Groth comments on the complete lack of foundation for this perspective noting that individuals from infants to senior citizens have been rape victims. "There is no place, no season, or time of day in which a rape has never occurred, nor any specific type of person to whom it has never happened."

Marital Rape

A 17th century British justice ruled that a husband cannot be found guilty of raping his wife "for by their mutual matrimonial consent and contract the wife hath given up herself in this kind unto the husband which she cannot retract." This logic prevailed in the United States until the 1970s. By 1993, marital rape was a crime in all 50 states under at least one section of the sexual offense codes. However, in approximately 30 states the treatment of a rape of a spouse differs from that of a non-spouse. These differences include reporting requirements (the time limit for a spouse to report a rape to authorities is shorter than for non-spousal victims) and that the spouse use force or the threat of force during attack.

Research indicates that between 10 and 14 percent of married women have been raped by their husbands. A nationwide study of violence against women concluded that about 7 million women have been raped by their intimate partners in the United States. A. Nicholas Groth noted that marital rape "may be the most predominant type of sexual assault committed." Victims of marital rape come from all social classes, racial and ethnic backgrounds, and geographic locations. One study found the rate of marital rape to be slightly higher for African American women than white, Hispanic, and Asian women.

Raquel Kennedy Bergen, one of the foremost experts on marital rape, notes that this crime generally falls into one of three categories. In force-only rape, husbands use only the amount of force necessary to sexually assault their wives. In battering rape husbands both rape and batter their wives, with the beating occurring either before, during, or after the sexual assault. Sadistic/obsessive marital rapists torture their wives and subject their wives to perverse sexual acts. They may force women to view pornography or enact what they see in pornographic photos or films. In one study, 18 percent of marital rape victims reported their children had witnessed the sexual attack, and 5 percent stated that children had been forced to participate in the sexual violence perpetrated by abusive husbands. Marital rape is often an ongoing crime, with many women reporting 20 or more attacks before the violence ends. As one observer noted: "When a woman is raped by a stranger, she has to live with a frightening memory. When she is raped by her husband, she has to live with the rapist."

As in non-marital offenses, spousal rape is a crime of power and domination committed by men who view their wives as sexual property. Battered women are more likely to be raped by a spouse than non-battered women, as physical and sexual abuse often occur together. Other risk factors include being pregnant, being ill or recently discharged from a hospital, and separation or divorce. In one study of marital rape, 17 percent of victims reported an unwanted pregnancy.

Acquaintance Rape

According to the United States Department of Justice (USDOJ) "rape is the most common violent crime on American college campuses." College women are more likely to be sexually assaulted and raped than women the same age who are not pursuing a university degree. The risk of victimization is not randomly distributed, as females at some schools have a greater risk of sexual assault than females at other institutions. Schools associated with frequent unsupervised parties, easily accessible alcohol, and single students living on their own are thought to have higher rates of sexual assault than institutions without these characteristics.

One study of two- and four-year schools discovered 35 rapes per 1000 female students over a 7-month period. This translates to 350 rapes per year on a campus with 10,000 females. Approximately 9 of 10 women college rape victims know their assailants. Rana Sampson, author of the USDOJ study, states that the frequently used term date rape is inappropriate for these crimes. Rather, college acquaintance rape is a more accurate designation, as these incidents typically occur when two people are in the same locale, rather than dating at a party or studying in a dorm room, for example.

Fewer than 1 in 20 college rape victims report the attack to campus police or local law enforcement personnel, although approximately two-thirds of victims confide in a friend. Among the reasons college women give for not contacting the police are embarrassment and shame, fear the police will not believe them, self-blame for being alone with the attacker, and their use of alcohol or drugs prior to the rape. Many rape victims mistakenly believe that if they did not physically resist their assailants, the sexual act was consensual. Victims of college acquaintance rape often drop out of school, fearing they will have to confront their assailants on a regular basis.

At least three drugs have been associated with sexual assault and acquaintance rape, substances that greatly reduce an individual's ability to resist or refuse sex. These drugs typically have no distinct color, smell, or taste and can be added to a victim's drink without his or her knowledge. So called date-rape drugs can affect victims within 10 to 20 minutes. Their potency may be facilitated by alcohol.

GHB or gamma hydroxybutyric acid can produce drowsiness, dizziness, slow heart rate, dream-like sensations, and in some cases, death. Rohypnol can cause sleepiness, a feeling of drunkenness, dizziness, confusion, problems speaking, and loss of consciousness. Individuals ingesting this substance often cannot remember what happed while they were drugged. Ketamine can produce hallucinations, a lost sense of time and identity, out of body experiences, dream-like feelings, and convulsions.

Women raped under the influence of these drugs may not have a clear memory of what happened to them, and as such, are less likely to report sexual assaults to police. Because these drugs effect cognition and memory, victims who do come forward make poor witnesses, significantly reducing the chances of convicting offenders. In addition, traces of these substances typically leave the body within 72 hours and are not detected with routine toxicology and blood tests. Medical personnel not specifically looking for date rape drugs will almost certainly miss them. Other than vigilance and trusting the people one is with, there is no defense against date-rape drugs.

Self Defense

Data suggest that individuals who resist their attackers are much less likely to be victims of a completed rape than non-resisting victims. In addition, most types of resistance are not associated with an increased risk of injury to the victim. However, there are exceptions to these findings. As one offender told Groth: "When my victim screamed, I cut her throat." The latter response is most likely to occur at the hand of a sadistic rapist whose primary pleasure comes from brutalizing women. An aggressive act of failed resistance and escape is likely to trigger an especially violent response that could result in a homicide. Five percent of Groth's subjects were sadistic rapists. Groth notes that "there is no defense strategy that will work successfully for all victims, against all offenders, in all situations, and the goal of survival is more important than the goal of escape."

While the criminal justice system has more vigorously pursued and prosecuted sexual assailants over the past 30 years, this system still has significant shortcomings. As recently as 2002, an estimated 180,000 to 500,000 rape kits across the country had not been processed by law enforcement agencies. Rape kits consist of the material—seminal fluid and pubic hair, for example—taken from victims after the attack. Obviously this material is crucial in obtaining sexual assault convictions. Speaking of the unexamined rape kits, Delaware Senator Joseph Biden noted that "The cause of this backlog seems to be pretty straightforward—woefully inadequate funding and understaffing in forensic laboratories. And the result of the backlog is clear—just delayed and sometimes denied."

ROBBERY

Robbery involves taking of something of value from an individual by force, by the threat of force, or leaving the victim in the state of fear. Unlike burglary, which requires an illegal entry but no contact with the victim, robbery is a crime of physical confrontation in which the perpetrator and target of the crime share the same physical space at the same time. Many robbery victims (especially those who resist) suffer severe personal injury. Robbery is predominantly an urban crime, with rates of this offense in metropolitan centers of more than one million inhabitants typically 25 to 30 times as high as rates in rural areas of the country. Almost 50 percent of the robberies in cities with populations over 250,000 were street/highway crimes or muggings. In rural America, muggings accounted for just over one-third of total robberies.

Robbery is overwhelmingly a young man's crime. In a typical year almost two-thirds of all those arrested for this offense are under 25 years of age, and 90 percent are males. African Americans account for just over 50 percent of total robbery arrests and whites between 40 and 45 percent of total arrests.

From 2005 to 2006, the national robbery rate increased almost 7.2 percent. Criminologist Richard Rosenfeld stated: "It strikes me that many of the cities that have experienced increases in poverty [in recent years] are the very same places experiencing increases in robbery." However, other urban areas show no correlation between poverty and robbery rates. Rosenfeld notes that economic conditions are not the sole reason for the upturn in robbery, and other factors must be considered. Contributing factors often cited are overextended police departments, released convicts reentering society, a change in the illegal methamphetamine market, and an increase in the number of males in the crime-prone 15-to-25-year-old age bracket.

Street Robberies

Jack Katz argues that what motivates many street robbers is something quite different than the need or desire for money. Indeed, the amount of money gained from muggings is relatively small and hardly worth the risk of eventually being arrested. For Katz, what drives many of these individuals is the need to establish and maintain a certain coveted identity—the badass. Each robbery, therefore, is an affirmation of the badass identity to the victim who is terrorized and intimidated, and the perpetrator who has thoroughly subjugated someone at his whim. Resistance to these individuals can be dangerous, because they are likely to view self defense as an intolerable affront to their identities.

In her study of 37 active street robbers (23 men and 14 women), sociologist Jody Miller discovered that the primary motivation for these crimes was economic, more specifically, to acquire status-conferring goods such as gold jewelry and spending money. Other motivating factors included the need for excitement, overcoming boredom, and revenge. Female street robbers commit this crime by targeting other females in a physically confrontational manner, targeting men by appearing sexually available, and/or by working with males in the robbery of other males.

Criminologists Richard Wright and Scott Decker asked what strategies offenders utilized to "compel the cooperation of would be victims" so that a successful robbery could be completed in the shortest amount of time with minimal risk to the offenders. By way of two informants, Wright and Decker contacted and interviewed 86 active street robbers in St. Louis, Missouri. A street robbery, they argue, can be broken down into four parts:

Approaching the Victim   Muggers use two approach strategies: 1) stealth and speed to sneak up on an unsuspecting victim, and 2) managing a normal appearance to get close enough for a surprise attack. Commenting on the first approach a robber stated:

> [Whoever I am going to rob, I] "just come at you. You could be going to your car. If you are facing this way, I want to be on your blind side. If you are going this way, I want to be on the side where I can get up on you [without you noticing me] and grab you . . . That's my approach."

A robber who is managing normal appearances approaches victims in a non-threatening manner.

> "Well if I'm walking and you got something that I want, I might come up there [and say], "Do you have the time?" or "Can I get light from you?" something like that. "Yeah it's three o'clock." By then I'm on you, getting what I want."

The approach method chosen by offenders is typically a function of the immediate situation as opposed to individual preferences, and most robbers are prepared to use either of these strategies. If the approach is successful, would-be victims have little chance to resist the attack, placing robbers in a position to dictate the unfolding crime.

Announcing the Crime   Offenders view announcing the crime as a make or break moment. Offenders are now committed to the crime and seek to establish dominance over the situation and victim as quickly as possible. This is usually accomplished by an "unambiguous declaration" of what is about to happen.

> [I tell my victims], "It's a robbery! Don't nobody move."

This declaration of intent is often followed by a warning of what will happen if the victim resists, or fails to do what he or she is told.

> [I say to the victim] "This is a robbery, don't make it a murder! It's a robbery, don't make it a murder!"

Wright and Decker noted the above declaration reassures victims who have been suddenly placed in what many no doubt perceive as a life-threatening situation; they believe if they submit to the robber's demands for cash and/or valuables, they will not be killed. At this stage of the crime, the robber is acting as a teacher. As one offender noted: "You have to talk to victims to get them to cooperate . . . They don't know what to do, whether to lay down, jump over the counter, dance, or whatever."

Transferring the Goods To avoid detection by passersby or the police, robbers attempt to take money and/or merchandise from their victims as quickly as possible. This task must be accomplished while keeping their prey under control. Wright and Decker found that offenders utilize two different strategies to transfer goods. The first is a command to relinquish their valuables.

> I tell [my victims], "'Man, if you don't want to die, give me your money! If you want to survive, give me your money! I'm not bullshitting!' So he will either go in his back pocket and give me the wallet or the woman will give me her purse."

This approach permits the robber to watch for any signs of danger.

Utilizing the second strategy, robbers simply take what they want from their targets. One offender noted a significant advantage (from his perspective) of this approach.

> "I don't let nobody give me nothing. Cause if you let somebody go in they pockets, they could pull out a gun, they could pull out anything. You make sure they are where you can see their hands at all times."

While most robbers have no intention of killing their victims, they (the robbers) will respond with severe-but non-lethal force if necessary. However, in a tense situation, a shot intended to wound may well prove fatal. As one respondent stated: "When you're doing stuff like this, you just get real edgy; you'll pull the trigger at anything, at the first thing that go wrong."

Making an Escape   Offenders can terminate the crime in two ways: they can leave the scene themselves, or force their victims to flee, with most robbers preferring the former. Before making their escape, offenders threaten victims with injury or death should they choose to pursue them. One robber stated:

> "I done left people in gangways and alleys and I've told them, If you come out of this alley, I'm gonna hurt you. Just give me five or ten minutes to get away. If you come out of this alley in 3 or 4 minutes, I'm gonna shoot the shit out of you."

Some offenders take a victim's identification and threaten to come to his or her home and kill them if they call the police. Other robbers rough up their victims, making pursuit difficult.

> [I hit my victims before] "I escape so as to give them less time to call the police . . . You hit them with a bat just to slow his pace."

Although most robbers wanted to leave the crime scene first, some opted for a strategy of ordering the victim to flee instead. This permitted offenders to depart in a leisurely manner without drawing attention to themselves. One robber who opted for this plan stated:

> "I try not to run away. A very important thing that I have learned is that when you run away, too many things can happen . . . Police could be cruising by and see you running down the street . . . I tell the victim to walk and not look back."

Wright and Decker conclude that street robbery is an "interactive event" with offenders attempting to control the situation via "tough talk, a fierce demeanor, and the display of a deadly weapon." While few robbers want to kill their victims, most will resort to violence, "and some are clearly prepared to resort to deadly force if need be."

While the overwhelming majority of robberies occur in the streets or places of business, about 10 percent of these crimes occur on private premises. This category of robbery involves breaking into homes or being admitted into a place of residence by an unsuspecting individual. Residential robberies are often committed by street gangs. Residents may be intimidated by the threat of force and left unharmed, or they may be physically assaulted, or, on occasion, murdered.

John Roman and Aaron Chalfin offer anecdotal evidence (that is, correlational rather than causal evidence) that an increase in the number of targets may contribute to a spike in robberies. Between the spring of 2004 and December 2006, the number of iPods Apple sold increased from 3.7 to almost 90 million units. During the first three months of 2005, major felonies rose 18.3 percent on the New York City subway. However, if cell phones and iPod thefts are excluded, felonies declined by 3 percent in that venue. Similar iPod-related robbery increases occurred in London and other cities. Roman and Chalfin argue there a four reasons why iPods were "crime creating" before the mini robbery boom related to this product ran its course: 1) these devices have no easily accessible anti-theft protection; 2) unlike cell phones, which require a subscription or contract to utilize, an offender can use a stolen iPod after the robbery has been reported; 3) because they are high status items, iPods were stolen for use by offenders and not resold; 4) as iPods transmit sound through both ears (rather than in one for cell phones), they make users less aware of their surroundings, and more vulnerable to victimization.

Just as high-priced, high-status basketball shoes and North Face jackets were stolen in the 1980s and 1990s, iPods were coveted from 2004 to 2006, and "unlike a jacket or a sneaker, one size fits all." It was the rapid proliferation of iPods rather than an increase in the number of motivated offenders that was primarily responsible for the spike in robberies. In other words, new, high-value, high-tech products can result in unexpected crime costs and risks for consumers. Roman and Chalfin conclude that the iPod crime wave:

> "was predictable and could have been prevented or mitigated, and yet little was done, or is being done, to slow the wave before it washes out on its own. U.S. crime policy is overwhelmingly focused on increasing the cost of committing crime for would-be offenders and pays little attention to the behavior of potential victims. As technology races ahead, we should expect to see more iCrime-like waves."

THE CHURCHES AND SEXUAL ABUSE

Prior to 2002, allegations against clergy members for sexually molesting children appeared periodically in newspapers. Church officials would invariably point out that if true, the priest, minister, or rabbi in question was an isolated case, a bad apple in an otherwise well-adjusted order of men doing God's work. Then in the spring of 2002, one story after another broke of sexual abuse on the part of Catholic priests in many of the church's 195 dioceses across the country. Over the next two years the Church was embroiled in arguably the worst scandal in its history in the United States.

The Roman Catholic Church commissioned the John Jay School of Criminal Justice in New York to investigate and prepare a report on the breadth of the scandal. Gerald Lynch, president of John Jay College stated the study was "accurate and comprehensive" noting that "This was not a sampling. We had an entire population." Principal investigator for the study Karen Terry said that "it is possible the bishops are not giving us everything." The following is a list of the major findings:

- Between 1950 and 2002, approximately 4400 of the 109,694 priests in this country (about 4 percent) were accused of sexually abusing children.
- For the entire 52-year period under investigation "the problem was indeed widespread and affected more than 95 percent of the dioceses and approximately 60 percent of religious communities."
- Just over 3 percent of accused priests had 10 or more victims, and of these, 149 priests accounted for almost 3000 victims, or 28 percent of allegations.
- Almost 70 percent of the allegations were made against priests ordained between 1950 and 1979, while priests ordained after 1979 accounted for less than 11 percent of the allegations. Two of every five accused priests were between 30 and 39 years of age.
- Victims, and typically the family of victims, were known to the priest prior to the molestation, with attacks occurring in the priest's home (41 percent), in church (16 percent), and other places (43 percent).
- Four of five victims were males, and almost two of three victims were between 9 and 14 years of age.
- Of the approximately 10,700 allegations, 63 percent were found to be true, 6 percent not true, and 31 percent were not investigated because the accused priest had died.

- The report found that only 3 percent of all priests against whom allegations were made were convicted and about 2 percent received prison sentences.

Victims' advocates were critical of the report, noting that there were untold numbers of wounded adults carrying the pain of sexual abuse who never reported their victimization. As is the case with other crimes, unreported victimizations cannot be determined with any degree of accuracy.

Abuse victims, the Catholic laity, and the public roundly condemned bishops and the church hierarchy for ignoring the problem. The John Jay report found that no action was taken in 10 percent of the allegations, while in 6 percent of cases the priest was reprimanded and returned to his ministry. In 29 percent of allegations the priest was suspended, and in another 24 percent of cases the priest was placed under administrative leave.

The extent and severity of the abuse, as well as the ongoing coverup, were painfully evident in a 418-page report issued by a Philadelphia grand jury in September, 2005. After sifting through 45,000 pages of documents from secret archdiocese archives, the report concluded that Cardinal Anthony Bevilacqua and the late Cardinal John Krol "excused and enabled the abuse by burying the reports they did receive and covering up the conduct . . . to outlast any statutes of limitations." While church documents referred to the sex crimes of priests in euphemistic terms, as "inappropriate touching," for example, the grand jury report had a different interpretation of these acts. "We mean rape. Boys who were raped orally, boys who were raped anally, girls who were raped vaginally." The report states that the "handling of the abuse scandal was at least as immoral as the abuse itself . . . the molestations, the rapes, the lifelong shame and despair . . . were made possible by purposeful decisions carefully implemented policies, and calculated indifference."

In addition to tremendous adverse publicity, the sex crime scandal also cost the Catholic Church (and insurance companies) a great deal of money. A survey conducted by the Center for Applied Research in the Apostolate at Georgetown University found that in 2006 almost $400 million was spent on settlements with victims, legal fees, and other costs. In July, 2007, the Archdiocese of Los Angeles paid $660 million to settle sex abuse claims. Although the exact figure is unknown, according to one estimate, clergy sexual abuse has cost the Catholic church more than $2.3 billion since 1950.

Since 2004, five dioceses (Tucson, Spokane, Davenport, Portland, and San Diego) have declared bankruptcy as a consequence of sexual abuse lawsuits. In February, 2007, the diocese of San Diego filed for bankruptcy protection only hours before a suit on behalf of 140 alleged victims was to be presented to a judge. As a consequence of a Chapter 11 (bankruptcy) reorganization in the Spokane, Washington diocese, the faithful of 82 parishes are being asked to contribute $10 million to settle claims brought by 177 sexual abuse victims.

Not Just Catholic Priests

Although the Catholic Church has received the bulk of the negative press coverage, scrutiny, and criticism regarding abusive priests, clergy members of other religions have also been accused of abusing children. According to Reformation.com, 838 ministers from various Protestant denominations, including Baptist (147), "Bible" Church Ministers (251), Episcopalian (140), Lutheran (38) Methodist (46) Presbyterian (19), and other faiths (197) were alleged to have abused children in recent years.

Based on a survey of about a thousand churches nationwide, Christian Ministry Resources (CMR), a tax and legal advice publisher serving more than 75,000 congregations, reported that "Despite headlines focusing on the priest-pedophile problem in the Roman Catholic Church, most American churches being hit with child sexual-abuse allegations are Protestant, and most of the alleged abusers are not clergy or staff, but church volunteers." A 2000 report to the Baptist General Convention in Texas stated that "The incidence of sexual abuse by clergy has reached horrific proportions." In the 1980s, the study notes, about 12 percent of ministers had "engaged in sexual intercourse with members" and almost 40 percent had "acknowledged inappropriate sexual behavior." According to the report, the incidence of clergy sexual abuse "exceeds the client-professional rate for physicians and psychologists." In their book *Ministerial Ethics*, Joe E. Trull and James E. Carter note that about "30 to 35 percent" of ministers from all denominations admit to having sexual relationships outside of marriage from inappropriate touching to sexual intercourse. Although many if not most of these encounters may be considered sexual relations between consenting adults, it's not clear to what extent clergy members took advantage of their status and the vulnerability of congregation members, someone in a psychological and/or spiritual crisis, for example.

Philip Jenkins, the author of *Pedophiles and Priests*, states, "You name me a denomination and I'll give you a case. Some [denominations] with huge problems include Mormons, Jehovah's Witnesses, Buddhists, Jews, Baptists, Pentecostals, Episcopalians—you name them."

Jenkins argues there is no evidence that Catholic priests have higher rates of child-related sexual abuse than any other group of clergy or non-clergy who work with children even though the Roman Catholic church received the most attention. "There's no evidence either way. If somebody says, 'Well it's obvious, they do,' I say, 'Fine, give me the evidence,' and the evidence isn't there."

The sexual abuse of children has not triggered the interest or outrage of the public to the extent this behavior has when committed by Roman Catholic priests. Jenkins cites the case of Pentecostal minister Tony Leyva who molested hundreds of boys in the 1980s. Yet few people have any knowledge of his longstanding criminal career.

New York Times reporter Laurie Goodstein writes that a Jehovah's Witness in Kentucky who attempted to form a group to monitor child sexual abuse allegations after a fellow elder abused a child was excommunicated from the church. Three fellow elders found William Bowen guilty of "causing divisions." His punishment was "disfellowshipping" or shunning. Bowen stated that church officials informed him of a database containing the names of 23,000 members and associates in the United States, Canada, and Europe who have been accused or found guilty of child abuse. Church officials stated that number is "considerably lower" but refused to divulge the figure.

In a three-month period in 2002, four other Witnesses who accused the church of covering up the sexual abuse of children were expelled. Carl Raschke, who has written about Jehovah's Witnesses, stated: "Groups that tend to be very tight-knit and in-grown historically have a higher incidence of sexual abuse and incest. That's an ethnological fact." According to Raschke, religious organizations claiming to be "thoroughly holy or godly" are loathe to admit that church members are not living up to the dictates of their faith.

Rabbi and law professor Arthur Gross-Schaefer believes the rate of sexual abuse among Jewish clergy is approximately the same as that found among Protestant clergy. According to Rabbi Shaefer, "Sadly, our community's reactions up to this point have been often based on keeping things quiet in an attempt to do 'damage control' . . . Victims tell me that no one really wants to listen to their story—not the board of directors, nor rabbinic colleagues, nor the Jewish community." In a paper entitled "Clergy Abuse: Rabbis, Cantors & Other Trusted Officials," the Awareness Center, the Jewish Coalition Against Sexual Abuse/Assault discusses clergy abuse in the Jewish community.

Halting the Abuse

In 2002, Minneapolis psychotherapist Gary Schoener, who has handled more than a thousand clergy sexual abuse cases, stated that mainline Protestant denominations have been the most responsive in adopting measures against clergy abuse. National policies have been implemented by Episcopalians, Presbyterians, Methodists, and Lutherans among others. Los Angeles Times reporter Teresa Watanabe notes that the Roman Catholic response to this crisis "has varied dramatically, in part because each of the nation's 195 American dioceses operates independently."

Roman Catholic Bishop Gregory M. Aymond, chairman of the Committee for the Protection of Children and Young People, recently stated that the decline in the number of victims and allegations in 2006 indicates that "what we are doing in creating safe environments is working." The church spent $25.6 million on child protection efforts in 2006, a 33 percent increase from 2005. Nevertheless, James Cobble, executive director of CMR, offers a more pragmatic and far less altruistic interpretation of the self-policing efforts by religious authorities of all persuasions. "What drove leaders to respond to this issue," he stated, "was not the welfare of the children. It was the fear of large, costly lawsuits."

KEY TERMS

first degree murder: The willful, deliberate, and premeditated killing of one human being by another, with malice aforethought; in states with capital punishment, this crime carries the death penalty

homicide: The killing of one human being by another; not necessarily a crime

manslaughter: Unlawful killings that are committed without malice aforethought, and, as such, are considered less severe and more explainable

mass murder: The killing of three or more people by one or more assailants within a single event lasting from a few minutes to several hours

rape: Forced sexual intercourse, including both psychological coercion as well as physical force

serial killing: The killing of three or more people in separate events with period between murders that can last for days, weeks, months, or years

robbery: Taking something of value from an individual by force or the threat of force

SUGGESTED READINGS

Brownmiller, Susan. 1975. *Against Our Will: Men, Woman, and Rape.* New York: Simon & Schuster.

Fox, James Alan, and Jack Levin. 2005. *Extreme Killing—Understanding Serial and Mass Murder.* Thousand Oaks, CA: Sage Publications.

Groth, A. Nicholas. 2001. *Men Who Rape: The Psychology of the Offender.* New York: Basic Books.

Katz, Jack. 1988. *Seduction of Crime.* New York: Basic Books.

LaFree, Gary. 2000. *Losing Legitimacy: Street Crime and the Decline of Social Institutions in America.* Boulder, CO: Westview Press.

Wright, Richard T, and Scott H. Decker. 1997. *Armed Robbers in Action: Stickups and Street Culture.* Boston: Northeastern University Press.

"Speed Bump" © Dave Coverly/Dist. By Creators Syndicate, Inc.

chapter five

Hookers, Dopers, and Corporate Crooks: Economic, Exploitive, and Consensual Crime

Violent street crimes—robbery, rape, and murder—share one distinguishing feature: offenders and victims share the same physical space during the commission of the crime. Most corporate crime is nonviolent, and offenders and victims are separated by time and space. Every year hundreds of thousands of Americans are victims of burglary and motor vehicle theft. Perpetrators of these crimes are often repeat offenders, many committing multiple crimes before they are apprehended, and many are never arrested. In this chapter we will examine the methods and motives of these career criminals.

Much has been written about the relation between technology and crime. From the law enforcement side, the use of squad cars and radio communication from the early years of the 20th century to high-tech crime scene analysis (DNA testing) in the 21st have been used to apprehend and prosecute offenders. However, the bad guys also make use of technology. Nowhere is this more evident then in the sophisticated use of computers in identity theft, one of the fastest growing and personally damaging crimes of the past decade.

Rates of illegal drug use in this country ebb and flow as a host of mind-altering substances become more or less popular and as different drugs become legal or illegal. We will analyze the relation between drugs and crime as well as examine the use and distribution of methamphetamine and cocaine. Our discussion of drug use in America concludes with an overview of the drug decriminalization debate.

The mostly consensual crime of prostitution has hundreds of thousands of participants in this country and is related to other criminal activity (robbery and drug use, for example) as well as a number of health issues, including the spread of sexually transmitted diseases, most notably HIV-AIDS. An especially perverse and exploitive dimension of this activity is child prostitution and international sex-tourism.

BURGLARY

Burglary is the unlawful entry of a structure to commit a felony or theft. The entry may or may not be forced (breaking a window, for example). Roughly two-thirds of burglaries known to police are residential crimes, with the remaining one-third involving commercial structures. Offenses for which the time of the crime was reported in 2006 reveal that 61 percent of residential burglaries take place in daylight hours, while 58 percent of non-residential burglaries occur at night. As is the case with robbery, burglary is overwhelmingly committed by young men. In a typical year, almost 9 out of 10 suspects arrested for burglary are males, and more than two-thirds have not reached their 30th birthday. Almost 70 percent of arrestees are white.

Why They Do It

In their study of 105 active burglars in St. Louis, Missouri, Richard Wright and Scott Decker discovered that these street criminals did not break into homes for money to realize some long-range goal. Rather, their crimes were motivated by more immediate financial problems. Money derived from criminal activity was spent in the following manner:

1. To keep the party going: Almost three of four subjects said they used money derived from burglary for good times, which typically involved purchasing drugs, most often cocaine.
2. To keep up appearances: Approximately half of the burglars who stole for financial reasons stated they used the money to buy status items, especially clothing. One subject noted, "See, I go to steal money and go buy me some clothes. See, I likes to look good. I likes to dress."

3. To keep things together: About half of the burglars interviewed said they spent a portion of the money they made from breaking into homes for daily living expenses such as food, shelter, and clothing for their children. A minority of the subjects spent all their illegally derived profits for this purpose.

The most important decision a burglar makes concerns target selection. In a study of 30 Texas burglars, sociologist Paul Cromwell and his colleagues concluded that these offenders consider three risk cues in the target selection process: surveillability, occupancy, and accessibility. Surveillability cues include location of the dwelling on the street, overall visibility of the target from the street and nearby homes, and, more specifically, the visibility of points of entry, such as doors and windows. If a burglar believes that he or she is not likely to be observed making an entry, the offender determines if the occupants are present. Of the 30 subjects in this study, 28 said they would never knowingly enter an occupied residence. Principal occupancy cues include a vehicle in the driveway, visible residents, and noises or voices coming from the home. Accessibility cues indicate how difficult it will be to enter the targeted home. Will entry be relatively simple, or is the home protected ("target hardening") by deadbolt locks, burglar bars, burglar alarms, wall, and/or dogs?

Criminologists often make a distinction between novice or amateur burglars and professionals. Novices are typically young, lower-, or middle-class males who will engage in only a few burglaries and then abandon this criminal activity, or are at the beginning of an extended criminal career. Terrance Miethe and Richard McCorkle note that "novices are not firmly entrenched in a burglar identity or criminal subcultures." The professional elite of the burglary world are relatively few in number. They differ from amateurs in terms of technical skills, organizational abilities, and the status accorded them by other criminals and police. These individuals also have well-established contacts for selling stolen merchandise quickly.

In many instances, skills learned in a noncriminal enterprise are effectively translated to illegal activities. Consider the following statement from a professional burglar.

The military taught me what I needed to know as a burglar. Planning, that's what I learned in the Army. Laying out a map in your head, getting it all together and knowing who you're going to unload the stolen goods on before doing anything is also important. I guess my training in the Special Forces taught me to be sneaky and rehearse things in my mind ahead of time because, you know, you're scared. The military taught me to have confidence in myself.

Nationwide, people attempting to protect their homes from burglars via alarm systems have become a serious drain on law enforcement resources. Annually, the time lost responding to false alarms is equivalent to the total work output of 35,000 officers, about 6 percent of the country's law enforcement personnel. Rana Sampson of the *Center for Problem-Oriented Policing* notes that "Each false alarm requires approximately 20 minutes of police time, usually for two officers. This costs the public hundreds of millions of dollars," money that in most jurisdictions is not recouped through fines. Burglar alarms are triggered because home owners do not understand their systems, the system was improperly installed, batteries are dead, or pets or insects move across motion detectors.

MOTOR VEHICLE THEFT

Motor vehicle theft is the attempted or completed theft of an automobile, truck, bus, or motorcycle. In the post-World-War-II era this has been one of the fastest-growing crimes in the United States, with the rate per 100,000 jumping from 108 in 1948 to 658 in 1990—a sixfold increase—before falling to just under 400 in 2006. Three of four vehicles stolen are cars, and almost all (93.5 percent) are stolen in cities. The highest rate of motor vehicle theft is in the western states, approximately three and a half times that of the northeastern states. In 2006, five of the ten metropolitan areas with the highest rate of this crime were in California—Stockton, Visalia-Porterville, Modesto, Sacramento, and Fresno. Las Vegas, Nevada had the highest rate of motor vehicle theft that year.

Arrest data indicate that the typical car thief is a white male under 25 years of age with a prior arrest record. However, because this profile is based on such a small percentage of car thieves (approximately 85 percent of motor vehicle thefts are not cleared by arrest) it must be viewed with caution.

Types of Auto Theft

In an effort to categorize the various forms of motor vehicle theft, criminologists and law enforcement officials have constructed a number of typologies. One of the most widely used is offered by the New York State Division of Criminal Justice Services. Researchers found that there are two fundamental motivations for this crime—unauthorized use, and profit—each of which has three subcategories.

The three types of unauthorized use are joyriding, transportation, and the commission of these crimes. Joyriding is the theft of a vehicle for short term enjoyment. These vehicles are recovered quickly, most often in the community where they were stolen. Transportation is the theft of a vehicle for personal for use, to transport the offender from one location to another. Stolen cars are abandoned at the final destination. Offenders also steal vehicles for the commission of crimes. For example, stolen cars are used by offenders who stage or create automobile accidents for the purpose of defrauding insurance companies.

Thefts for profit account for the overwhelming number of motor vehicles stole in the United States. Vehicles may be stolen for parts, for resale, or for export.

In some cities, 50 percent or more of stolen vehicles are taken to chop shops and disassembled. The value of individual auto parts may be much greater than the value of an intact vehicle. A car with a resale value of $6,000 can be worth $15,000 to $20,000 when dismantled and sold as parts. This is especially true of automobiles such as the 1995 Honda Accord, the vehicle with the highest theft rate in 2006 according to the National Insurance Crime Bureau. Because there are so many cars of that make, model, and year on the road, there is a high demand for parts to keep them running. Stolen parts are sold at a reduced rate to the customer or an insurance company paying full price. This criminal enterprise depends on illegal parts distributors and thousands of unscrupulous auto parts store owners who buy stolen goods. (While owners may not know for sure that a part is stolen, the heavily discounted prices are a reasonable indication.) Thieves also sell stolen auto parts at swap meets. Speaking of stolen Honda parts in his city, a Dallas police officer noted, "Would you rather pay $500 in a store or $50 on the black market?" Thieves also use the Web to sell stolen auto parts.

Professional thieves working the lucrative buy back or strip and run scheme strip a car completely and discard the frame, which the police eventually recover. Declared a total loss by the insurance company, the car frame (no longer considered stolen property) is eventually resold at an auction. Car thieves track the targeted vehicle (frame), and when it comes up for sale, buy it for next to nothing. Now legal, the car is reassembled with the original stolen parts and sold.

There is a growing number of thefts for small auto components, especially radios, DVD players, high-end rims, air bags, and, more recently, electronic control modules or ECMs, auto computers that command numerous vehicle functions. In addition, xenon gas headlights and expensive leather seats are in demand and will increasingly be targeted by car thieves.

A much-used strategy for reselling stolen vehicles is title washing, wherein the title of a stolen car is quickly transferred from one state to another "in order to disguise the history of the vehicle and confuse the ownership trail." In some cases, four or five titles are obtained in less than a week, with the final clean title used in the sale of a stolen vehicle to an unsuspecting customer. The New York state report notes that car thieves utilizing this scheme rely on loose and inconsistent vehicle title laws in this country.

Stolen cars are increasingly appearing for sale on the Web. In some case the same vehicle is sold to numerous individuals, with thieves keeping the initial payment (or the entire cost of the vehicle) without delivering the car. In other instances the car is sold with improper or fraudulent ownership papers.

Cars also may be stolen for export. Organized crime groups steal luxury cars, late-model mid-size sedans (especially Hondas and Toyotas), and parts for export to eastern Europe, the Caribbean, and Asia. As most countries in these regions have no laws prohibiting the sale of vehicles stolen in the United States and Canada, this is a very lucrative business. In some cases thieves will lease or buy a new vehicle, putting down the least amount of money necessary to complete the transaction. Once the title is obtained, they will remove the lienholder from the document, or forge a lien release letter and obtain a clean title. Presenting the now apparently legal title to customs officials, the vehicle is cleared for export.

Although anti-theft alarms and devices such as steering wheel locks or bars, have contributed to declining rates of auto theft, these mechanisms are no match for a skilled offender. A thief will gain entry to a locked vehicle by breaking the driver's side window. If the alarm goes off, he opens the hood and disables the alarm. If the thief encounters a steering wheel lock or bar, he can cut through the steering wheel, and if necessary, break the steering column. A New York City police detective noted that it takes auto thieves usually less than 20 seconds to start a car. A vehicle disabled during the course of a crime may be towed away by well-prepared offenders.

Sophisticated electronic devices use a hidden transmitter to help law enforcement personnel track a stolen vehicle. According to the Insurance Information Institute, electronic tracking devices "not only help police find individual stolen vehicles, but they lead them to chop shops, which thwart the export of stolen motor vehicles and lead to the recovery of expensive construction vehicles as well as passenger cars." To date car thieves have not neutralized transmitter type devices as successfully as they have other anti-auto-theft products.

CORPORATE CRIME

There are two common misconceptions about corporate crime. First, most people believe significantly more money is lost via street crimes such as robber, burglary, and auto theft. In fact the opposite is true. Criminologist Jeffrey Reiman estimates that in 1997 the cost of corporate crime in this country was $338 billion dollars, while the combined losses from all motor vehicle thefts in 2006 was $7.9 billion. The General Accounting Office of the federal government estimates that healthcare fraud alone costs Americans about $100 billion annually. Corporate crime is the most cost-effective criminal activity; it makes the illegal drug trade look like a nickel and dime operation.

Second, most people believe that corporate crime is not violent. Nothing could be farther from the truth. According to the National Commission on Product Safety, approximately 20 million Americans are injured each year as a consequence of faulty products. (According to one estimate 467 faulty products were recalled in 2006.) In addition, as many as 110,000 people are permanently disabled and 30,000 are killed. The latter number is considerably higher than the 15,000 to 20,000 homicide victims in the United States annually over the past 10 years.

Gilbert Geis notes that corporate offenses against the environment or environmental crimes result in the "compulsory consumption of violence" on the part of the public. Whereas people may be able to distance themselves from street predators through home and auto alarm systems, in most instances they can do little if anything to protect themselves from the harmful effects of environmental crimes. For example, the Rockwell International Corporation pleaded guilty to illegally disposing of radioactive waste near the Rocky Flats nuclear weapons site a short distance from Denver. Company employees dumped highly toxic and radioactive waste in streams that flowed through the industrial complex and then falsified records submitted to state and federal health officials.

The four most common forms of environmental crime are water pollution, air pollution, the unlawful storage and transportation of hazardous materials, and the illegal trafficking of wildlife. The FBI notes that as environmental laws become more restrictive and the cost of properly disposing hazardous material increases, the incentive for the illegitimate disposal of these materials also increases. At any one time the FBI is investigating approximately 450 environmental crimes, about 50 percent of which are violations of the Clean Water Act. Environmental crimes are increasingly coming to the attention of local law enforcement officials. The Los Angeles County Environmental Crimes Strike Force has permanent representatives from 20 state and local agencies. However, as federal and state law enforcement officials have launched a more aggressive campaign against environmental criminals, these individuals are becoming more sophisticated in the methods they utilize to circumvent the law. For example, some firms shield their involvement in these crimes by the use of intermediaries and dummy corporations. In addition, many corporate environmental attorneys are former prosecutors with the expertise to protect their clients.

Marshall Clinard and Peter Yeager argue that historically, three industries seem to violate government regulations and laws more than any others. The oil industry has been involved in restrictions of independent dealers, contrived shortages, misleading advertising, and interlocking directorates. The auto industry has engaged in deceptive advertising, unreliable warranties, unfair dealer relations, and violations of safety standards. The pharmaceutical industry has been guilty of false advertising, inferior product quality, improper research and inspection, and excessive markups.

In his investigation of the pharmaceutical industry, Australian criminologist John Braithwaite found evidence of "substantial bribery" in 19 of the 20 largest American drug companies. Bribes were paid to government officials who might work for the benefit of these corporations, to [Australian] cabinet ministers to get the drugs approved for marketing, to health inspectors, "to customs officials, hospital administrators, tax assessors, political parties, and others." Braithwaite argues that as serious as these crimes are, they pale in comparison to the corruption found in the testing of drugs: "Rats die in clinical trials on new drugs and are replaced with live animals; rats which develop tumors are replaced with healthy rats; doctors who are being paid $1,000 a patient to test a new product pour the pills down the toilet, making up the results in a way which tells the company what it wants to hear."

Former U.S. Attorney General John Aschcroft stated: "Corrupt corporate executives are no better than common thieves." No better, but much wealthier, and much less likely to be imprisoned than street criminals. Despite the high-profile convictions of Dennis Kozlowski (former CEO of Tyco), John Rigas (former CEO of Adelphia), and the late Ken Lay (former Chairman and CEO of Enron), corporate crime in the United States goes largely unpunished. The Multinational Monitor, an organization that examines the behavior of international corporations, reports that between 2000 and 2005, district attorneys in this country chose not to prosecute corporate crime, but to make deals with offenders and let them escape conviction.

Used mostly in securities and banking fraud, these deals are known as non-prosecution agreements and deferred prosecution agreements. "With a non-prosecution agreement, the prosecutor agrees not to charge a company criminally, in exchange for fines, cooperation and changes in corporate structure. With a deferred prosecution agreement, the prosecutor files a charge, but agrees to drop the charge if the company abides by promises to the prosecutor." The article notes that this practice establishes a double standard for criminal justice: individuals who commit crimes are arrested, convicted, and punished (often with lengthy prison sentences), while corporations having committed crimes that can cost the public billions of dollars escape with modest fines and promises not to repeat their crimes.

Criminologist Henry Pontell, co-author of *Profit Without Honor: White-Collar Crime and the Looting of America*, states that "There are too many ways for [white-collar criminals] to escape the ultimate sanction." These offenders hide behind high-priced legal talent and the many diffuse layers of corporate responsibility.

IDENTITY THEFT

The many scholarly, government, and popular press articles on identity theft routinely begin by noting that this is the fastest growing crime in the United States. Unfortunately, estimates of how fast it is escalating and how much money consumers lose each year vary widely. For example, estimates of the number of identity theft victims annually range from 700,000 to 10 million while consumer losses from this offense are reported to be between $745 million to $5 billion, with the average victim losing from $1000 to $17,000. Whatever the true cost, identity theft may already be, as noted by a U.S. Justice Department prosecutor, "the single greatest type of consumer fraud."

Identity theft has been called an equal opportunity crime, affecting people of all incomes, races, and ages. Not only do people lose money to identity thieves, but offenders often commit crimes using the names of their victims. According to Robert Richardson of the Computer Security Institute "ID theft usually occurs not because of the carelessness of individual consumers, but because of the carelessness or vulnerability of organizations they deal with, including the government." Businesses, credit bureaus, and government agencies do not safeguard their databases and often use social security numbers (the most valuable piece of information in stealing someone's identity) when other identifiers would suffice. In addition, these organizations do not notify consumers when their databases have been breached by hackers, and typically fail to help victims of identity theft.

How They Do It

Consumer Reports magazine noted the most common ways identity thieves steal information:

1. *Stealing company data:* In 1993, Visa, MasterCard, and American Express stated that an unknown hacker had accessed as many as 8 million credit card records. A financial services organization estimated that approximately one percent of those records, or 80,000 consumers, would likely become victims of identity theft.

2. *Pretexting:* Telemarketers (phone solicitors), store clerks, and computer spammers use a false pretense to gather personal information from consumers. This can include credit card numbers and other account information.

3. *Dumpster diving:* Criminals dig through trash for discarded mail that includes credit card bills, social security numbers, medical statements and other personal information that can be used to obtain credit in your name and/or access existing accounts.

4. *Mail theft:* Criminals and criminal organization steal mail from unlocked mailboxes to find letters with personal information and preapproved credit cards.

5. *Account takeover:* Thieves use stolen or fake identity cards and drivers' licenses to take over existing credit card accounts. They avoid detection by filling out change of address forms at the post office and having the victim's mail sent to a private mailbox or new address.

6. *Skimming:* Using hand-held magnetic card readers that can be purchased on the Web, identity thieves obtain personal information that is encoded on the cards' magnetic strips. Culprits in this scam include waiters, gas station attendants, and store clerks who are paid per card by

organized crime rings. Some automatic-teller machines have been rigged to skim personal identification numbers.

7. *Phishing expeditions:* By way of emails and pop-ups that sound and look official—banks, credit institutions, and recognizable corporations—con artists have tricked millions of people into sending them personal information such as social security numbers, credit card numbers, and bank account information. Allan Paller, research director of the SANS Institute, and organization that trains security personnel states that over the past few years organized crime has figured out that phishing "is a better way to make money than selling drugs." In spear-phishing, the sender appears to be someone inside the company for which the victim works, or informs a potential victim that his or her bank account has been compromised, and instructs the victim to email personal information to the institution via a Web site that looks legitimate. Microsoft has identified four common email phishing come-ons:

- "Verify your account." No legitimate business should ask for passwords, social security numbers, or other personal information through email.
- "If you do not respond within 48 hours your account will be closed." These messages convey a sense of urgency in the hope that individuals will send personal data without thinking.
- "Dear Valued Customer." This message implies that you are a customer, although the salutation does not contain your first or last name.
- "Click below to gain access to your account." This message appears to come from a legitimate business but is masked, that is, the link will take you, and your personal information, to a phony Web site.

8. *Robot networks:* One of the most difficult forms of identity theft to combat is the work of "bot-herders" who surreptitiously install malicious software on computers—viruses, worms, and Trojan horses—by way of what appears to be a harmless program. Once the software has been loaded, hackers can control the computer remotely. Users unwittingly grant access to bot-herders when they click on an advertisement, open an email attachment, or provide information via a phishing Web page that appears to be a legitimate site. Assistant Director of the FBI James Finch notes "The majority of victims are not even aware that their computers have been compromised or their personal information exploited." The Bureau estimates that as of November 2007, approximately 2.5 million computers had been victimized by robot-controlled botnets. FBI Director Robert Mueller considers botnets the Swiss army knives of cyber crime as this criminal activity is so versatile. "A botnet could shut

down a power grid, flood an emergency call center with millions of spam messages, or disable a military command post."

9. *Raiding your old computer:* Graduate students at the Massachusetts Institute of Technology were able to recover sensitive data from the hard drives of about one-third of the computers they tested. Consumer Reports bought hard drives on Ebay and attempted to recover data using an easily obtained, inexpensive software program. They found "a Microsoft Word tax document that included annual salary information, Quicken files filled with income and expense data; email from Outlook Express; love letters; photographs; and lists of favorite Web sites."

Victims of identity theft find that putting their financial lives in order is a time-consuming nightmare of making seemingly endless phone calls and writing dozens of letters. According to the Identity Theft Resource Center, the average victim spends 607 hours—the equivalent of almost 15 full-time work weeks—attempting to clear his or her credit records.

In a fast-paced, information-driven society, it is likely the number of identity thefts annually will only increase. To date there are no coordinated attempts by local, state, or federal agencies to deal with this crime. Former New York State Attorney General Elliot Spitzer stated that if identity theft is to be stopped, "Anyone who stores information needs to do more" and "federal legislation is going to be necessary."

Medical Identity Theft

This most pernicious form of identity thievery is usually defined as the "use of a person's name or insurance information without consent to obtain medical goods or services, or to make false claims for goods and services." The World Privacy Forum, a non-profit public interest and research group, states that medical theft is "deeply entrenched in the healthcare system . . . and may be done by criminals, doctors, nurses, hospital employees, and increasingly, by highly sophisticated crime rings." Victims of this crime may receive inappropriate medical treatment, discover that their health insurance benefits have been exhausted, and/or be unable to secure health and life insurance. They may fail job-related physicals due to the recording of diseases in their health records they do not have. According to the Federal Trade Commission there are approximately 200,000 instances of medical identity theft annually in this country.

Anndorie Sachs received a phone call in April, 2006, from the Utah Social Services Department noting that her hospitalized infant son had tested positive for methamphetamines. The mother of four was both surprised and stunned as she hadn't given birth for years. The next day authorities showed up at her home and threatened to remove her children. "As much as I denied it," she stated, "they just kept insisting that yes, I was the mother of this child . . ." The matter was eventually resolved when officials discovered that someone had stolen Sachs' driver's license, checked into a local hospital and had a baby. The Salt Lake City resident was forced to hire a lawyer to untangle her medical records and ward off debt collectors wanting her to make good on $10,000 worth of bills.

A psychiatrist in Boston made false entries in the charts of individuals who were not his patients. He gave these "patients" diagnoses of drug addiction and severe depression over the course of several billed sessions these individuals never had, then used this information to submit false claims in insurance companies. A Colorado man whose Social Security number and other personal information were stolen was confronted with a $44,000 bill for a surgery he never had. Medical identity theft can be life threatening as a Florida woman discovered. Someone made false entries in her health records including changing her blood type. Victims of these offenses have also had their medications changed as well as their underlying medical conditions.

Medical identity theft may go undiscovered for years, and in many cases, will never be detected. And when this crime is discovered it may be attributed to sloppy bookkeeping or the errors of a hospital employee. One of the most frightening aspects of medical identity theft is that victims do not have the same rights and resources to fight this crime as do victims of financial identity theft. With the passage of the Health Insurance Portability Accountability Act, the health records of tens of millions of Americans are more difficult to access, giving consumers an additional layer of protection and privacy. However, as a consequence of HIPAA, it can be extremely difficult for individuals to examine their own records. This makes it difficult to determine if one's health history and insurance information are accurate, and to rectify incorrect information in the aftermath of medical identity theft.

ILLEGAL DRUG USE

Some have argued that drug use is an inevitable consequence of human nature. Psychopharmacologist Ronald K. Siegel states, "I have come to the view that humans have a need—perhaps even a drive—to alter their state of consciousness from time to time." Donald X. Freeman, an expert on substance abuse policy, thinks that drugs have such a powerful allure simply because they make people feel so good. Nevertheless, a purely biological explanation does not tell us why the rate of drug use varies so dramatically both within and between cultures and from one historical period to the next. Any biological drive toward mind-altering substances would have to be extremely malleable and susceptible to social pressures, values, and norms to account for the fundamental observation that tens of millions of individuals do not use drugs.

Hundreds, if not thousands of studies have been conducted in an effort to determine why people use and abuse mind-altering drugs. Although the results of this research have been varied, and at times, contradictory, one general fact has emerged. Drug use, at least initially, is learned behavior. Even if there is such a thing as an addictive personality, individuals must acquire the knowledge where to buy drugs, how to use them, as well as how to rationalize such use to themselves and others, and we know this behavior can be learned almost anywhere. With users from all social classes and every racial and ethnic background and occupation, it is evident that people use and abuse for a number of reasons: because they do not have enough money or they have too much, because they are attempting to escape painful situations or seek adventure, because they are depressed or overjoyed, bored or have too much to do. Regardless of whether people are pushed by emotional problems, or pulled by the allure of money and excitement, drug use is learned and reinforced in a social context.

For whatever reason they initially use illegal drugs, the ingestion of these substances can lead to drug dependence (formerly called drug addiction), a condition that has three basic characteristics:

1. The user continues to ingest the drug over an extended period of time.
2. The user finds it difficult (if not impossible) to quit, and may take drastic measures, such as stealing, dropping out of school or work, abandoning family and friends to continue using the drug.
3. An increasing tolerance for a drug's effects requires people to use more and more of the substance to create the same experience they had during initial usage. If the user stops ingesting the substance he or she may experience painful physical or psychological symptoms.

Over the past 30 years the notion of an "addictive personality" to explain in part or whole addiction to drugs as well as other "destructive" behavior (overeating and gambling, for example) have become popular. A 1983 study by the National Academy of Sciences concluded that while there is no one group of psychological characteristics that explain all addictions, a number of "significant personality factors" can contribute to addiction. These characteristics include impulsive behavior, difficulty in delaying gratification, an antisocial personality, a high value on nonconformity, a sense of alienation, a sense of heightened stress, and a general tolerance for deviance.

Some researchers are of the opinion that one or more of these addictive characteristics are in-born, that is genetically inherited personality traits. While few researchers doubt there is a personality component to addiction, the evidence for a genetically determined addictive set of behavioral characteristics is inconclusive at best. Even if a genetic component to addiction is eventually isolated, environmental components of addiction cannot be discarded. Medical geneticist Elizabeth Simpson notes that "There are almost no examples where genetics work in exclusion of the environment." The University of British Colombia researcher argues that environmental factors "are important not just in the disease itself, but the course of the disease, the severity of the disease and whether it is actually a significant event in a person's life or not." Genetics alone cannot predict whether someone will develop an addiction. At most an individual's genetic blueprint will identify risk factors. However, genetic mapping may determine what intervention therapies have the greatest chance of success.

How Many Drug Users?

The latest annual survey of drug use in this country conducted by Substance Abuse and Mental Health Services Administration (SAMHSA) contained both good news and bad news. The good news is that use of illicit drugs among youths ages 12 to 17 continues to decline, dropping from 11.6 percent of individuals in that group admitting use in the past month in 2002, to 9.9 percent in 2005. The bad news is that drug use among people between 50 and 59 years of age increased from 2.7 percent in 2002 to 4.4 percent in 2005. Marijuana is the drug of choice, with 14.6 million Americans reporting past-month use of this substance. In 2005 approximately one percent of the juvenile and adult population used cocaine, while no more than one in a thousand individuals reported using heroin. While the number of individuals admitting to methamphetamine has declined in recent years, the number of users admitting to being dependent on these substances increased from 164,000 in 2002 to 257,000 in 2005.

Drug use in the 12-to-17-year-old cohort varies significantly by race and ethnicity. The 2004 SAMHSA survey found that 26 percent of American

Indian and Alaska Native youths used drugs, compared to 12.2 percent of youths reporting a heritage of two or more races, 11.1 percent of white youths, 10.2 percent of Hispanic youths, 9.3 percent of African American youths, and 6.0 percent of Asian youths. A survey published in the *Archives of General Psychiatry* found that drug abuse and dependence are higher among certain subpopulations, including males, individuals 18 to 44 years of age, and people who have never married. A 2004 *Journal of the American Medical Association* article reported the number of deaths and causes of death in this country in 2000. Among the findings: 435,000 deaths were tobacco-related, 85,000 a function of alcohol consumption, and 17,000 a consequence of illicit drugs. There were no marijuana-related deaths that year. That is, tobacco causes 25 times as many deaths as illegal drugs each year, and alcohol causes 5 times as many deaths.

Drugs and Crime

Drug use can and does lead to criminal behavior. However, the association between drug use and crime is not a simple, straightforward cause-and-effect relationship. James Q. Wilson and Robert Herrnstein remind us that drug use as a cause of crime is only one of four feasible connections between these variables. To begin, the relation between drugs and crime may be spurious, meaning there is no causal connection. For example, one or more factors, such as an individual's personality or extreme poverty, are responsible for both drug use and crime. Second, direct causality (i.e., drug use causes crime) has three components. The pharmacological effects of drugs may produce behavioral changes, such as increased impulsivity and/or higher levels of aggression, that result in the commission of crime; the prohibitive cost of illegal drugs may cause people to engage in criminal activity to support their drug taking; or in what has been called the enslavement theory of addiction, drug users become hooked on one or more illegal substances and require ever-increasing amounts of money to support their habits. Unable to sustain their addiction via legitimate employment, they turn to crime. That is, the addiction precedes and causes the criminal behavior.

According to criminologist James A. Inciardi, 30 plus years of research indicates that the enslavement theory "has little basis in reality."

"All of these studies of the criminal behavior of careers of heroin and other drug users have convincingly documented that although drug use tends to intensify and perpetuate criminal behavior, it usually does not initiate criminal careers. In fact, the evidence suggests that among the majority of street drug users who are involved in crime, their criminal careers were well established prior to the onset of either narcotics or cocaine use." That is, the criminal behavior preceded drug use or addiction.

In 2004, 17 percent of state prisoners and 18 percent of federal inmates stated that they committed their current offenses to obtain drug money. During that same year, 55 percent of all inmates in federal prisons were incarcerated for drug offenses. A 2002 survey of jails concluded that 52 percent of incarcerated women and 44 percent of men met the criteria for alcohol or drug dependence. Over the past 20 years between 4 and 7.4 percent of homicides in which circumstances of the crimes are known to police were drug related. In 2005 law enforcement agencies made approximately 1.8 million drug-related arrests (just over 1 in 8 arrests), more than for any other crime.

The drug/crime relationship may be conditionally causal; that is, drug use may cause criminal behavior providing some other condition(s) exist. Drug use may result in criminal behavior among adolescents only if they are in the company of other young males and subject to peer pressure. Although the final alternative is rarely considered, the ingestion of certain drugs may reduce crime. Narcotics such as heroin produce drowsiness and reduce both tension and sexual activity.

The monetary cost to society for drug abuse is staggering. In 2002, this cost was over $180 billion of which $107 billion was a result of drug-related crime. Because drugs are widely and readily available, whatever impact they have on crime rates as well as the financial loss to individuals and societal institutions is likely to continue in the foreseeable future.

The Cocaine Distribution System

The simple facts that citizens of the United States have a voracious appetite for mind-altering drugs, and that most of these substances are illegal, have led to a complex system of drug distribution that spans five continents. Unprecedented in terms of their economic might and ability, international drug cartels influence policymakers and the politicians of numerous countries. Just as organized crime in the United States met the demand for illegal alcohol during Prohibition in the 1920s, today an assortment of criminal organizations in a number of countries provide a similar service in the much more lucrative enterprise of producing and distributing illegal drugs. The following is a classification and breakdown of the international drug distribution system as outlined by Michael Lyman and Joseph Albini:

Growers plant, nurture, and harvest coca plants and opium poppies. By the early 1990s, well over 520,000 acres were under coca cultivation in Bolivia, Peru, and Colombia, with growers employing between half a million and 1.5 million people. According to the *United Nations 2007 World Drug Report*, about 68,000 Colombian households in 23 of the nation's 32 provinces are producing coca, usually on less than one hectare (2.2 acres) of land. Inasmuch as growing coca plants is labor intensive, requiring approximately 225 days of work from the beginning of the growth process until the plants reach maturity, hundreds of thousands of people earn much if not all of their income from this initial stage of the illegal drug business.

Manufacturers turn raw materials, coca plants and opium poppies, into finished products, the street-saleable drugs of cocaine and heroin. With cocaine, mixers, pit laborers, and pasters are engaged in a chemical process that transforms coca leaves with a cocaine content of 0.5 percent into a paste that may have a cocaine content as high as 90 percent, although the drug that reaches the consumer is usually no more than 40 percent pure. This procedure takes place in rain forests and in villages on the outskirts of towns in the central Andes mountains (Peru). In some cases, a powerful individual will serve as both grower and manufacturer, and possibly, the importer as well.

Albini states that the next step in the distribution process requires the interaction and cooperation of three players: the importer, the smuggler, and the trafficker/primary distributor. The importer devises a plan by which the newly refined drug now ready for shipment will reach its final destination, most often the United States, Europe, or Japan. This may entail, in part, bribery and intimidation of officials ranging from rural police officers to officials at the highest level of government. Because importers often have a significant financial investment in these illegal substances, they frequently make enormous profits.

Inasmuch as an estimated 75 percent of all illegal drugs consumed in the United States come from other countries, the role of the smuggler is vital to the overall distribution system. Lyman notes that smugglers may use private and commercial aircraft, boats and ships, ground vehicles (cars, trucks, trailers, etc.), and body packing. Private planes with their cargos of cocaine and marijuana fly across the U.S./Mexico border and land on lightly traveled highways, country roads, swamps, even fields. Once on the ground, they are met by trucks (which often illuminate the landing point with their headlights) that will transport the illicit cargo to distribution points throughout the country. A large ocean-going vessel with a shipment of drugs will rest in international waters off the coast of the United States, where it is met by numerous smaller boats. These latter vessels will carry reduced drug pay-loads to the mainland. Almost 90 percent of all coca destined for the United States is transported via the Central America/Mexico corridor.

Ground vehicles transport drugs to the United States as they enter states that border Mexico (California, Arizona, New Mexico, and Texas). Smug-glers hide their illegal cargo in false-bottom gas tanks, hidden compartments, and other inventive places of concealment.

Body packers attempt to conceal drugs (usually high-priced substances like heroin and cocaine) on their persons or in body cavities as they enter the United States by plane or train. Female smugglers sometimes fill condoms with drugs which are then inserted into their vaginas prior to passing through customs. Other smugglers put drugs into balloons that are swallowed and later retrieved after vomiting or sifting through feces upon defecation. Albini notes that "One can safely conjecture that a substantial amount of cocaine has been brought into the United States from Colombia using this method." On occasion, however, digestive acids will eat through the balloon unleashing the contents and killing the smuggler in short order from a massive drug overdose.

While the Colombian cartels dominated the cocaine trade in the 1970s and 1980s, they have given way in large measure to Mexican mini-cartels used previously as sub-contractors by Colombian drug organizations. The competition for billion-dollar drug profits has resulted in violent turf wars among Mexican drug gangs. United States officials estimate that in 2006 alone as many as 2500 Mexicans were killed in these drug battles.

The U.N. Report states that the global drug trafficking system is becom-ing more complex than the highly structured and centralized organizations of the 1980s. Newer groups are smaller, more flexible, and more temporary than their predecessors. These new organizations are also more likely to comprise individuals from multiple racial and ethnic backgrounds.

The trafficker/primary distributor allocates the illegal drugs (often in significant quantities) once they have reached the port of entry by way of smugglers. Found in virtually every major city in this country, distributors deal in multi-pound lots of cocaine, heroin, and/or marijuana. These individ-uals (some of whom have legitimate front occupations such as banker or attorney) regularly resell the drugs they receive from traffickers. Distributors are usually well-known to local law enforcement authorities. In addition, gangs from minority neighborhoods may be contracted to do the dirty work of drug traffickers.

Dealers sell drugs to individual consumers. Because they are much more visible than those higher up on the drug distribution ladder, dealers are more likely to be observed, arrested, and convicted for their activities. Small-time dealer-users are often their own worst enemy. Frequently under the influence of their own products, these individuals exercise poor judgement and end up selling drugs to undercover officers. Fearful of arrest and incarceration, many of these individual are turned into informers and used by police in their pursuit of distributors and traffickers.

Although drug consumers are not, by definition, part of the distribution chain, they are, as Albini correctly notes, "the most vital structural-functional component of syndication." Without the constant, heavy demand of consumers in the United States and other rich, industrially developed nations, all of the individuals involved in the illegal trade from field hands in Peru and Colombia to a small army of urban dealer-users would be searching for other means of employment.

Dealing Drugs: Gangsters and Wannabee "Gangstas"

The stereotypical drug dealer is a young, inner-city black or Latino gang-banger whose fast and furious, often short-life is immersed in the seedy world of illicit drugs and violence. While this image has some basis in fact, it is much more myth than reality. Using data originally compiled by gang leaders to help manage their organizations, Steven D. Levitt and Sudhir Venkatesh discovered that only a few individuals (the leaders) earned between $50,000 and $130,000 annually in the drug trade. A second tier of sellers (officers) earned approximately $12,000 a year—slightly more than minimum wage—while the lowest level street dealers made less than $2,500 a year for their average 20 hours a week of drug work, an hourly rate significantly below minimum wage. Earning so little money via this illegal activity, it is hardly surprising, as Venkatesh notes in his book *Gang Leader for a Day*, that the vast majority of drug selling gang members live at home with their mothers.

Studies of street level drug dealers indicate that individuals involved in this criminal enterprise move in and out of illegal employment markets. Research in Washington D.C. concluded that 67 percent of those charged with selling drugs were also gainfully, legally employed at the time of their arrest. As Ryan Kings of the *Sentencing Project* notes, for these individuals "drug selling was not a specialized industry, and was a complement to, rather than a substitute for legitimate employment." Similarly, Levitt and Venkatesh found that 75 to 80 percent of street level dealers held legitimate, low-paying jobs at some time during the year. These findings suggest that for at least some low-level street dealers, their illegal activity is a way to make ends meet in an inner-city environment where good paying jobs are few and far between. John Hagedorn who studied street level drug dealers in Milwaukee, argues that these individuals hold many conventional values, including the core value of attaining the American way of life. For Ryan King "drug dealing can be seen not as the cause of urban decline, but rather as a response to the evaporation of a sustainable employment market with an exodus of manufacturing and commerce from urban areas in the 1960s."

On the other side of the economic divide, typically far removed from the concerns and watchful eyes of law enforcement personnel, is another group of drug dealers. Sociologists A. Rafik Mohamed and Erik Fritsvold examined a drug dealing network at a high-tuition, private, Southern California university. The economic, "rational choice" model of crime, as the authors note, suggests that illegal behavior is the result of a careful cost/benefit analysis on the part of offenders. This is why poor black and Latino kids sell drugs, they need the money to survive. If the economic model of crime falls short in accounting for drug dealing on the part of college students from affluent, upper middle-class families, then how can this behavior be explained? Mohamed and Fritsvold offer five "motivations" for this illegal activity.

Motive #1: Underwrite Costs of Personal Drug Use   Most of the student dealers interviewed smoked a great deal of marijuana and viewed selling the "herb" as a mechanism for not having to pay retail prices for their personal supply. When asked why he sold drugs one individual stated: "I don't know, I really just do it to smoke, that's the only reason I sell pot. Then you just start selling tons of pot. Then, like now, I get to smoke tons of pot."

Motive #2: Underwrite Other Incidental and Entertainment Expenses   Although most of the dealers interviewed could be classified as "heavy" marijuana users, others rarely smoked the drug. However, these individuals viewed dealing marijuana as way to supplement entertainment related activities. One dealer noted that "when I was a freshman, I drank a lot, so you know the alcohol budget was extensive. And just other stuff; you know shoes, clothes, whatever. I just had money around, so why not?"

Motive #3: The Spirit of Capitalism   Inasmuch as campus drug dealing was often a mechanism for subsidizing one's good times income, for more entrepreneurial students, selling marijuana and cocaine was a cash-making opportunity that could not be ignored. "It was an easy way to make money. Because in the [dorms] ... my spring semester, a bunch of people got busted and there was nobody dealing and there was a demand." Another student dealer moved from simple marijuana sales to offering a menu of illicit drugs to his customers.

Motive #4: Ego Gratification and the Pursuit of Status   Mohamed and Fritsvold asked campus drug dealers why they chose a university setting for their illegal activity. All of the dealers responded in terms of ego-gratification, status enhancement, and the power they wielded over others in terms of supplying them with drugs, or withholding these substances from their customers. On a campus known for its crass materialism, even a dealer who sported a $50,000 car and designer clothes could not easily differentiate himself from other students until he began selling drugs. However, as one campus drug dealer noted: "If you said, where'd you get pot, where can I get pot? I'm sure my name would be mentioned."

Motive #5: Sneaky Thrills and Being a Gangsta   For many campus drug dealers a fundamental motivation for their criminal activity was wannabee gangstaism coupled with the thrill of getting away with something. Mohamed and Fritsvold argue that "rich white college drug dealers are not motivated by material need as much as they are the idea that they can outsmart those in formal positions of authority, both within and outside the university setting." And if student drug dealers are caught their "status," more accurately, the status of their parents, will keep them from being held fully accountable for their transgressions.

When asked if he was fearful of the campus police detecting his criminal activity one dealer stated that they (the campus police) "can kiss my ass. They can't touch me. They can't do anything to me . . . What's [the campus police] going to do, take my weed away?" The illegal activities of this brazen individual were known to university administrators, with one official relating to the researchers that "He must have some very influential parents." This official stated that because of the parents' status, even though the administration was cognizant of his campus drug dealing, the investigation was being handled with extreme caution.

Methamphetamine and the Making of an Epidemic

Over the past few years government statistics indicate that 0.2 percent of American 12 years of age and older have used methamphetamine. Individuals in this category are classified as "current users" of the drug. Surveys indicate that four times as many Americans ingest cocaine on a regular basis, and 30 times as many use marijuana, as use methamphetamine. Of the 12 specific drugs or categories of mind-altering substances the federal government tracks only LSD, sedatives, and heroin (all at 0.1 percent) have lower rates of regular use. From 1999 to 2005 the lifetime prevalence of high school seniors reporting methamphetamine use declined by 45 percent from 8.2 percent to 4.5 percent.

So how did this relatively little-used drug turn into a national epidemic? In a superb presentation and analysis of the data, the answer is provided by the Sentencing Project's Ryan S. King. To begin, most observers have selectively focused on data supporting the epidemic perspective while downplaying, if not completely ignoring information that undermines this position.

The Arrest Drug Abuse Monitoring (ADAM) data set reports the percent of persons arrested in 19 cities who test positive for drugs at the time of arrest. In 2003, six cities were responsible for most of the overall increase in methamphetamine positive tests from 1990: Los Angeles (from 5.7 to 28.7 percent), Omaha (from 0.6 to 21.4 percent), Phoenix (from 6.7 to 38.3 percent), Portland, Oregon (from 10.9 to 25.4 percent), San Diego (from 27.3 to 36.2 percent) and San Jose (from 8.9 to 36.9 percent). The remaining 13 cities in the survey had a negligible average increase of 1.5 percent in that 13 year span. Methamphetamine use appears to be a highly localized phenomenon. In addition, ADAM is likely to overreport methamphetamine usage as it focuses only on persons arrested, an unknown number of whom may have been arrested two or more times during the year in question.

King notes the spike in meth ingestion among arrestees in the six high-use cities does not represent an increase in overall drug use, but a shift in drug preference. In San Diego, for example, between 1998 and 2003 the number of positive cocaine tests in the arrest population decreased by 46 percent. Similarly, positive cocaine tests dropped 25 percent in Phoenix and 44 percent in Los Angeles. Ryan makes the important policy observation that enforcement and treatment strategies in cities characterized by high methamphetamine use are very different from locales where cocaine and heroin dominate the market.

The public perception that the nation is in the throes of a methamphetamine epidemic has been fueled by the media, especially newspapers and television. King argues that "A general lack of critical analysis coupled with widespread reporting masquerading as facts have resulted in a national media that has been complicit in perpetuating a 'myth of a methamphetamine epidemic.' " Angela Valdez of the *Willamette Week* (Oregon) examined the 261 stories *The Oregonian* newspaper ran on methamphetamine from October, 2004 through March, 2006. Valdez argues the conclusions reached by reporters of that publication were largely unsubstantiated by reliable, credible evidence. For example, *The Oregonian* reported that methamphetamine use fuels about 85 percent of the state's property crimes without attributing the source of that statistic. Valdez quotes Portland State University criminology professor Kris Henning about that figure: "If meth causes property offenses, and meth use has been going up, then property offenses should have gone up. And they haven't. It's either that, or all the people who commit property crimes have disappeared and been replaced by a small number of meth users."

The U.S. Department of Justice reports the production of methamphetamine in Mexico has increased sharply since 2002 as organized criminal groups in that country have been able to acquire large quantities of pseudophedrine and ephederine—key ingredients of the drug—in bulk quantities from China. The Mexican conduit was probably facilitated by the number of methamphetamine laboratory seizures in this country that increased annually from 1999 (6777) to 2003 (10,182). While the demand for this drug has remained relatively constant over this period, reduced production in this country resulted in a supply vacuum that was filled by Mexican crime groups. Frank Owen notes that Mexican methamphetamine is significantly more powerful than the drug made in the United States. This high-potency "ice" methamphetamine is usually smoked and can lead to a rapid onset of addiction.

Methamphetamines are dangerous drugs and the use of these substances should be discouraged, especially among adolescents. However, attempting to frighten the public via bogus epidemic stories can only be counterproductive. Ryan King argues that:

". . . it is our national responsibility to use prevention and educational techniques that are honest and portray the consequences of drug use and efficacy of treatment in a realistic fashion. Exaggerating these points only undermines the credibility of the source of information and increases the likelihood that the recipient will discount future warnings."

Is Decriminalization the Answer?

Discussions about solutions to the country's drug problem usually end up in a debate over decriminalization—the reduction or elimination of penalties for using drugs. For example, decriminalization might involve the imposition of fines or medical treatment for using drugs instead of arrest and criminal prosecution. Advocates of this strategy argue that decriminalization would result in the following:

- a significant reduction in drug-related street crime and violence
- a dramatic decrease in organized and criminal drug-trafficking activity in this country
- reduction of the number of young people in prison during some of the most productive years of their lives
- pronounced savings in criminal justice expenditure (especially prison costs) including tens of millions of dollars given to Colombia each year to disrupt the cultivation of coca and smuggling of coca
- reduction of the number of police officers, prosecutors, judges, and government officials corrupted by drug money
- increased government revenue, as drugs will be taxed in a manner similar to cigarettes and alcohol
- hundreds of thousands of extra hours for police to pursue violent criminals
- elimination of the flow of tens of billions of dollars to Colombian crime syndicates
- a reduction in AIDS cases and other sexually transmitted diseases contracted via injection with contaminated needles

Opponents argue that the fundamental problem with drugs is not there legality or illegality, but that they destroy people's lives. In 2001, Rafael Lemaitre, a spokesman for the Office of National Drug Control Policy stated that "Decriminalization ignores the fact that drug use effects the brain, may lead to addiction and causes untold misery to the user and costs society $110 billions annually in health and social costs." From this perspective decriminalization is an illogical, unworkable policy and would send the wrong message to people—especially adolescents—that drug use was not only permissible but morally acceptable. If drug laws were relaxed or abandoned, rates of drug consumption would certainly escalate.

Perhaps the most well known experiment in decriminalizing drugs was undertaken by the Netherlands beginning in 1976. With the passage of the Opium Act, the Dutch government made a distinction between "soft drugs" such as marijuana, and "hard drug" including heroin, cocaine, and metham-phetamines. Soft drug health risks to the individual were considered low in comparison to hard drug related health problems as well as drug related social problems (drug related crime, for example). In some cities cannabis is sold in cafes or coffee houses and is monitored by the government and pro-hibited from being sold in large quantities and to minors. These soft drug venues are tolerated by the government in the belief that it reduces the chance or risk of soft drug users becoming exposed to hard drugs via their interaction with dealers. In 2001, 25 years into the Dutch experiment, Robert Keizer, Drug policy advisor of the *Ministry of Health, Welfare, and Sports of the Netherlands* offered a summary of the results of his country's drug policy:

• Cannabis use has increased in recent years and the age of first use decreased as it has in other European countries. "There are also signs that cannabis use is stabilizing ..." The rate of cannabis use in the Netherlands is similar to rates of use in the surrounding countries of France, Germany, and Belgium, and lower than rates in the United Kingdom, and the United States.
• There is no indication that the "soft drug" cannabis policy has led to an increase of "hard drug users."
• The rate of hard drug use (notably heroin and cocaine) stabilized in 1991with the Netherlands along with France and Germany the three nations in the European Union with the smallest number of problem addicts.
• The population of hard drug users is comprised more or less of the same group of people as indicated by the fact that each year the average age of this group (40 in 2001) increases by almost a year. "Not many young peo-ple are taking up heroin and crack."
• The number of drug-related deaths as well as addicts infected with HIV is low.

In July, 2001, Portugal reduced penalties for drug use and possession toward the eventual end of decriminalizing all illegal mind-altering sub-stances. Drug law violations that resulted in imprisonment from three months to a year now result in the confiscation of the banned substances and a referral—not a trial—to a three person commission comprised of a physi-cian, lawyer, and social worker to determine the individual's level of addiction and recommend treatment options. Drug users deemed not addicted to an illegal substance are subject to monetary fines and other penalties. It is unclear what impact these new laws have had on the rates of drug use and drug trafficking in Portugal.

Proponents of decriminalization counter the argument that scrapping penalties for drug use would lead to increased drug use as completely unsubstantiated. They cite studies from the Netherlands indicating that marijuana use declined in that country after cannabis was decriminalized. Opponents of legalization counter by noting that rates of alcohol use increased with the end of Prohibition and speculate that use of drugs more exciting and addictive than marijuana could rise if those drugs were legalized.

PROSTITUTION

In its earliest form, prostitution—the sale of sex for money or some material goods—was associated with temple orgies and fertility rites. At the time of Christ, the city of Rome had 36,000 registered prostitutes. This activity increased rapidly in money-based urban economies, and by the Middle Ages, prostitution was firmly entrenched in European societies. Today, prostitution flourishes in rich and poor countries throughout the world. As is the case with numerous types of behavior, it is difficult to determine the number of people who make all or part of their living via prostitution. According to some estimates, there are between 100,000 and 500,000 prostitutes (female and male) in this country. Data from the FBI's *Uniform Crime Report* reveal that each year between 85,000 and 100,000 individuals are arrested for prostitution and commercialized vice.

Streetwalkers

Prostitutes ply their trade as streetwalkers, bar girls, call girls, escorts, masseuses as well as hotel and convention hookers. Most research in this area has focused on streetwalkers, as these individuals are highly visible and accessible. Sociologist Craig B. Little notes that the street hooker "lives in a world of pimps, roach-infested hotels, vice officers, 'pros wagons,' night courts, bail bondsmen, prostitution lawyers, and day-to-day existence with a criminal record." To this list we could add that streetwalkers are at risk for contracting a host of sexually transmitted diseases (STDs) including AIDS. These women and men are also frequent victims of violence at the hands of customers and pimps. Working in pairs, streetwalkers may rob clients and or be involved in a number of hustles, including shoplifting, con games, and passing bad checks.

A study of 130 San Francisco streetwalkers found that 82 percent had been assaulted since entering the profession, with 55 percent of these individuals having been attacked by customers. Almost 7 in 10 had been raped and 48 percent were raped 5 or more times in their role as a streetwalker. Sixty-eight percent met the criteria for post traumatic stress disorder (PTSD)—a psychological reaction to extreme physical and emotional trauma—and their mean score was higher than a group of 123 Vietnam war veterans seeking PTSD treatment. Melissa Farley, who has done extensive research on prostitution, notes that streetwalkers afflicted with PTSD suffer "acute anxiety, depression, insomnia, irritability, flashbacks, emotional numbing and physical hyper-alertness."

By way of police and health department records, John J. Potterat and his colleagues identified 1969 women who worked as prostitutes (more than 90 percent as street prostitutes) in Colorado Springs, Colorado between 1967 and 1999. Of those who died during this period, 19 percent were homicide victims and 18 percent of the deaths were a consequence of drug ingestion. The workplace homicide rate for women in this study was 51 times higher than for females who worked in liquor stores, the legal occupation with the highest female homicide rate. The authors note that "The high homicide and overall mortality rates observed in our cohort probably reflect circumstances for nearly all prostitutes in the United States where prostitution is illegal."

Research indicates that the majority of streetwalkers come from dysfunctional families and broken homes where they were physically and/or sexually abused. Many were runaways or abandoned by their parents. Living on the streets, lacking even a high school education and marketable job skills, they turn to prostitution for money. Income is often used to support drug use, a frequent pathway into prostitution. Studies find that up to 80 percent of street prostitutes use drugs before entering "the life." Some observers believe that the average age of entrance into prostitution is dropping, with girls as young as 13 or 14 working the streets.

Anthropologist Jennifer James divides prostitutes into two categories, true prostitutes and part timers. True prostitutes include:

- *Outlaws:* prostitutes who work on their own without pimps
- *Rip-off artists:* thieves posing as prostitutes who steal from johns; prostitution is not their main source of income
- *Hypes:* females who enter the profession primarily to support their drug habit
- *Ladies:* prostitutes identified by their class, finesse, and professionalism

- *Old-timers:* veteran prostitutes who lack the class of ladies
- *Thoroughbreds:* young, professional prostitutes

Pimps

The role of the pimp—an individual who derives some or all of his income from the earnings of prostitutes—figures prominently in the literature on prostitution, although it is unknown how many streetwalkers work for these individuals. Pimps often control women through violence and drug dependence while taking up to 70 percent of a streetwalker's income. The relationship between a woman and her pimp has been described as similar to that of an abused wife and her husband.

Criminologist Larry Siegel characterizes the pimp as a man who steers customers toward women in his "stable," protects his girls from johns who would abuse them, and posts bail when they are arrested. The pimp is the most important person in the life of many prostitutes (especially streetwalkers and bar girls), as he serves as surrogate father as well as husband and lover. Pimps may pick up women with previous experience, or "turn out" women who have never worked as prostitutes.

The prostitute-pimp relationship can be mostly businesslike or "shockingly brutal." Pimps control the women in their stable by a variety of means. If emotional ploys do not yield the desired results, most pimps will not hesitate to use violence as a mechanism of control. Violence may well be used by pimps in controlling females who make a decision to leave.

While pimps may offer women in their stable some measure of protection from other pimps, and are quick to help get them back on the street upon arrest, they can do little to spare these females from violence at the hands of their clients. Streetwalkers in the employ of pimps may be more vulnerable to street violence as they tend to work in more dangerous neighborhoods.

Call Girls

At the top of the prostitution hierarchy, call girls are usually well-educated and very attractive. Contacts are typically made via telephone by customers who can afford the hundreds if not thousands of dollars for a single date. Sociologist David R. Simon notes that some call girls are corporate prostitutes, women "frequently retained to help close business deals, including bidding on contracts, mergers with other corporations, political lobbying, undercutting competitors, gathering stock holder proxy votes and securing oil leases."

Many call girls work out of escort services found in the yellow pages of any big-city phone directory. *New York Times* reporter Andrew Jacobs notes that a New Jersey-based escort service boasts a roster of 2200 clients who are carefully screened by the owner. Prior to a date, customers can check out the merchandise (including reviews) on an independently run Web site. Women in this agency are in their early 20s to early 40s and keep two-thirds of the fees.

The now-infamous Emperors Club VIP, the prostitution ring that supplied women to former governor Elliot Spitzer, served clients in New York, Washington D.C., Miami, London, Paris, and Vienna. Clients paid between $1000 and $3100 an hour for women. Day rates, "dawn to dawn" packages ranged from $10,000 to $31,000. Provocative images of the women (with faces obscured) were available on a Web site with a diamond rating under each photo. The Web site noted that "We act for a select group of educated, refined, and successful international clients who give their best in all they do and who, in return, only wish to receive the best." Clients paid for services by credit card, wire transfer, or cash sent by mail (the latter described by Emperors Club organizers as "packages"). Gift certificates were available upon request.

As high as the Emperors Clubs fees appear to most people, sociologist Sudhir Venkatesh argues that these prices are on the low end of an elite, three-tier call-girl market.

Tier 1 fees range from $2000 to $4000 per session. Women in this tier come from all age and ethnic backgrounds; most have a high-school education and limited work experience. They are employed by escort services that promise discretion but may not always deliver on that guarantee as they are often under surveillance. "In practice, this means buyer beware."

Tier 2 sex workers charge up to $7500 per session. These women are usually white, enrolled in college or have a university degree. They have a small, exclusive clientele (as few as a dozen men) and only take new customers by way of referrals. Most of these women do not work for escort services and keep all of the money.

Tier 3 women charge $10,000 or more per session. They typically have no more than four or five clients and may charge an additional monthly service fee.

Venkatesh notes that in the world of high priced call-girls, Elliot Spitzer got a bargain, and not paying enough for sex may have been the biggest mistake he ever made. If the former governor had coughed up the money that more expensive prostitutes command, he likely would have been beyond the radar of law enforcement agencies that monitor escort services.

What do men obtain for thousands of dollars per hour? Venkatesh states that after interviewing hundreds of sex workers, "approximately 40 percent of trades in new York's sex economy fail to include a physical act beyond petting and kissing . . . Flush with cash, these elite men routinely turn their prostitute into a second partner or spouse," sometimes persuading these women "to take on a new identity, replete with a fake name, a fake job, and a fake life history."

In recent years the Internet has become a major venue for linking prostitutes and johns, with prostitutes and organized sex rings selling their services in much the same manner as any legitimate business. The Miami Herald examined three Web sites where escort services typically place advertisements and counted 804 listings for female prostitutes and 80 for male prostitutes in the South Florida area. Procuring prostitutes via the Internet the Herald noted, "is almost as easy as buying a book on Amazon." If former U.S. Attorney Marcos Jimenez is correct, escort service/call-girl operations have little to fear from federal law enforcement agencies as prostitution busts are "relatively rare." According to Jimenez, combating prostitution rings "fits in some FBI priority list somewhere, but I can assure you it's not near the top." The owner of a legal brothel in Nevada stated that "When the FBI gets interested in an escort agency, it's because of money laundering, not prostitution."

Child Prostitution

In their research, Richard Estes and Neil Alan Weiner estimate that approximately 293,000 children in this country are at risk of becoming victims of commercial sexual exploitation. More than 50 percent of girls living on the street engage in formal prostitution, with 3 of every 4 child prostitutes working for a pimp. In a television interview, Richard Estes stated that "Child sexual exploitation is the most hidden form of child abuse in the U.S. and North America today." Most of these children are runaways and homeless youth who trade "survival sex" for food, shelter, and clothing. In addition, children living at home trade sex for money or expensive clothing as well as a variety of consumer goods. A significant number of men who solicit sex from children are married men with families of their own.

Approximately 1 in 5 child prostitutes are ensnared in organized crime networks with some transported from one city to another to provide fresh faces for clients as well as to make pornography. According to the FBI, Atlanta, Las Vegas, Tampa Bay, Miami, and Washington DC are among 14 cities with the highest incidence of child prostitution in the United States.

Children also number among the estimated 17,500 individuals trafficked to the United States from foreign countries each year. Most of these individuals come from Southeast Asia, Russia, and Mexico, with the youngest often kidnap victims earmarked for prostitution.

The U.S. State Department estimates that globally each year over 1 million children are exploited via child sex tourism (CST). The International Labour Organization states that sex tourism accounts for between 2 and 14 percent of the gross domestic product of Malaysia, the Philippines, and Thailand, with CST a significant component of this revenue. Clients of CST are primarily men from the United States, Germany, Sweden, Australia and Japan.

Tourism per se is not the cause of child sexual exploitation as extreme poverty, high rates of illiteracy, few occupational opportunities, and the status of women are the background for this activity. Sex tourism is a mechanism for males from developed countries to exploit children from poor countries with who often lead a hand to mouth existence.

Perpetrators of CST often concoct one or more rationalizations to justify this activity to themselves and others. Perhaps the most common is the "I'm helping these destitute children and their impoverished families" explanation. Consider this statement from a retired U.S. schoolteacher. "On this trip, I've had sex with a 14 year-old girl in Mexico and a 15 year-old girl in Colombia. I'm helping them financially. If they don't have sex with me, they may not have enough food. If someone has a problem with me doing this, let UNICEF feed them." Others employ bizarre (and untrue) anthropological explanations including the children in these countries are less "sexually inhibited" argument, and the host country does not have taboos against adults and children having sex. Still others employ a straight out racist perspective considering the welfare of children in poor countries as unimportant, or less important than their sexual gratification.

Peter Piot, the Executive Director of UNAIDS summed up the morality of child prostitution: "the commercial sexual exploitation of children is an atrocity. It has rightly been called the ultimate evil. It denies children their fundamental rights. It has devastating consequences for them. It is a perversion of the natural order—adults should be there to protect and nurture children, not take advantage of their emotionally and physically vulnerable state."

Clients

An estimated 10 to 20 percent of the male population is estimated to engage female prostitutes at some stage in their life, but only 1 percent pay for sex regularly. Men who frequent prostitutes come from all social classes, racial/ethnic groups, and occupational backgrounds. A study of 101 males in New Jersey who frequented streetwalkers found that the average customer was 40 years of age, had lived in the area approximately 20 years, was currently or previously married, and had been a client of local prostitutes for about 5 years. Another study found that 80 percent of men apprehended for soliciting prostitutes reported that their marriage or steady relationship was sexually satisfying. These findings and those of other researchers belie the conventional wisdom that most johns are lonely, frustrated men unable to establish more conventional sexual relationships.

Among men who purchase sex acts there is a wide range of how often they engage in this behavior. Donna M. Hughes notes that findings from several studies indicate that a group of hard core, habitual users accounts for a significant percentage of the demand for illicit sex. She cites one survey's findings that 22 percent of the men questioned had purchased sex up to 4 times, 19 percent between 5 and 10 times, 14 percent between 11 and 25 times, and 11 percent more than 100 times. Hughes notes that the more often men buy sex the more likely they are to accept the idea that sex is a commodity.

Hughes cites an unpublished research paper by Steven Grubman Black, who interviewed 92 self-identified purchasers of sex acts between 22 and 65 years of age. Many of these men believed that paying women for sex was normal, natural male behavior and a number were introduced to this activity at a young age, by a father, uncle, or older male identity figure. Grubman Black found that the motivation for these johns could be classified as follows:

- Lonely, shy, awkward males. Unable to establish long-tem relations with women; for these individuals, sex is a secretive, shameful yet necessary activity.
- Desirous of sex acts they cannot have with primary partners. These men want oral, anal, or a threesome and are willing to experiment.
- Changing the rules. These clients are males who force women to perform sex acts they had not agreed to and keep her beyond the allotted time against her will.

- Roughing her up. These misogynists physically and sexually assault prostitutes, particularly when they learn the "woman" is a transsexual.
- Ultimate control. These clients believe they own the woman by virtue of paying her and are entitled to do whatever they want with and to them.

Prostitution is often described as a "victimless crime." Counselors who try to rehabilitate streetwalkers, police officers who see the violence perpetrated upon prostitutes, and the women themselves know this is anything but true.

KEY TERMS

burglary: Unlawful entry of a structure to commit a felony of theft

cocaine: An addictive central nervous system drug extracted from the leaves of the coca plant that produces a sense of euphoria, excitement, and restlessness.

corporate crime: Illegal activity such as price fixing, restraint of trade, dumping hazardous waste, and corruption committed by company officials to maximize corporate profits

identity theft: A crime wherein an imposter obtains key pieces of personal information in order to impersonate someone else

methamphetamine: An addictive central nervous system stimulant that increases energy and decreases appetite; legally used to treat narcolepsy and some forms of depression

motor vehicle theft: Attempted or completed theft of automobiles, trucks, buses, and motorcycles

prostitution: The exchange of sexual access for money or goods; drugs, for example

SUGGESTED READINGS

Felson, Marcus. 1998. *Crime in Everyday Life*. Thousand Oaks, CA: Pine Forge Press.

Flowers, R. Barri. 1998. *The Prostitution of Women and Girls.* Jefferson, NC: McFarland & Company.

Geis, Gilbert. 2006. *White-Collar and Corporate Crime.* New York: Prentice Hall.

Musto, David. 2002. *Drugs in America: A Documentary History.* New York: NYU Press.

Owen, Frank. 2007. *No Speed Limit: The Highs and Lows of Meth.* New York: St. Martin's Press.

Simon, David R. 2002. *Elite Deviance.* Boston: Allyn & Bacon.

Wright, Richard T., and Scott H. Decker. 1996. *Burglars on the Job: Street Life and Residential Break-Ins.* Boston: Northeastern University Press.

"Speed Bump" © Dave Coverly/Dist. By Creators Syndicate, Inc.

chapter six

Victims and Victimization: Will You Be Next?

In the colonial United States, the criminal justice system and the philosophy of that system were very different from what they are today. Victims or their agents were responsible for apprehending criminals and bringing them to justice. Crimes were thought of as injuries to the victim as opposed to injustices committed against the state. In other words, a crime was considered a conflict between two individuals, and not an attack on society. Convicted offenders were required to compensate the victim up to three times the amount of property and/or cash that was lost as a result of the crime. Offenders who could not pay were legally required to work as servants for a period of time equal to the amount of money owed. Victims also had the option of selling offenders so as to recoup their monetary loss. Individuals who chose this latter option had one month to find a buyer; incarceration costs past 30 days were to be paid by the victim or the offender was released.

In contemporary society, crime victims have been reduced to little more than witnesses for the prosecution or state. The criminal justice system no longer exists to redress the wrongs suffered by victims; rather, its primary focus is on the offender and his/her relation to the state. William McDonald notes that the primary purpose of the justice system is to deter crime as well as rehabilitate and punish criminals. Once the criminal has been punished, his/her debt to society has been paid, regardless of the social, economic, and/or medical condition of the victim. This is what justice means not only in the United States but in most modern societies. Many social critics argue that the victim has been forgotten if not completely disregarded by the criminal justice system.

Criminologist Ezzat Fattah no doubt echoes the sentiments of countless crime victims when he states, "It is difficult to understand how an act that hurts, injures, or harms a human being, an act that might thoroughly affect the person's life and disrupt his or her existence, can be considered an offense not against that individual but against society." For Fattah, it makes no sense that if the source of a physical injury or monetary loss is a criminal offense, the fine paid by the person who caused it goes to the state, whereas if the cause of the wrongdoing is defined as a civil matter (medical malpractice or negligence, for example), the damages go to the person who has suffered.

For most of its history, criminology focused almost exclusively on offenders. Leading textbooks in the field did not even have a listing for victims in the index, much less devote a chapter to this subject. Sociologist Andrew Karmen notes that until the 1940s and 1950s "criminology could be characterized as 'offenderology.' " Researchers focused on offenders: "who they were, why they engaged in unlawful activities, how they were handled by the criminal justice system, why they were incarcerated, and how they might be rehabilitated." The central problem in offender-driven criminology was the search for the origins of criminal motivation.

Approximately 50 years ago a number of academicians and researchers began considering the victim in the criminal event, and by 1970 victimology became a part of criminology. An interdisciplinary field that includes social scientists, medical practitioners, and members of the criminal justice system, victimology explores a number of issues, including how some individuals contribute to their victimization; the financial, physical, and psychological consequences of being a crime victim; and how the police, courts, and various social agencies deal with victims. In this chapter we will briefly examine each of these issues, including the victim's movement and the successes and failures this movement has had in gaining additional legal and financial rights for crime victims.

VICTIM CHARACTERISTICS

Not every resident of the United States has an equal chance of being a crime victim. Overall, some individuals are more likely to become targets of offenders than others, and the likelihood of being victimized for particular offenses, such as robbery and rape, is even more skewed. The National Crime Victimization Survey, as well as other measures, indicates that becoming a crime victim is correlated with a number of characteristics.

Age

In general the younger a person is, the more likely he or she will be the victim of a violent crime, with teens and young adults experiencing the highest rates of violent victimization. In recent years, persons age 12 to 24 were 22 percent of the population but were 35 percent of murder victims and 49 percent of victims of all serious crimes (murder, robbery, rape, sexual assault, and aggravated assault). Beginning with the 25-to-34 age group, as people get older, the likelihood declines. The widely held perception that older persons are disproportionately victims of violent crime is false. Between 1993 and 2002, Americans 65 years of age or older experienced less violence and fewer property crimes than younger persons. Property crimes, not violent victimizations, account for most offenses against the elderly.

Gender

With the exception of rape and sexual assault, males have higher rates of violent victimization than females. For example, in any given year approximately four of five murder victims are males. The crime victimization gender gap exists at every age group and diminishes as people get older. When males are violent crime victims their assailants are likely to be strangers. However, about two of every three attacks against women are perpetrated by someone the victim knows, typically a husband, boyfriend, family member, or acquaintance.

Estimates regarding the number of females victimized by boyfriends and husbands each year range from just over a million to as many as 18 million. This wide discrepancy is explained in part by the type and purpose of the survey, the definition of violence used, and the political context of the research. Abused women are slapped, punched, kicked, burned, and occasionally shot. A significant number of these victimizations do not come to the attention of the police, and, therefore, are not reflected in the annual report, *Crime in the United States.* Abused women may be too embarrassed or frightened to inform crime victimization interviewers of their treatment, or believe that only physical assault at the hands of stranger is a "real" crime.

Race and Ethnicity

For every 1000 persons in that racial group, 27 African Americans, 20 whites, and 14 persons of other races were violent crime victims in 2005. Although only 13 percent of the population is African American, in any given year approximately 49 percent of murder victims are African American (49 percent are white and 2 percent are Asians, Pacific Islanders, and Native Americans). Between 1992 and 2001, Native Americans were victims of violent crimes at rates more than twice that of African Americans, two and a half times that of whites, and four and a half times that of Asian Americans.

A significant amount of violent street crime involves victims and offenders of the same race. For example, between 1976 and 2004, 86 percent of white murder victims were killed by whites, and 94 percent of African American murder victims had their lives ended by African American assailants. The exception to this victim-offender relationship is Native Americans. One study found that 7 of 10 violent victimizations of Native Americans involved an offender of a different race. While African Americans are less likely to be victims of sex-related homicides, workplace killings, and homicides by poisoning, they are greatly overrepresented in killings involving drugs.

In 2005, property crime victimizations were slightly lower for African Americans (145 per 1000 African American households) than for whites (156 per 1000 white households). Property crime rates for Hispanics were significantly higher (210 per 1000 households) than for non-Hispanics (148 per 1000 non-Hispanic households). Hispanics suffered motor vehicle thefts at nearly three times the rate of non-Hispanics.

Immigrants

Research indicates that immigrants are victimized at rates similar to that of the larger population but are less likely to report crimes to law enforcement authorities. A national survey of 92 police chiefs, prosecutors, and court administrators concluded that the crimes least likely to be reported by immigrants are domestic violence, sexual assault, and gang violence. Explanations for low reporting rates include:

- Fear of becoming involved with authorities
- Fear of embarrassing their families
- Difficulties speaking English

- Cultural differences in conceptions of justice
- Lack of knowledge of the U.S. criminal justice system

Income

Individuals with low incomes face an additional problem of high crime victimization rates. As household income increases, rates of crime victimization decrease. In 2005, individuals 12 years of age and older from households earning less than $7500 annually were more than twice as likely to be victims of violent crime and personal theft than Americans with household incomes of more than $75,000. As a number of criminologists have noted, the poor have the highest rate of street crime victimization because they live in neighborhoods that produce a significant number of street crime offenders. Rates of sexual assault and rape were approximately the same among all income groups.

Marital Status

The risk of crime victimization is a function in part of one's marital status, with rates of violent crime and personal theft per 1000 individuals age 12 or older as follows:

Never married	37.4
Divorced/separated	31.7
Married	10.3
Widowed	6.1

One possible explanation for these findings is that never married and divorced/separated individuals spend more time away from home, and, therefore, are more likely to be victims of street crime. Although unattached, widows and widowers are more likely to be older individuals who stay home making them less vulnerable to violent crimes and personal theft.

Region, Urbanization, and Home Ownership

Rates of property crime per 1000 households in 2005 were highest in the West (206.5) and lowest in the Northeast (103.9) with the Midwest (155.8) and South (146.8) in the mid range. The West also had the highest rate of motor vehicle theft. Rates of burglary, motor vehicle theft, and household theft were highest in urban centers, followed by the suburbs and rural areas of the country. People who owned their homes had a significantly lower rate of property crime victimization (137 per 1000 households) than individuals who rented their homes (192 per 1000 households).

THE COST AND CONSEQUENCES OF VICTIMIZATION

Economic and Medical Costs

Criminal victimization results in the loss of significant amounts of money annually on the part of hundreds of thousands of individuals as well as the larger society. Nearly 18 million violent and nonviolent criminal victimizations in 2002 (about 77 percent of all victimizations) resulted in economic losses. The cost of medical expenses, lost earnings, and assorted victims services totals about $105 billion each year. When intangible factors such as pain and suffering of victims and their families and a reduced quality of life are considered, the total cost to the nation of criminal activity is approximately $450 billion annually. When factors such as loss of time by victims, criminals, and prisoners is factored into the equation, economist David Anderson calculated the annual cost of crime in this country at $1.7 trillion. He notes that to date economists have "only looked at the tip of the iceberg."

Based on a nationally representative survey of 168 mental health professionals, Mark Cohen and Ted Miller estimated that between 20 and 25 percent of the client population of these practitioners are crime victims. More than half of this criminally victimized population are thought to be adults who suffered physical and/or sexual abuse as children.

Psychological Consequences

One of the major contributions of victimology has been an increasingly thorough examination of the psychological impact of being a crime victim. This is especially important in light of the fact that each year approximately one in four households is victimized by one or more acts of violence or theft. Numerous studies have indicated that the psychological consequences of victimization can be much more disruptive to people's lives than the loss of property and/or physical injury. In addition, while material losses often can be replaced in short order, the emotional pain resulting from criminal victimization can persist for years, and in some cases a lifetime. Victims of robbery, burglary, and non-sexual assaults may suffer post traumatic stress disorder (PTSD) that is, the persistent reexperiencing of the traumatic event by way of intrusive memories (flashbacks) and dreams. PTSD may be associated with a host of anxiety-related emotional and behavioral symptoms.

Victims of robberies, physical assaults, and sexual assaults may be stigmatized because they failed to resist their attacker forcefully. Morton Bard and Dawn Sangrey argue that "Submission in our culture is viewed as cowardly and we tend to feel contemptuous of the victim who does not fight back." Some people may resist an assailant even against their better judgment (resistance may trigger an especially violent response on the part of the offender) simply because they fear being labeled a coward.

VICTIMS AND OFFENDERS

The Geography of Crime

Crime is not randomly distributed in time and space. It follows, therefore, that victims are not randomly distributed in these dimensions. In other words, people are more or less likely to be victimized depending on where they live and in what historical period they reside in these locations. Recall from our discussion of the routine-activities perspective that three elements are necessary for a crime to occur: motivated offenders, a suitable target, and the absence of capable guardians of these targets. In their formulation of this theory, Lawrence Cohen and Marcus Felson state that the routine activities of Americans have changed significantly since World War II. People now spend significantly more time away from home (both husbands and wives working, more frequent and longer vacations, for example) and interact with a larger group of people outside the family network. Time spent away from home means that one's residence is increasingly devoid of "capable guardians," while interacting with nonhousehold members makes people viable targets for face-to-face victimizations such as physical assault, sexual assault, and robbery.

While the routine activities of tens of million of people have changed so as to increase their chances of being victimized, these individuals do not have the same likelihood of becoming crime victims. Michael Gottfredson makes a distinction between the absolute and probabilistic exposure to risk. Absolute exposure simply means that the victim (a person or property) must come into contact with an offender if a crime is to occur. However, when potential victims are exposed to offenders, it does not follow that a crime will occur on each and every such occasion. The offender must believe that he or she can complete the crime successfully, that is, obtain a reasonable reward and not be apprehended. This is probabilistic exposure.

Probabilistic exposure is correlated with residence and lifestyle. In their analysis of all 115,000 addresses and intersections in Minneapolis over a 12-month period, Lawrence Sherman and his colleagues discovered that 50 percent of all calls to police for assistance came from just 3 percent of places. All of the calls regarding the predatory crimes of robbery, rape, and auto theft came from 2.2 percent, 1.2 percent, and 2.7 percent of locations respectively. Every one of the almost 25,000 domestic disturbance calls were recorded at just 9 percent of addresses (although each address sometimes included many apartments). These intensely high-crime areas are often referred to by law enforcement personnel as hot spots. Even in high-crime neighborhoods some addresses and locations had no calls to the police for assistance. High-crime areas, therefore, may contain hot, and one might say, cold spots. Obviously living in and/or frequenting a high-crime area significantly increases one's chances of becoming a crime victim.

What makes a neighborhood (or portion thereof) a hot spot, a zone of exceptionally high victimization? In their study of Cleveland, Ohio, Dennis Roncek and Pamela Maier found that "the amount of crime of every type was significantly higher on residential blocks with taverns or lounges than on others." Locations with drinking establishments generate crime and victims for a number of reasons. Patrons of these establishments typically have cash on them and make for attractive targets, especially when they are intoxicated. Taverns and lounges are likely to have cash on hand, making them desirable robbery targets as well. Intoxicated individuals are likely to engage in behavior they would not exhibit when sober, such as fighting and sexual assault. Bars often attract a significant number of people (especially on weekends and happy hour periods) from other neighborhoods. People may be more likely to engage in criminal behavior when they are in the presence of strangers and separated from local, informal agents of social control such as family and friends. Businesses such as bars, lounges, and liquor stores may attract a clientele with a disproportionate number of offenders.

To the extent that individuals who live in neighborhoods that are also home to a significant number of offenders and potential offenders, their chances of becoming crime victims increase proportionately. Numerous studies have concluded that most perpetrators of street crimes commit a significant number of offenses close to home. Murders, rapes, and assaults often transpire within a few blocks of the offenders' residence, while crimes such as burglary and larceny occur further from the transgressors' residences. The reduction of criminal activity as the distance from one's residence increases is called distance decay. As Patricia and Paul Brantingham note: "It takes money and effort to overcome distance." Lack of money for transportation to another neighborhood on the part of offenders explains to some degree why poor people of color have such high rates of street crime victimization. They are victimized by the offenders among them (other poor people of color) who do not have the financial means to rob and burglarize individuals in more affluent neighborhoods. In addition, a racial/ethnic minority member is likely to arouse the suspicion of guardians in a locale primarily or exclusively comprised of members of a different racial/ethnic group.

Victim Precipitation

Criminologists have long known that some people contribute to their own victimization. The phenomenon wherein an individual starts or triggers a chain of events that culminates in his or her victimization is called **victim precipitation**. In 1958 Marvin Wolfgang was the first researcher to study victim-precipitated homicides. He examined 588 homicides in Philadelphia between 1948 and 1952 and discovered that 26 percent of these crimes could be classified as victim precipitated. In these killings, the victim was the first to exhibit deadly force, "the first to strike a blow or fire a shot." The following are examples of this category of homicide from Wolfgang's research:

- A drunken husband, beating his wife in the kitchen, gave her a butcher knife and dared her to use it on him. She claimed that if he should strike her one more time she would use the knife, whereupon he slapped her in the face and she fatally stabbed him.
- A victim became concerned when his eventual slayer asked for money the victim owned him. The victim grabbed a hatchet and started in the direction of his creditor who pulled out a knife and stabbed him.
- A drunken victim with a knife in hand approached his slayer during a quarrel. The slayer showed a gun and the victim dared him to shoot. He did.

Wolfgang found that whereas about a third of all homicide victims are stabbed, in victim-precipitated killings, more than half are stabbed. Consuming alcohol was also more prevalent in victim-precipitated murders, with the victim most likely to have been drinking. Numerous investigators, including Wolfgang, have speculated that some of these homicides were thinly masked suicides, that is, the victim wanted to die and used the killer to perform the act for him or her.

In one of the most controversial pieces of crime research conducted, Menachem Amir reported that19 percent of the rapes he studied in Philadelphia were victim precipitated, that is, a victim first agreed to have sexual intercourse and then changed her mind. Many people were outraged by Amir's findings, arguing that he was engaging in a classic example of blaming the victim. Criminologists Terrance Miethe and Richard McCorkle note that "Although studies of victim precipitated rape are widely criticized for blaming the victim, it is undeniable that lifestyle changes by some potential victims can dramatically alter their risks of sexual victimization." That female hitchhikers are at increased risk for being robbed, raped, and/or physically assaulted is self evident, an observation that is not lost on police who are apt to consider the rape of a young woman thumbing a ride as a victim-precipitated crime. To the extent that law enforcement views such an offense as victim precipitated, it is conceivable that officers will not investigate the crime with as much enthusiasm as a rape wherein the female is thought to "really be a victim."

Choosing a Victim and Repeat Victimization

Imagine that you desperately needed money as a result of some unexpected emergency. After selling your car and other possessions, and borrowing money from family, friends, and lending institutions you were still short. Trapped in a financial corner, you reluctantly decide that engaging in criminal behavior is your only way out. Upon deciding to commit residential burglary you would be faced with the question of what home or homes to target. As a novice thief you might decide to skip any home that advertises the presence of a burglar alarm. Neither amateur nor professional criminals are likely to burglarize the first house they see. Victims (property or people) are rarely chosen haphazardly or randomly. Some amount of planning, from a few minutes to a few months, is needed to produce a crime target.

So how do offenders decide whom or what they will victimize? Regarding street crimes such as physical and sexual assaults and muggings, David Finkelhor and Nancy Asdigian argue that the physical characteristics of individuals may increase a person's chances of being victimized, independent of his or her activities, "because these characteristics have some congruence with the needs, motives, or reactivities of offenders." In other words, certain offenders are drawn to certain types of victims they perceive as particularly vulnerable. The matching of offender's strengths and victim's weaknesses is called "target congruence." This sort of congruence helps explain the high rate of youth victimization in three ways:

1. *Target vulnerability:* These characteristics include small stature, physical weakness, and/or psychological problems indicating that individuals can be overpowered easily and will offer little or no resistance.
2. *Target gratifiability:* These individuals have something offenders want, possessions for example, that are trendy, relatively expensive, and portable, or in terms of female victims, a sexual outlet. "Femaleness itself is a risk attribute."
3. *Target antagonism:* These are characteristics of the victim that arouse anger, resentment, jealousy, or destructive impulses in the offender. Victim characteristics in this category would include sexual orientation and one's racial/ethnic identification.

Target vulnerability is a common theme when offenders are questioned about victim selection. In their study of 65 arrested and/or incarcerated female robbers, Ira Sommers and Deborah Baskins recorded the following responses from these women:

- Yeah, I would look for easy people. People who looked timid, who ain't gonna put up much of a fight.
- You know, I felt like if somebody looked soft, or you know young people they would have got stuck up.
- I would look for a certain type of person and I would plan what I was gonna say and when I was gonna make my move. I was looking for, uh, alcoholics, people who were coming out of bars drunk.

In what has been dubbed the law of repeat victimization, Graham Farrell and Ken Pease note that the likelihood of being a crime victim is lower for individuals who have never been victimized, and higher for those who have been victimized on one or more occasions. Individuals victimized three or more times have an even greater likelihood of falling prey to offenders in the future.

By way of victim surveys, interviews with offenders, and crime reports we know repeat victimization (RV) is a common occurrence. RV is more likely for violent crimes such as assaults (including domestic violence) and robberies than it is for property offenses. For example, a study in Indianapolis found that 32 percent of establishments victimized by commercial robberies had been robbed before.

Research also indicates additional victimizations often occur quickly, sometimes within a week of the first offense. After an initial period of increased risk, the chances of RV diminish rapidly until victimization risk is approximately the same as that of individuals or property that have never been victimized. However, researchers have found that the period of declined risk may be punctuated by a slight resurgence or bounce when risk of RV increases somewhat. Deborah Lamm Weisel notes that the bounce is likely associated with the replacement of stolen property via insurance money. "It seems likely," Weisel notes, "that some repeat offenders may employ a 'cool down' period, perceiving victims to be on high alert immediately after an offense but relaxing their vigilance after a few months."

There are two primary explanations for RV. According to the flag interpretation, some targets are especially attractive to offenders and/or viewed as particularly vulnerable. For example, apartments with sliding glass doors are susceptible to break-ins, as are homes located on corners where offenders can more easily determine if anyone is home. Convenience stores open 24 hours a day have a heightened exposure to crime, especially in the middle of the night. Some jobs such as taxi driver or delivery person put employees at a higher risk than individuals in other lines of work. Popular products, such as the latest electronic equipment and vehicles considered desirable for joy riding, have a higher risk of being stolen.

From the boost perspective, RV is a function of the successful outcome of an initial crime. Some offenders gain important information about a target, and on the basis of their experience and this knowledge, decide to victimize the same individuals or steal similar property. For example, offenders may gain information about defeating specific types of home and/or auto burglar alarms and target dwelling and vehicles with these security systems.

During the course of the initial burglary offenders might not be able to carry away all of the merchandise, so they return later for the goods left behind. Offenders may tell others about their crime, allowing these individuals to victimize the home or business yet again. Some victims may not be able to protect themselves from future crimes. A broken window or door left unrepaired from the first burglary allows easy access for another break-in. Victims of domestic violence who cannot defend themselves, or leave the location of the initial attack (usually the home) are vulnerable to additional assaults.

Interviews with offenders suggests that much RV may be a function of boost factors. Seasoned criminals are adept at calculating both the rewards and risks of offending. Boost and flag explanations can overlap, with the specifics of these perspectives varying by offense type. Weisel notes that bank robberies are most likely to recur if the initial crime yields a significant take. When the amount of money stolen from one of these establishments is minimal, the risk of a second robbery is low. According to Weisel, "most offenses are highly concentrated on a small number of victims while the majority of targets are never victimized."

According to the report on repeat victimizations, there are four primary types of repeat victimization:

- *True repeat victims* are the same targets initially victimized, such as apartment occupants burglarized multiple times within a year.
- *Near victims* are individuals or targets physically close to the initial victims that are victimized in a similar manner. Offenders may determine that apartments close to a burglarized unit have a similar layout, contain the same type of merchandise, and are subject to the same vulnerability that made the initial break-in possible.
- *Virtual repeats* are repeat victims identical to the initial victims in one or more important ways (for example, a chain of restaurants or convenience stores having the exact layout and/or management practices regarding the number of personnel in the store or cash handling and depositing procedures).
- *Chronic victims* are individuals who suffer from the commission of multiple crimes, burglary, robbery, and domestic violence, for example, over time.

Repeat victims of violent crimes in the United States are frequently admitted to urban trauma centers. Three studies concluded that between 33 and 44 percent of individuals who go to these facilities for medical care have been victims of violent crimes in the past, often in the past few weeks or months. All of these studies found that violent crime victims admitted to urban trauma centers are especially likely to be young, male, African American, living in poverty, with low levels of education, and histories of alcohol and/or drug abuse. Many of these individuals had psychiatric problems and few had medical coverage. One study found that victims of violent crime admitted to trauma centers were more likely to carry a gun or a knife, and to put up a fight when attacked.

Victims as Offenders and Offenders as Victims

We tend to think of criminals and victims as two distinct categories of individuals—the good guys and the bad guys. Victimologist Ezzat Fattah rejects this dichotomy and considers "the roles of 'victim' and 'victimizers' as neither fixed, assigned or predetermined." Rather, these roles are "interchangeable and may be assumed simultaneously or consecutively . . . with many individuals moving alternatively between the two . . . as yesterday's victims become today's offenders and today's offenders becoming tomorrow's victims." A number of studies have found that crime victims are much more likely than non-victims to report a history of deviant behavior.

In their study of active armed robbers in St Louis, Richard Wright and Scott Decker found that six of every ten offenders specializing in street robbery stated they usually targeted individuals involved in criminal pursuits, especially drug dealers. Comments of two of these robbers are as follows:

- [I like robbing] them drug dealers [because] it satisfies two things for me: my thirst for drugs and the financial aspect. [I can] actually pay my rent, pay my car, [and things like that too.]
- [Dope men are perfect victims] cause they have money on them . . . They carry all they money, jewelry, and all that on them, and all they drugs."

Janet Lauritsen and her colleagues argue the delinquent lifestyle of many adolescents places them at greater risk of criminal victimization. For example, gang members are often victimized by rival gang members in fights and drive-by shootings. Drinking and drug use at bars and parties increases the risk of physical and sexual assault on the part of adolescents using mind-altering substances. If Lauritsen is correct, the victimization patterns of many young people cannot be adequately understood apart from their criminal and deviant activities.

Offender-victims are especially vulnerable to other street criminals because they are viewed as less likely to call police than nonoffender-victims. Offender-victims may believe that if they contact law enforcement officials, their past or current criminal behavior will be discovered, thereby making an unfortunate situation even worse. One might speculate that the more heavily involved an offender is in criminal activities, the less likely he or she is to notify the authorities regarding a victimization. Offender-victims known to local officers may have less credibility, and, therefore, are less likely to be believed by police and prosecutors.

Crime victims often become criminal offenders themselves. For example, property crime victims sometimes attempt to recoup their losses by stealing from others. Dutch criminologist Jan van Dijk notes that in a country where bicycles are important and common vehicles of transportation, one bicycle theft may trigger a number of similar events as victims turn into offenders in a crime, victim, crime, victim sequence. "First A steals B's bike. Then B steals C's bike. Next C steals D's bike. Finally, D steals E's bike, but E is left out in the cold." Not only do victims shortly become offenders in this victim/offender chain, but the original victimization activates a chain of offenses that significantly contributes to both crime and victimization rates. Criminologist Marcus Felson speculates that "van Dijk chains" may contribute to the theft of certain items in this country, notably auto parts and compact disc players. Children whose school books or other belongings are stolen may steal similar articles from their classmates.

Employees who rightly or wrongly believe their employers are cheating them (for example, not receiving adequate compensation for the number of hours worked) may retaliate by stealing from the company or engaging in workplace sabotage. A multitude of research on family violence has made it painfully clear that abused children are more likely to be aggressive and delinquency-prone teenagers and abusive adults than children not similarly victimized. It is not uncommon for rapists to have a history of childhood brutalization, and young victims can become lifelong offenders in a self-perpetuating cycle of violence.

Crime Displacement and Victimization

Displacement is the transfer or movement of a crime from one place, time, or kind to another, place, time, or kind as a result of a change in the social and/or physical environment. When displacement occurs, the likelihood that one category of people will become crime victims decreases while the probability that another category of individuals will be victimized increases. There are at least four types of crime displacement: spatial, temporal, target, and type-of-crime displacement.

Spatial displacement occurs when offenders who are deterred by increased police presence in one neighborhood move to another neighborhood where the risk of being apprehended is lower. Temporal displacement occurs when offenders change the time of day, or the day, or the season, to one in which it is safer to commit a crime(s). Target displacement occurs when offenders seek new victims (people or property) because existing targets have become more difficult to penetrate (target hardening). Victimologist Andrew Karmen refers to this manner of displacement as a "valve theory of crime-shifts." That is, when and where one area of illegal opportunity has been shut off, offenders look for replacement targets. For example, if burglars discover that certain home alarms are difficult to disarm, they avoid these households and concentrate on dwellings with alarms that can be disarmed readily, or homes without anti-theft devices. If a bus company only accepts riders holding a prepurchased bus pass, offenders may shift their attention to cab drivers or pizza delivery employees.

Type-of-crime displacement occurs when offenders abandon what has become a high-risk crime and turn to offenses considered less risky. For example, bank robbers who realize their chances of being apprehended have increased dramatically as a consequence of enhanced technological surveillance begin to rob supermarkets instead of financial institutions.

Because this type of "crime spillover" is very difficult if not impossible to measure, the displacement of victims from one time, location, and type of crime to another is unknown and will never be more than conjecture.

THE VICTIMS MOVEMENT

In December 1993, Mary Byron of Jeffersontown, Kentucky, was murdered by Donovan Harris, her estranged boyfriend. Harris, who raped Byron at gunpoint three weeks earlier, had been released from jail on bond. On the day of her death Byron did not know that Harris was no longer in police custody, because the Jeffersontown police department, which she had asked to notify her upon Harris' release, failed to do so. Mary Byron's estate subsequently sued a detective and the police department. When the suit was dismissed, Byron's estate appealed the decision. The appellate court ruled that neither the city of Jeffersontown nor the detective had an obligation to notify Byron of Harris' release from custody.

The case of Mary Byron was an especially tragic example of what many people believe is the shameful way the criminal justice system treats crime victims. While not as poignant as the Byron incident, there are numerous examples of how people who have suffered at the hands of criminals are, according to many observers, victimized a second time by the police, the courts, and/or the penal system. Whereas the Sixth Amendment to the Constitution guarantees the defendant's right to remain in the courtroom during a trial, crime victims have no such privilege. In most jurisdictions, victims appear in court only to give testimony. When the defendants in the Oklahoma City bombing were tried, the judge ruled that family members of fifteen victims be excluded from the courtroom. In 1983, Alabama became the first state to permit victims to sit at the prosecutor's table during the trial, and in the following years a number of other states permitted crime victims to remain in court at the judge's discretion. The plight of crime victims was rediscovered by some members of the media, "enlightened" criminal justice practitioners, social scientists, and advocacy groups beginning in the late 1950s and early 1960s. Television, newspapers, and news magazines brought to the nation's attention stories focusing on the suffering of victims of crime. None of these cases was more chilling than that of a young Bronx, New York City woman named Kitty Genovese. Upon returning to her Bronx apartment after work at 3:15 a.m., the 30-year-old Genovese was assaulted by Winston Mosely, who stabbed her twice in the back. When she cried for help, a man in her building opened the window and shouted, "Let that girl alone." Mosely fled in his vehicle, returned 10 minutes later and found Genovese, barely conscious lying in a hallway at the back of the building. He stabbed her several more times, than raped the screaming woman before leaving the scene. Approximately a dozen of Genovese's neighbors saw the attack—that lasted more than 30 minutes—before notifying authorities. Kitty Genovese died en route to a local hospital.

The get-tough-on-criminals and victim's rights movement advanced a campaign to convince the public that they had more to fear from becoming crime victims than being falsely accused of violating the law. In addition to fighting for equality in the workplace and the political arena, the most recent phase of the women's movement focused on females as victims of domestic violence and rape. From the first crisis centers for female victims that began in Berkeley, California and Washington, D.C. in the early 1970s, shelters for abused women have spread across the country. Groups like Mother's Against Drunk Driving and Parents of Murdered Children also focused on victims of specific crimes. By the end of the twentieth century, a multifaceted victim's movement had helped bring about significant changes in the nation's perception of crime victims and, more importantly, how these individuals are treated by the criminal justice system. In this section some of these changes will be examined.

Victims and Offender: Restitution and Restoration

We noted at the beginning of this chapter that during the colonial era offenders were required to compensate victims for up to three times the amount of property and/or cash that they had lost. The victims movement resurrected this sentiment and had some measure of success in bringing about programs by which individuals convicted of property crimes make restitution to their victims, even if so doing results in economic difficulties for the offender. This philosophy was summed up in the 1982 *President's Task Force on Victims of Crime*:

> It is simply unfair that victims should have to liquidate their assets, mortgage their homes, or sacrifice their health or education or that of their children while the offender escapes responsibility for the financial hardship he has imposed. It is unjust that a victim should have to sell his car to pay bills while the offender drives to his probation appointments. The victim may be placed in a financial crisis that will last a lifetime. If one of the two must go into debt, the offender should do so.

That criminals should make monetary restitution for their transgressions is part of a more comprehensive philosophy regarding offender/victim relations called restorative justice. From this perspective, the emphasis shifts from punishing offenders (retribution), to restoring the situation and well-being of victims prior to the crime. As in the colonial era, crime is viewed as a transgression against a person as opposed to an offense against the state. It follows from this position that offenders owe a debt not to society, but to the individuals they have harmed. Restorative justice also emphasizes the possibility of reconciliation between offenders and victims.

The first victim offender reconciliation programs (VORPs) in North America began in Ontario, Canada, in 1974 and Elkhart, Indiana four years later. To date there are well over 300 such programs in the United States and Canada, and approximately 800 in Europe, Australia, and New Zealand. Several North American VORPs receive nearly a thousand case referrals a year from local courts.

VORPs would never have succeeded in this country if people did not embrace the philosophy of restorative justice. While the punitive, get-tough, law-and-order perspective is part and parcel of many political campaigns, Mark Umbreit, one of the foremost experts on restorative justice, argues there is a "growing body of evidence to suggest that the general public is far less vindictive than portrayed and far more supportive of the basic principles of restorative justice than many think, particularly when applied to property offenders." For example, in a survey of adults in Minnesota, nearly 75 percent of respondents said that having the offender compensate them for their losses was more important than a jail sentence for burglars. Over 80 percent of the respondents stated that they were interested in taking part in a face-to-face mediation with offenders. These sentiments make VORP possible.

VORPs are dialogue-driven sessions that bring the offender and victim together in the presence of a trained mediator. The emphasis on these sessions or meeting is upon victim healing, offender accountability, and the restoration of losses. Most VORPs are run by nonprofit community agencies and consist of a fourfold process, the centerpiece of which is the mediation session. The mediator facilitates a dialogue between the victim and the offender, focusing on how the crime has affected their lives. In most cases, a written, mutually agreeable settlement is reached at this stage. The agreement may also take the form of monetary payments to the community or service to the victim and/or community. The offender's behavior is monitored to insure that he or she fulfills all the terms of the agreement. Additional sessions are scheduled if problems occur.

Judges, probation officers, prosecutors, police, and on occasion, defense attorneys and victim advocates refer both juveniles and adults to VORPs. Referrals are usually made at the pre- or post-adjudication level, although they can be made at any point in the criminal court proceedings. The settlement can become part of the sentence or some component of the court's overall decision in a case. Regarding court-referred cases, offenders are often on probation while fulfilling stipulations of their contracts.

Based on his evaluation of VORPs in four cities (Albuquerque, Austin, Minneapolis, and Oakland) Umbreit reached a number of conclusions:

- These mediation programs result in a high degree of satisfaction for victims (79 percent) and offenders (89 percent). Over 80 percent of respondents in both groups perceived the mediation process as fair. Offenders made the following comments: "I liked the fairness of it." "To understand how the victim feels makes me different . . . I was able to understand a lot about what I did." "I had a chance of doing something to correct what I did without having to pay bad consequences." Victims were overwhelmingly positive about VORP, as indicated by their comments. "I was allowed to participate and I felt I was able to make decisions rather than the system making them for me." "The mediation made me feel like I had something to do with what went on . . . that justice had been served."
- These mediation programs significantly reduce the fear and anxiety experienced by crime victims. Prior to their involvement in VORP, 25 percent of victims feared they would be victimized again by the same offender. That figure dropped to 10 percent after the mediation.

- Juvenile offenders do not perceive their participation in VORP as getting off easy or beating the system. Rather, they view the settlement as a demanding response to their law violating behavior, as severe as other options available to them via the courts.
- Judges and probation staffs support VORPs enthusiastically and increasingly integrate them into the juvenile court system.
- Significantly fewer and less-serious additional crimes were committed by juveniles who participated in VORP-style programs over a 12-month period when compared to a group of offenders who committed similar crimes but did not participate in any form of mediation.

The effectiveness of VORPs on reducing recidivism is not conclusive, although it appears to support the mediation philosophy. A number of researchers have compared recidivism rates of offenders who participated in VORP to those individuals processed by the courts. For example, a study of six California VORPs found that at five sites recidivism rates of VORP groups were between 21 and 105 percent lower than the non-VORP group while at the sixth site the recidivism rate was 46 percent higher. One large meta-analysis found lower recidivism rates of 619 juvenile offenders who participated in mediation programs compared to a 679 juvenile offenders who did not. Another meta-analysis of 14 studies with over 9000 juvenile offenders found a 26 percent lower recidivism rate among mediation participants, and less-serious offenses among those who broke the law again. Howard Zehr, one of the founders of the restorative justice movement, argues that "Restitution programs, especially intentionally structured programs such as VORP, do indeed appear to reduce recidivism rates and by measurable amounts."

Similar to VORPs are victim-offender conferencing programs, including family group counseling (FGC). This program usually includes a conference facilitator, the victim, the offender, members of their respective families, and support groups. Fundamental goals of FGCs are to allow the victim to express the impact of the crime on him or her to the offender, to help offenders own up to their criminal behavior by way of shaming them, and to reach some agreement on a plan to compensate victims.

Victims of Crime Act

Deeply dissatisfied with the treatment of crime victims in her country, British magistrate and prison reformer Margaret Fry launched a campaign to have the state compensate crime victims in 1957. Although this initial attempt was unsuccessful, victim compensation programs were eventually introduced in New Zealand (1963), England (1964), and in California and New York (1965 and 1966 respectively). By the mid 1990s, all 50 states, the District of Columbia and the Virgin Islands had established victimization programs. The 1996 Antiterrorism Act also provides compensation to U.S. residents who become victims of terrorist attacks within or outside this country. Established with passage of the 1984 Victims of Crime (VOCA), the Crime Victims Fund is a major source of revenue for state compensation programs. Victims programs are subsidized by criminal fines, forfeited bail bonds, penalty fees, and special assessments collected by federal courts, the U.S. Attorney's Office, and the Bureau of Prisons. No funds come from federal tax dollars. Money deposited in state crime funds increased steadily from 1985 through 1996 before declining. The first $10 million placed in these funds each year is used to improve the investigation and prosecution of child abuse cases, with the remaining money distributed in the following manner: 48.5 percent to state compensation programs, 48.5 percent to state assistance programs, and 3 percent to support training and improve the delivery of services to federal crime victims.

Victim compensation provides direct reimbursement to crime victims for medical costs, lost wages or loss of support, funeral and burial costs, and mental health counseling. Every state has its own crime victims compensation program. The majority of states place a ceiling on awards distributed to victims in the $10,000 to $25,000 range. California caps awards at $46,000, while New York and Maryland have no specified limit. To be eligible for compensation, most states require the victim to notify law enforcement officials within three days of a crime and file a claim within a designated period of time, usually two years. These time limits can usually be extended for good cause. In addition, states have a number of program requirements that must be met before awards are granted. For example, most states closely examine the relationship between the offender and victim. Relatives, sexual partners, or residents of the same household as the offender are typically not eligible for compensation. These individuals are disqualified for fear that the husband who has badly beaten his wife, for example, will be enriched by the money the woman receives from the victim's fund. Compensation programs only pay crime victims to the extent that other forms of reimbursement such as medical insurance do not cover the loss.

Victim assistance funds provide, but are not limited to, the following services: crisis intervention, counseling, emergency shelter, criminal justice advocacy, and emergency transportation. Nationwide, thousands of organizations provide services to nearly 200,000 crime victims and their families.

From 1986 through 1998 the Office for Victims of Crime (OVC) distributed over $700 million in VOCA compensation funds, and approximately $3.2 billion between 1999 and 2006. In recent years compensation programs have been expanding across the country. Although a significant amount of money has been distributed to crime victims, it is almost inconsequential when compared to the tens of billions of dollars these individuals lose each year. In addition, a number of states still cap compensation awards at the exceedingly low amount of $10,000.

Government compensation programs focus on victims of violent crimes for at least three reasons. First, victims of property crimes often have insurance that covers losses (or a portion of these losses) incurred by way of burglary, auto theft, etc. Second, the loss of goods does not have as strong an emotional appeal as offenses that result in bodily injury or death. This being the case, using public funds to compensate property crime victims would not be supported enthusiastically by citizens. Finally, there are so many more property crimes than violent victimizations that an all-inclusive compensation program could not possibly help the vast number of individuals who suffer crime-related property losses annually without expending billions of dollars.

CRIME VICTIMS' RIGHTS ACT OF 2004

In 2004, the Congress passed and President Bush signed into law, the Crime Victims' Rights Act (CRVA), legislation that granted additional rights to victims of federal crimes in federal court proceedings. However, CVRA is vague about some of these rights including a victim's right to be "reasonably heard" in proceedings addressing pleas, sentencing, or the release of defendants. An early test of the CVRA occurred in California involving a father and son who swindled dozens of victims. After the defendants pled guilty to wire fraud and money laundering, more than 60 people submitted victim impact statements. At the father's sentencing numerous victims spoke about the impact the crimes had on their lives. During the sentencing phase of the son's hearing the judge refused to allow victims to speak. In January, 2006, the Ninth Circuit Court of Appeals ruled that the judge's decision was wrong and made the following three rulings about CVRA:

- In passing this act it was the intention of Congress to allow victims to speak at sentencing hearings, not just to submit a victim impact statement.

- Victims have a right to speak even if there is more than one criminal sentencing.
- If a crime victim is denied the right to speak at a sentencing hearing, the sentence shall be vacated and a new sentence hearing held wherein the victim does have the opportunity to speak.

Victims and Sentencing

One segment of the victims movement advocates longer prison sentences, harsher treatment of convicted offenders, and speedy executions; it has been called (rightly or wrongly) the vengeance rights lobby. This faction of the movement scored a victory in 1974 when the criminal courts in Fresno, California permitted victims to introduce a victim impact statement (VIS) at the sentencing phase of the trial. These statements allow victims to tell the court in detail how they have been medically, financially, and emotionally affected by the crime **prior** to the offender's sentencing. In some jurisdictions, prosecutors send out VISs when charges are filed. Every state permits some form of impact information at sentencing. In a few states victims have input at crucial junctures of the criminal court process, such as bail hearings, pretrial release hearings, and plea bargain hearings. In addition, the majority of states also permit input from victims at the parole hearings of offenders. A number of states include the original impact statement in the offender's file, with these documents reviewed during parole hearings. Some states allow victims to state their opinion about an appropriate sentence for offenders via a **victim statement of opinion**.

In the federal justice system as well as many state systems, the VIS is prepared by a probation officer as part of the presentence investigation (which describes the offender's background and the criminal offense) that is forwarded to the judge. The judge gives the VIS as much or as little weight as he or she chooses in the sentencing determination.

Critics of the VIS argue that allowing victims to affect sentencing decisions makes for a dual track system of punishment. That is, tougher sentences are likely dealt to offenders whose victims submitted a VIS. However, several studies have concluded that VISs have had little impact on sentencing decisions primarily because victims do not exercise their right to use them. In many cases victims are not aware that they have the right to participate at this stage of the proceedings. If a trial has had many continuances and/or has been particularly long, victims cannot always be located at the sentencing stage. Some people are aware of the VIS but refrain from exercising their right to address the court because they are dissatisfied with the

criminal justice system, or fear their emotional well-being may be jeopard-
ized if they have to explain everything in detail again. Still other people may
fear retaliation from friends and family members of the offender if they push
for a more punitive sentence.

In their study of victim impact statements and sentencing, Mary Lay
Schuster and Amp Popen cite a victim's advocate who stated that "very
rarely does an impact statement affect sentencing because it's already been
decided." In one case a judge was handed a 10 page impact statement, took a
recess from the hearing, read the entire document, then rejected a plea
agreement stating she could not support the negotiated plea (and sentence
recommendation). Commenting on this case, a victim advocate noted it was
a rare event, the ".01 percent" of sentences handed down by judges.

In response to the murder of Mary Byron, Jefferson County, Kentucky,
established an automated victim notification system (December 1994). In
1996, the Kentucky Board of Corrections implemented the Victim Information and Notification System (VINE) making Kentucky the first state to provide this service across all jurisdictions. Via this computerized system,
information is available 24 hours a day about inmates housed in local jails,
adult correctional facilities as well as some juvenile offenders. By way of
computer-generated telephone calls, registered persons are contacted when
an inmate is scheduled for release or if an inmate escapes. This is but one of
the many services that has come about since the victims movement began
almost 40 years ago.

KEY TERMS

crime displacement: The transfer or movement of a crime from one place, time,
or kind to another place, time, or kind as a result of change in the social and
physical environment

Crime Victims Fund: Established with the passage of the VOCA in 1984, it is a
major source of revenue for state compensation programs

van Dijk chain: The chain of events that is triggered with the first victimization;
for example, if A has a bicycle stolen, he may steal B's bike, who in turn may
steal C's bike, and so on

victim impact statement (VIS): A statement that allows victims to tell the
court in detail how they have been medically, financially, and emotionally
affected by a crime prior to the offender's sentencing

victim offender reconciliation programs (VORPs): Programs that bring the
offender and victim together in the presence of a trained mediator for the pur-

pose of reconciliation, which usually results in the offender making some form of restitution to the victim or community

victimology: The study of the role of the victim in the criminal event and the relation between crime victims and the criminal justice system (the police, the courts, and the corrections officials)

victim precipitation: The chain of events started by the victim that culminates with his or her victimization

victim statement of opinion: A subjective declaration by the crime victim telling the court what sentence he or she believes the offender should receive

Victims of Crime Act (VOCA) of 1984: Legislation that authorized federal funding for state-administered victim compensation programs; these funds are typically limited to victims of violent crimes with no other resource to money or services

SUGGESTED READINGS

Jerin, Robert A. and Laura J. Moriarty. 2007. *Victims of Crime: Understanding Victimology, Victimization and Victim Services.* Los Angeles: Roxbury Publishing Company.

Karmen, Andrew. 2006. *Crime Victims: An Introduction to Victimology.* Belmont, CA: Wadsworth/Thompson Learning.

McDevitt, Jack and Judith M. Sgarzi. 2002. *Victimology: A Study of Crime Victims and Their Roles.* Englewood Cliffs, NJ: Prentice Hall.

Meadow, Robert. J. 2006. *Understanding Violence and Victimization.* Upper Saddle River, NJ: Prentice Hall.

Ruback, R. Barry and Martie P. Thompson. 2001. *Social and Psychological Consequences of Violent Victimization.* Thousand Oaks, CA: Sage Publications.

Zehr, Howard and Barb Toews. 2004. *Critical Issues in Restorative Justice.* Monsey, NY: Criminal Justice Press.

"Speed Bump" © Dave Coverly/Dist. By Creators Syndicate, Inc.

chapter seven

Crime and Criminal Law: Order, Liberty, and Justice for All?

The criminal law is very complex, and most Americans are naturally unfamiliar with many of its basic concepts and provisions. TV shows about police and lawyers might tell us something about the criminal law, but the terms and content can be confusing and the TV characters obviously do not stop to explain them to the vast viewing audience.

ORDER VERSUS LIBERTY IN AMERICAN DEMOCRACY

This chapter outlines some of the key elements of the criminal law and also discusses the rights of suspects and defendants. As Chapter 2 pointed out, a basic challenge in the U.S. criminal justice system is to balance the need for order and safety with the need to respect civil liberties. The founders' emphasis on liberty reflected the colonial experience. England's abuse of the legal system—for example, its denial of jury trials for certain offenses— in its struggle with the colonists, was fresh in the minds of the writers of the Constitution and Bill of Rights. They saw the need to limit the ability of government to take away individual freedom and, more generally, to limit the ability of government to abuse its powers. Thus, the Bill of Rights specified not only the freedoms of speech, press, and religion but also the rights of individuals charged with criminal offenses and limits on government actions against such individuals.

Many U.S. Supreme Court rulings have tried to define the proper powers of the police and other law enforcement agents under the Constitution. Many people believe Court rulings have handcuffed the police and made it more difficult for them to control crime, while many other people are afraid that the police may abuse their powers if appropriate restrictions are not placed on their behavior. Thus the question remains: What is the proper balance in American democracy between order and safety on the one hand and civil liberties and individual liberty on the other hand?

Crime Control Since September 11

This question, so difficult to answer in normal times, became even more difficult and urgent after the terrorist attacks of September 11, 2001. Only six weeks after the attacks, Congress passed the USA Patriot Act with almost no debate. As Chapter 2 discussed, critics condemned the Patriot Act and other government actions for violating many protections guaranteed by the Bill of Rights. Government officials and other parties strongly defended these actions as necessary to protect the citizenry in an age of terrorism. This debate continues until this day and no doubt will continue long after you finish reading this book. We hope this chapter's discussion of the criminal law and of important Supreme Court decisions on criminal justice will give you a basic understanding that will help you understand the dilemma between crime control and due process in American society and also the continuing debate over the Patriot Act and other legal actions taken in the wake of 9/11.

TYPES OF CRIME: FELONIES AND MISDEMEANORS

The criminal law distinguishes between serious crime and minor crime. **Felonies** are serious crimes, such as homicide and rape, that are punishable by at least one year of incarceration, while **misdemeanors** are minor offenses, such as disturbing the peace, that are punishable by less than one year of incarceration. Someone incarcerated for a felony is ordinarily sent to a state prison for a state offense or a federal prison for a federal offense; someone incarcerated for a misdemeanor is ordinarily sent to a local or country jail. In many areas, misdemeanors are further divided into gross and petty misdemeanors, the former punishable by more than 30 days in jail and the latter punishable by fewer than 30 days in jail. The most minor offenses, such as many kinds of traffic offenses, are called **violations** or **infractions** and are punishable only by fines and do not result in a criminal record.

WHEN IS AN ACT A CRIME AND WHEN IS IT NOT A CRIME? THE ELEMENTS OF CRIMINAL BEHAVIOR

Six distinct elements characterize an act as a crime. If these elements are not present, a defendant should not be found guilty.

Prohibited by Criminal Law: Now, Not Later

The first element is that the behavior must be prohibited by criminal law. This element seems so obvious that it should go without saying. In fact, however, it is one of the most important cornerstones of criminal law in a democratic society. It would certainly be unfair for someone to be arrested for committing an act that was not prohibited or that was prohibited only after the act occurred.

Guilty Act: Watch What I Do, Not What I Think

A second element is ***actus reus***, a Latin term meaning "guilty act." This refers to the idea that an actual act, not just the thought of doing the act, is required before something can be considered a crime. Thus, you cannot be found guilty of merely thinking that you would like to shoplift or break some other law; instead you must actually commit some illegal act. This concept also implies that a law must actually prohibit an act before it can be considered a crime.

Criminal Intent: Where There's a Will, There's a Way?

The third element is criminal intent, or ***mens rea***. This refers to the idea that the person must have intended a criminal act before it can be considered a crime. Another way of saying this is that the offender must have had criminal intent. In practice, this means the offender wanted to do some harm and knew what he or she was doing, and the act was not just an accident or something that occurred under duress. Thus, a defendant should not be found guilty of a crime unless it is established that he or she committed the crime knowingly and willfully. For example, if someone gives you something to drink that, unknown to you, contains an illegal drug, you should not be found guilty of using this drug because you did not know you were drinking it. The law also assumes that a criminal act has occurred even if it did so from extreme negligence or recklessness. Thus, if a parent accidentally leaves an infant inside a car on a hot, sunny day and the infant becomes very ill or even dies, the parent may be charged with a crime even though the parent did not intend to harm the child. In practice, guilty act and criminal intent are the key elements that must be proven to justify a guilty verdict.

Concurrence: Not Just a Coincidence

A fourth element is **concurrence**. This means that a criminal act and criminal intent must occur at roughly the same time, with the intent preceding the act. Of course, this is how most crimes occur. But suppose a man is planning to kill a business colleague by pushing him onto a subway platform. The day before he plans to carry out this act, he accidentally kills the colleague in a car accident. Even though the colleague is dead, as the would-be offender was intending, no crime occurred because the actual act of murder that was intended did not occur.

Causation: An Act and an Impact

A fifth element is causation. This means that the criminal act actually caused the harm suffered by the victim. Suppose a defendant tries to rob someone with a knife. During the robbery, the victim suffers a heart attack and dies. The prosecution might argue that the victim had the heart attack only because of the robbery and that the defendant is thus guilty of a homicide. The defense will probably look for any evidence that the defendant had a weak heart and, if such evidence exists, argue that the defendant should not be blamed for a death resulting from a weak heart, not from the robbery itself, especially if the defendant was not intending to harm the victim physically.

Harm: Something Actually Happened

The idea of causation implies that the sixth and final element of a crime is **harm**. This means that some harm must actually occur to an individual or property before a crime can be established. This element raises the question of whether harm occurs in so-called victimless crimes, such as prostitution and illegal drug use, which involve willing participants and no actual victims as that term is usually defined. Even though people committing these behaviors participate willingly, the law assumes that they are harmed whether or not they agree with this assessment, and it also assumes that these behaviors harm society itself.

WHEN IS A DEFENDANT NOT RESPONSIBLE FOR COMMITTING A CRIME? LEGAL DEFENSES TO CRIMINAL CHARGES

All the elements just discussed must be proven before a defendant should be found guilty of a crime. It is more difficult to prove these elements for some defendants and criminal acts than for others, and several legal defenses to criminal responsibility challenge the existence of one or more elements. Most of them challenge the existence of criminal intent. All these defenses involve many complexities but can be summarized here briefly.

Self Defense

Self defense is probably the legal defense with which you are most familiar. If someone is about to rob or attack either you or a companion and you use a weapon in self defense or otherwise injure the offender, you should not be found guilty of a crime as long as you can prove that the actions you took were necessary to prevent harm to yourself, your companion, or your property. One key issue here is the degree of imminent peril the offender was actually posing. If the offender was a big man holding a gun, your claim of self defense might be more credible than if the offender was a small, slender youth with no weapons. Another key issue is the amount of force you are allowed to use. If you shoot an unarmed burglar, your force may exceed the amount that the principle of self defense permits. If you shoot someone who was assaulting you and wound them but then keep on shooting, your force may again exceed the legal amount. The criminal justice system would then need to determine whether your claim of self defense against criminal responsibility is legally valid.

Entrapment

Entrapment is another familiar defense. If you commit a crime only because a law enforcement agent induced you to do so, you may claim you were entrapped. Such entrapment defenses are difficult to prove but ordinarily have the best chance of succeeding if the defendant is otherwise a law-abiding citizen who committed a crime only after repeated requests by law enforcement agents and/or after the use of particularly strong inducements by them. Thus, if an undercover agent asks you to buy some illegal drugs for a small reward of $50 and you quickly agree, your entrapment defense might not succeed. But if you agree only after many requests to buy the drugs and the promise of a reward of several thousand dollars, your entrapment defense might carry more teeth. If a man is cruising a red-light district filled with prostitutes and is arrested after agreeing to have sex with an undercover police officer posing as a prostitute, an entrapment defense probably will not work since it would be difficult to argue that the man would not have engaged the services of a prostitute if he had not been induced by the officer.

Accident and Mistake

One set of legal defenses is that a defendant committed a harmful act by accident or mistake. For example, suppose someone gives you a package to mail that, unknown to you, contains a bomb. If you had no reason to suspect a bomb was in the package, you should not be held criminally responsible. Of course, that decision would be up to a jury if the case ever went to trial. The prosecution would probably try to prove that a reasonable person should have been suspicious, while the defense would try to establish that you had no knowledge of the package's contents and no reason to believe that it contained anything dangerous.

Duress and Necessity

Duress means that a defendant was forced to commit a criminal act out of imminent threat to his/her safety or that of loved ones. If someone threatens to shoot you if you do not help him commit a robbery and if you have no realistic chance of escaping his reach, you should not be found guilty if you then help him commit the robbery. A key question when this defense is used is whether the defendant had no reasonable alternative. A related defense is necessity, which means that someone commits a crime to prevent a greater harm from occurring. Suppose a child is seriously injured and the child's parent, not wanting to wait for an ambulance, carries the child to the family car and then rushes to the hospital. Along the way, the car exceeds the speed limit and runs a red light or two. Technically, the parent has broken several traffic laws, but, if caught and cited by police, could argue that this was necessary to help the injured child. A key question again is whether the defendant had a reasonable alternative; another is whether the harm prevented by the offense was greater than the harm it caused.

Insanity

The most controversial criminal defense is undoubtedly the insanity defense, which argues that a defendant cannot not by reason of mental defect appreciate the consequences of his or her actions and thus should not be held criminally responsible. The insanity defense is very complex but generally means that a defendant does not know right from wrong and/or was not able to control his or her behavior.

American courts actually use several standards of insanity, with the specific standard depending on the jurisdiction in which a case is heard. The oldest standard is called the M'Naghten rule, named after the case of Daniel M'Naghten, who assassinated Edward Drummond, the private secretary to the British Prime minister, Edward Peel, in 1843. During his trial, M'Naghten claimed he had been delusional, and he was found not guilty by reason of insanity. The case led to the development of the M'Naghten rule, which says that defendants cannot be held criminally responsible if they have a mental defect that prevents them from understanding the act of which they are accused, or, if they did understand it, from understanding that their act was wrong. The M'Naghten rule, or the "knowing right from wrong" rule is the standard in several states.

Some individuals may have a serious mental illness that drives them to commit a criminal act even though they do understand that the act is wrong. As a result, many states have developed the irresistible impulse test. Under this standard, individuals are not held criminally responsible if they had a serious mental condition that compelled them to commit a criminal act, even if they did understand that the act was wrong.

A fourth standard is the substantial capacity test, used by several other states. This rule states that a defendant cannot be held responsible for a criminal act if, because of a mental disease or defect, he or she lacked substantial capacity to understand that the act was wrong or to obey the law.

Federal law uses part of the substantial capacity test, as it declares that defendants are not criminally responsible if they lack the ability to understand their actions were wrong. Because this test demands that they lacked this understanding, it is more stringent than both the irresistible impulse test and the substantial capacity test; it allows for defendants to be criminally responsible as long as they understand their act was wrong, even if a mental condition compelled them to commit the act.

The insanity defense is quite controversial and often receives much media attention when it is used in cases involving serious crime. Perhaps the most controversial use of the insanity defense occurred in the trial of John Hinckley, Jr., who was arrested in 1981 for shooting President Ronald Reagan. Hinckley said he shot Reagan in order to capture the interest of actress Jodie Foster. When the jury found him not guilty by reason of insanity, this verdict was condemned by many observers and much of the public. Despite this controversy, in practice very few defendants present an insanity defense, and its use does not unduly impede the prosecution of criminal cases.

Age or Developmental Disability

Underlying the insanity defense is the concept that defendants must be able to comprehend their criminal acts for them to have the criminal intent required for a guilty verdict. Two other statuses, age and developmental disability, also raise the question of whether defendants have criminal intent.

We can all agree that some individuals are too young to be able to understand the consequences of their actions and thus to have the required criminal intent. For this reason, if, say, a five-year-old gets angry at a friend, finds and retrieves his parents' loaded and unlocked gun, points it at the friend, says "I'm going to shoot you," and then pulls the trigger, our criminal law holds that the child was too young to understand what he was doing, however tragic the death that results.

How old must defendants be to understand their actions? There is no clear answer to this question, but historically the criminal justice system has considered age 17 or 18 the age at which a teenager turns into an adult. However, many states in recent years have begun to prosecute offenders as young as 13 or 14 as adults and to have the (adult) criminal justice system, rather than the juvenile justice system, handle their cases. Proponents say this shift is necessary to reduce juvenile crime and to punish serious juvenile offenders severely. Opponents say children that young do not understand their actions sufficiently to justify being considered adults and that their criminal tendencies will simply be reinforced if the criminal justice system handles their cases. Studies find that juveniles who are prosecuted as adults have a higher recidivism (repeat offending) rate than juveniles accused of similar offenses whose cases are handed by juvenile court.

We can also all agree that some individuals have developmental disabilities that make it difficult and perhaps impossible for them to understand the consequences of their actions. Whether such defendants should be held criminally responsible has been the subject of much debate, especially in cases involving possible execution upon conviction. In June 2002 the U.S. Supreme Court ruled 6–3 in *Atkins v. Virginia* that executions of mentally retarded individuals violated the Eighth Amendment's ban on cruel and unusual punishment. This decision reversed the Court's 5–4 ruling in a case just 13 years earlier that such executions were constitutional.

THE RIGHTS OF SUSPECTS AND DEFENDANTS: CONTROVERSY AND THE U.S. CONSTITUTION

As noted earlier, a key dilemma of the U.S. criminal justice system is to strike the proper balance between public safety and the rights of those accused of crimes. Many Supreme Court decisions since the 1960s have defined the rights that suspects and defendants have under Constitution. Here we summarize these rights and discuss some of the controversial Court decisions in this area, and we then discuss whether these decisions have hampered the ability of the police to control crime and made our country less safe.

What Rights Do Suspects and Defendants Have?

The Constitution and Bill of Rights list several rights and protections for suspects and defendants that the U.S. Supreme Court has interpreted.

Habeas Corpus This right from Article 1, Section 2 of the Constitution requires that defendants be brought before judges to determine whether their arrest and detainment is lawful. The reasoning behind this guarantee is that defendants should not be allowed to languish in jail. The Patriot Act restricted *habeas corpus* for non-citizens suspected of terrorism.

Probable Cause and Protection Against Unreasonable Search and Seizure These related rights come from the Fourth Amendment and mean that the police and other law enforcement agents may not search someone's person or property, seize that property, or arrest the person without adequate justification. Such restrictions on the police are critical to protect personal privacy and to prevent abuse of police power. The Patriot Act restricted these protections for persons suspected of terrorism.

Double Jeopardy   The Fifth Amendment specifies that no one can be placed in double jeopardy by being prosecuted twice for the same crime. This provision was meant to limit prosecutorial harassment. If someone is acquitted of a crime, they may not be prosecuted again even if new evidence emerges of their guilt.

Self-Incrimination   The Fifth Amendment protects defendants from being forced to testify against themselves and, in so doing, to incriminate themselves. This provision was again meant to limit police abuse and also to require prosecutors to prove a defendant's guilt without any help from the defendant.

Speedy and Public Trial   The Sixth Amendment requires that all defendants "shall enjoy the right to a speedy and public trial." Both these features were designed to protect defendants from government abuse in the form of unjustifiably long detainment and secret legal proceedings.

Trial by Jury   Article 3, Section 3 of the Constitution and the Sixth Amendment both guarantee defendants a jury trial. During the colonial period, juries often came to the aid of colonists unfairly prosecuted by England, and the inclusion of the right to jury trials in both these documents reflected the importance that the new nation placed on the jury.

The Right to Question Witnesses and to Be Told of the Charges   Again from the Sixth Amendment, these rights seem essential in any democracy's criminal justice system.

Right to Counsel   The Sixth Amendment also guaranteed the representation by counsel. For many years, defendants who could not afford to hire an attorney were not able to enjoy this right if their state did not provide them counsel. In 1963, the Supreme Court ruled in *Gideon v. Wainwright* that states must provide free counsel to felony defendants too poor to afford their own;

the court later extended this right to defendants accused of misdemeanors. In practice, many poor defendants today do not get effective counsel, as their lawyers' caseloads are far too heavy to permit adequate legal representation.

Excessive Bail and Cruel and Unusual Punishment   The Eighth Amendment prohibits excessive bail and fines and, more important, "cruel and unusual punishment." During the 1960s and early 1970s, the U.S. Supreme Court interpreted the latter provision as requiring various reforms of prison and jail conditions.

Suspect and Defendant Rights and Controversial Supreme Court Rulings

As the Supreme Court has interpreted various Constitutional provisions, it has imposed restrictions on the police that, according to some critics, have unduly hampered police efforts to control crime and maintain public safety. The most controversial restrictions are the **exclusionary rule** and the **Miranda warning**. Another restriction concerns police use of deadly force.

The Exclusionary Rule   The exclusionary rule arose from several Supreme Court rulings and prohibits the prosecution from using illegally obtained evidence to help convict a defendant. This rule reflects several protections in the Bill of Rights, including the Fourth Amendment prohibition of unreasonable searches and seizures, the Fifth Amendment prohibition of self-incrimination, and the Sixth Amendment's guarantee of counsel, which the Court has said applies once a suspect is arrested.

The key Supreme Court ruling for the development of the exclusionary rule was *Mapp v. Ohio* (1961). This case arose after three Cleveland police officers arrived at the home of Darlene Mapp and demanded entrance, stating that a person wanted in connection with a recent bombing was believed to be hiding at that location. Mapp telephoned her attorney, who advised her not to let the police enter the home without a search warrant. The police forcibly entered the home a few hours later and confiscated "obscene material." At her trial, Mapp was convicted of possessing this material even though the prosecutor never produced a search warrant. The Supreme Court reversed Mapp's conviction because, it said, the police lacked probable cause to arrest her and had no search warrant for her home. As a result, the Court said, any evidence the police found was obtained illegally and thus inadmissible in a court of law.

Before this decision, the exclusionary rule applied only to federal law enforcement officers. The *Mapp* decision significantly extended the exclusionary rule because most criminal cases occur at the state level, not the federal level.

The Right to Counsel after Arrest   We mentioned earlier a defendant's right to counsel. In 1964, the Supreme Court extended this right to the immediate post-arrest period in *Escobedo v. Illinois*. Danny Escobedo had been arrested and interrogated for 15 hours regarding the fatal shooting of his brother-in-law. Eleven days later, he was arrested a second time and taken to a police station for further questioning. Escobedo's attorney arrived a short time later, and, after repeated requests to see his client, he was told this was not possible. When Escobedo asked to see his lawyer, he was informed that the attorney did not want to see him. Police then told Escobedo that he could not see his attorney until they were finished with the interrogation. He then made self- incriminating comments that were later used against him in court, and he eventually was convicted for his involvement in his brother-in-law's murder. Upon appeal, the Supreme Court ruled that Escobedo had been denied the right to the assistance of legal counsel at an important stage of a criminal proceeding. Refusal of the police to permit Escobedo access to his attorney constituted a denial of his Sixth Amendment rights.

The Miranda Warning   Anyone who has seen television police shows has heard the *Miranda* ruling: "You have the right to remain silent. Anything you say can and will be used against you in a court of law. You have the right to speak to an attorney, and to have an attorney present during any questioning. If you cannot afford a lawyer, one will be provided for you at government expense."

This warning resulted from a 1966 Supreme Court case, *Miranda v. Arizona*. In 1963, Ernesto Miranda was arrested in connection with the forcible abduction and rape of an 18-year-old girl in Phoenix, AZ. Two hours after being identified by the victim in a police lineup, Miranda signed a confession admitting his guilt. During the course of the trial, Miranda's attorney asked one of the police officers who took his client's statement if Miranda had been told that "he is entitled to the advice of an attorney before he made it." The officer responded that he did not. The court admitted the confession into evidence and Miranda was convicted. On appeal, the Arizona Supreme Court affirmed the conviction and noted that, since Miranda had been arrested on two previous occasions (and convicted one time), he was familiar with criminal legal proceedings and knowingly waived his rights.

The Supreme Court reversed the Arizona high court's judgment in a close 5-4 decision. The majority decision recognized the need to provide procedural safeguards to protect a suspect's constitutional protection against self-incrimination. The decision specifically declared that suspects must be warned that they have the right to remain silent, that any statement they make may be used against them, and they have the right to consult with an attorney. As with evidence obtained after an illegal search and seizure, any statements or evidence obtained in violation of the *Miranda* rule are not admissible in criminal court. The Supreme Court reaffirmed its *Miranda* decision by a 7–2 vote in a 2000 case in which the majority decision stated that the *Miranda* warning has become part of our national heritage.

Police Use of Deadly Force   The most serious use of police power obviously involves the possible use of deadly force. For many years, the police were fairly free to use deadly force to stop fleeing suspects. In a 1985 case, *Tennessee v. Garner*, the Supreme Court addressed this issue. This case began in Memphis, TN, where police officers who were searching for a prowler spotted a teenager, Edward Garner, near a chain-link fence in a residential backyard. The officers did not see a weapon and believed Garner was unarmed. One officer yelled, "Police, halt," and moved forward. When Garner began to climb the fence, an officer fired his gun at him. The bullet struck Garner in the back of the head, and he died at a hospital a short time later. When this incident occurred, Tennessee was one of 32 states that permitted police to use deadly force against fleeing felons. The Supreme Court decided that the officer in the Garner case could not have reasonably believed that the suspect posed any threat; if someone has committed a burglary, that does not automatically mean that he or she is dangerous. It ruled that police may not shoot fleeing suspects who are otherwise unarmed and not dangerous.

Have These Rights Hampered the Police and Prosecutors?

Suspects or defendants who believe they have been denied any legal right may challenge their prosecution. For example, they may believe they were arrested without probable cause and ask a judge to dismiss the charges. Or they may believe that their homes were searched without a proper warrant and ask a judge to exclude any evidence gathered during the search. They may also challenge their prosecution if they believe that either the police or the prosecutor engaged in various kinds of misconduct. In recent years, disclosures of police misconduct in cities like Los Angeles and Philadelphia forced the dismissal of many charges and the reversal of convictions in cases already decided. In these cities, the police fabricated evidence against drug and other suspects and lied on the witness stand.

If some prosecutions may be dropped because police or officials violated the rights of suspects and defendants, it is natural to wonder whether these rights have unduly hampered police and prosecutors and made our nation less safe. Not surprisingly, both police officers and citizens criticized the Supreme Court decisions of the 1960s (including *Mapp v. Ohio*, *Escobedo v. Illinois*, and *Miranda v. Arizona*, discussed in the previous section) that expanded the rights of suspects and defendants. Regarding the *Miranda* case, a police chief in Texas stated, "Damnedest thing I ever heard, we might as well close up shop." In 1968, President-elect Richard Nixon promised to appoint justices to the Supreme Court who were more receptive to the desires of the law enforcement community than they were to the arguments of criminal defendants. Nixon blamed the "liberal" Warren Court (Chief Justice Earl Warren was a member of the court from 1954 to 1969) for the nation's rising crime rates.

Despite its initial reception, *Miranda* has been embraced by an increasingly large number of police personnel. Police support this ruling because it provides them with clear guidelines for interrogating suspects and does not eliminate their ability to deceive suspects who waive their Miranda rights. The *Miranda* ruling may discourage people from making outright confessions of guilt, but it also encourages suspects to talk to police and often incriminate themselves indirectly. Because they know they do not have to talk to police, they may feel more at ease if they do decide to talk and more likely to say something that will later be used against them.

Reflecting these possibilities, several studies find that the *Miranda* ruling has not unduly hampered the police. One study of 118 interrogations found no evidence that *Miranda* reduced the number of confessions resulting from police interrogation of suspects; this finding has been replicated in other research. Other research finds that the majority of suspects choose to waive their *Miranda* rights after hearing the *Miranda* warning from the police. For example, a Salt Lake City study found that almost 84 percent of suspects chose this alternative.

In sum, the *Miranda* warning does not hamper the police, as the majority of suspects waive their *Miranda* rights and answer police questions anyway, with many confessing. They do so partly because they feel safe with the *Miranda* warning, partly because the police already have a lot of evidence against them, and partly because the police promise them a more lenient outcome or even lie about the strength of the evidence. The law enforcement community as a whole eventually discovered that *Miranda* was a positive step. A California sheriff stated that, "When *Miranda* came down all of law enforcement thought the bad guys had won again. But after some reluctance and suspicion we began to work harder. We became more professional. Instead of relying on outwitting somebody in interrogation, we went and got good evidence." This increased emphasis on gathering good evidence has improved the quality of detective work and led to technological advances to obtain this evidence.

What about the exclusionary rule that suppresses illegally obtained evidence? Although critics initially feared that the police would be unduly shackled in gathering evidence, opposition to the rule has lessened considerably. For example, all 26 Chicago narcotics officers interviewed in one study stated that the exclusionary rule should be retained along with a tightly worded "good faith" exception, meaning that, if an officer intended to comply with rules of gathering evidence but inadvertently made a mistake, the evidence would not be suppressed. One officer stated, "I would not do anything to the exclusionary rule . . . it is not a detriment to police work. In fact the opposite is true. It makes the police department more professional . . . Throughout this department the majority of cases are not hurt by the exclusionary rule." Another officer said, "In the old days if we knew something was in the house . . . we would just knock the door down. Now we use a search warrant."

Studies of the exclusionary rule find it does not hamper police and prosecutors from carrying out their jobs. Two reasons seem to account for this fact. First, conviction in most cases does not depend on evidence gathered from searches of a suspect's home or person. Second, the exclusionary rule is used very infrequently; motions to suppress evidence based on the exclusionary rule are made in only about 5 percent of all cases and only rarely granted. A study of more than 500,000 California cases found that prosecutors rejected less than one percent of the cases because of illegally gathered evidence.

Trends in the Rights of Suspects and Defendants

The major Supreme Court cases that defended the rights of suspects and defendants occurred during the 1960s under the leadership of Chief Justice Earl Warren. Since that time, several Supreme Court rulings under Chief Justices Warren Burger, William Rehnquist, and now John Roberts have restricted these rights as the Court has taken a more conservative turn. Evidence of this trend is seen in a number of Court's decisions involving police searches.

In the 1984 case *U.S. v. Leon*, the Court weakened the exclusionary rule by permitting the prosecution to introduce illegally obtained evidence if police officers obtained this evidence by relying on a search warrant that they reasonably, but erroneously, believed to be valid. In this case, they had arrested Alberto Leon for drug possession after obtaining a search warrant that relied on outdated information and was thus invalid. Because the police believed the search warrant was valid, the Court's ruling established what is commonly referred to as a good faith exception to the exclusionary rule.

In another case, *Ohio v. Robinette* (1996), the Supreme Court also defined police powers to search during traffic stops in a way that made it easier for police to conduct such searches. A police officer had pulled over Robert Robinette for speeding, issued him a warning, and returned his driver's license. The officer then asked Robinette if he was carrying any contraband, illegal drugs, or weapons. When Robinette said he was not carrying any of these things, the officer asked if it would be okay to search his car. Robinette later testified that he "automatically" consented to the search. The officer found a small amount of marijuana and a methamphetamine pill and arrested him for drug possession. The Ohio Supreme Court ruled that the officer had extended the stop beyond its traffic enforcement purpose and used this coercive setting to gain consent for a search from Robinette in violation of his constitutional protection against undue search and seizure. This court ruled that the police are required to inform individuals they are free to go after a lawful detention but before the officer attempts to gain permission for a search.

The U.S. Supreme Court reversed this ruling in an 8-1 vote, with the majority decision declaring that the Fourth Amendment does not require that a legally detained individual be informed that he/she is free to go before an officer requests a search. The Court said it would be unrealistic to expect police officers to routinely inform detainees that they are free to go before a consent to search may be considered voluntary.

In another 1996 case, *Whren v. United States*, the Supreme Court permitted searches conducted after the police used traffic violations as a pretext for stopping motor vehicles in order to search for evidence of other illegal activity. In this case, Washington, DC police had spotted two young males in a new sport utility vehicle lingering at a stop sign in a neighborhood known for drug activity. Suspicious that the two men were engaged in drug offenses but having no evidence of this, the police decided to detain and question them for waiting too long at the stop sign and headed their car toward the vehicle. At this point the suspects sped away. The police caught up to and stopped the vehicle and found a bag of cocaine in each hand of the passenger, Michael Whren. Whren's attorney attempted to have the cocaine evidence suppressed on the grounds that the police traffic stop was merely a pretext for a drug search (which the police admitted). The Supreme Court's unanimous decision stated that pretextual detentions (stopping a vehicle for some minor traffic violation when the real objective was investigating something else for which there was no cause to detain) are lawful.

In 1999, the Supreme Court again expanded police search powers in *Wyoming v. Houghton*. A car occupied by David Young, his girlfriend, and Sandra Houghton was stopped for speeding and having a faulty brake light. A police officer noticed a syringe in Young's pocket and ordered him to step out of the car and produce the syringe, which Young then admitted he used to take illegal drugs. His two passengers were then ordered out of the car and searched for drugs and weapons. The police next searched the car for drugs and found a purse belonging to Houghton. Because the purse contained illegal drugs, Houghton was arrested (while Young and his girlfriend were free to go). During a pretrial hearing, Houghton's attorney asked the judge to suppress the drugs found in her purse on the grounds that the police had no probable cause to search it. The judge disagreed, saying that in view of the driver's syringe, the police did have probable cause to search the car, and thus the purse. Houghton was convicted and sentenced to three years in prison.

The Wyoming Supreme Court reversed her conviction after concluding that the purse search violated Houghton's Fourth Amendment right against illegal search and seizure. The U.S. Supreme Court concluded in *Wyoming v. Houghton* that the police can search a motor vehicle passenger's possessions even if the passenger had nothing to do with the alleged traffic violation and had done nothing to suggest involvement in criminal activity. The Court reasoned that police officers with probable cause to search a vehicle may also inspect passengers' belongings that might conceal the object of that search.

Taken together, these police search rulings and other Court decisions from the last two decades have restricted the rights of suspects and defendants. Many people in the law enforcement community have been overjoyed that the Supreme Court has moved in this direction. After the *Whren* decision one California Highway Patrol Officer noted that the "the game was over. We won." Critics said the Court had crossed the line between public safety and individual liberty. The proper extent of police power in American democracy will undoubtedly remain an important debate for many years to come, with the attacks of September 11 only intensifying the controversy this debate represents between public safety and individual freedom.

KEY TERMS

actus reus: As an element of crime, an actual act committed

concurrence: As an element of crime, the correspondence of criminal intent and a criminal act

crime: A behavior deemed so harmful to the public welfare that it should be banned by criminal law

criminal law: The body of law that prohibits acts that are seen as so harmful to the public welfare that they deserve to be punished by the state, and that governs how these acts are handled by official state procedures

exclusionary rule: The principle that evidence obtained by the police in violation of a suspect's constitutional rights may not be used to prosecute the defendant

felony: A serious crime punishable by a sentence of at least one year in prison

harm: As an element of crime, the requirement that some injury is done to an individual or property by a criminal act

infraction: A minor violation of the law punishable by only a fine

mala in se: Offenses that are inherently evil; serious crimes

mala prohibita: Offenses that are crimes only because the law prohibits them; often applied to "victimless" crimes

mens rea: As an element of crime, a guilty mind or criminal intent

misdemeanor: A minor offense punishable by a sentence to jail of less than one year

violation: Also an infraction, a minor offense punishable by only a fine

SUGGESTED READINGS

Cole, David, and James X. Dempsey. 2002. *Terrorism and the Constitution: Sacrificing Civil Liberties in the Name of National Security.* New York: W.W. Norton.

Davenport, Anniken. 2006. *Basic Criminal Law: The United States Constitution, Procedure and Crimes.* Upper Saddle River, NJ: Prentice Hall.

Friedman, Lawrence M. 2004. *Law in America: A Short History.* New York: The Modern Library.

Head, Tom (ed.). 2004. *The Bill of Rights.* San Diego: Greenhaven Press.

Lewis, Anthony. 1964. *Gideon's Trumpet.* New York: Random House.

Rogers, Richard. 2000. *Conducting Insanity Evaluations.* New York: Guilford Press.

"Speed Bump" © Dave Coverly/Dist. By Creators Syndicate, Inc.

chapter eight

Why They Break the Law

Why does crime occur? Why do some people commit crime and other people not commit it? Are parents too permissive? Is the criminal justice system too lenient? Do criminals have defective genes? Are poverty and overcrowding to blame? As we stressed on the first pages of this book, in order to know how to reduce crime, we need to know what causes it. If the main problem is that the criminal justice system is too lenient, then we should crack down on crime and criminals with more arrests, more convictions, and longer prison terms. If we think parents are too permissive, then we should encourage parents to be stricter. If instead we blame poverty and overcrowding, then we should implement social programs to address these problems.

In this chapter we discuss the major explanations for crime and the research evidence for and against these explanations. We think the evidence is clear that crime stems much more from various social and economic problems than from a lenient criminal justice system. To the extent this is true, American public policy and government spending should focus much more on addressing these problems than on putting more people behind bars and building more prisons. Before we get too far ahead of the game, let us turn to the many explanations of crime to see what they say and to examine the evidence for and against their assumptions and beliefs.

RATIONAL CHOICE AND DETERRENCE

Rational Choice Theory: Where There's a Will, There's a Way

The idea that crime is a freely made choice reflects the more general view in the social sciences, and especially in economics, that people act rationally by carefully weighing the potential rewards and risks before engaging in any behavior. This view is called **rational choice theory**. In the study of crime, the rational choice perspective assumes people decide to commit a crime by carefully weighing the potential rewards and risks.

However, many criminologists question the validity of this assumption for the majority of criminals. They have at least two reasons for their skepticism. First, as we pointed out in Chapter 4, many violent crimes, including most homicides and assaults, are very emotional and rather spontaneous. As such, people who commit these crimes do not pause to assess the consequences of their actions before trying to injure and even kill someone. Instead, they often act without thinking. Other crimes such as robbery, burglary, and motor vehicle theft are often at least somewhat planned and fairly unemotional, and the people who commit them might indeed weigh the consequences of their action as the rational choice view assumes.

That brings us to a second reason to be skeptical about the rational choice view when applied to criminal behavior. Studies of convicted offenders find that about half of them were drunk or on drugs when they committed their offense. Because they were drunk or on drugs, it is obviously very doubtful that they carefully weighed the consequences of their actions before breaking the law. After assessing the evidence on whether criminals act in the way rational choice assumes, criminologist Samuel Walker concluded, "Actual offenders do not appear to make their decisions about criminal activity on the basis of a rational and carefully calculated assessment of the costs and benefits." Many and perhaps most readers of this book might assess the costs and benefits of our intended behavior, but criminals do not.

Deterrence Theory: The Risk Isn't Worth It

An offshoot of rational choice theory is **deterrence theory**, which assumes that potential criminals can be deterred from breaking the law by increasing the threat of arrest and punishment. As we discussed in the opening pages of this book, this assumption has guided criminal justice policy in the United States during recent decades, as our nation has followed a get-tough strategy involving longer prison terms, more prisons, and increases in the number and powers of the police.

Although deterrence theory sounds quite plausible, criminologists are again skeptical for at least two reasons. First, for potential criminals to be deterred by the threat of arrest and punishment, that implies that they weigh their risk of arrest and punishment carefully before committing their offense. Yet, as we have just seen, many offenders do not sit down and weigh this risk, because they are either acting emotionally and/or because they are too drunk or high. And if they do not do weigh their risk of getting caught, then they cannot be deterred by the prospect of arrest and punishment. Second, while robbers, burglars, car thieves, and other such criminals do plan their crimes to at least some degree, as both rational choice and deterrence theories assume, studies find that these offenders still do not give much thought to their risk of getting caught or, when they do give it some thought, decide they will not be caught or resign themselves to being caught. Once they make sure that no police or bystanders are nearby, they go ahead with their crimes. To the extent this is true, they are not deterred by the threat of arrest and punishment.

If studies find that criminals do not act in the way that rational choice and deterrence theories assume, then we would not expect that the threat of arrest and punishment significantly deters criminal behavior. It should come as no surprise, then, that most research fails to find that increases in the risk of punishment have a deterrent effect. In particular, it fails to find that increases in the penalties for a crime reduces the rate of that crime.

For example, although states passed firearm sentence enhancement (FSE) laws during the past few decades that require minimum prison terms or longer prison terms for people committing gun crimes, the rates of these crimes generally have not gone down. Thomas B. Marvell and Carlisle E. Moody studied the effects of FSE laws, in all 44 states that passed such laws, beginning in the 1960s. In some states gun crimes decreased, while in other states gun crimes increased. This inconsistent pattern suggested to Marvell and Moody that "on balance the FSE laws do little nationwide to reduce crime or gun use." The so-called three-strikes laws that require a life sentence or extremely long prison term for anyone committing a third felony (in some states a second felony) also have not reduced crime rates in the states that implemented these laws. Some scholars even think that three-strike laws may have increased the homicide rate because felons did not want to let a witness live who might testify that the felon had committed his third felony.

Two other kinds of evidence suggest that deterrence does not work. First, as the number of prisoners has increased steadily since the 1970s, one would expect the crime rate to decrease steadily if the increased threat of imprisonment was having a deterrent effect. Yet this steady decrease did not occur; in particular, violent crime rose sharply during the mid to late 1980s. Second, if the threat of imprisonment deters crime, then states with higher incarceration rates (the number of prisoners per 100,000 residents) and longer prison terms should have lower crime rates than states with lower incarceration rates. Yet states with higher incarceration rates and long prison terms often have very high crime rates.

If deterrence does not work for potential offenders in the general public (this type of deterrence is called *general deterrence*), perhaps it might still work for prisoners once they are released back into society (this type of deterrence is called *specific deterrence*). In this way of thinking, exprisoners dislike the prison experience so much that they should not want to risk another prison term by committing additional crime once they are back in society. If specific deterrence does work in this manner, then inmates who serve longer prison terms should have lower recidivism (repeat offending) rates once they are back on the streets than inmates who serve shorter prison terms.

Once again, however, most research finds that the threat of punishment does not deter ex-prisoners' criminal behavior. If anything, the research finds an opposite effect: More severe punishment makes prisoners *more* likely to commit new offenses, perhaps because they become more embittered and have more time in jail or prison to be influenced by other inmates.

At least two kinds of evidence support this pessimistic conclusion. In the last few decades, increasing imprisonment rates have forced criminal justice officials to release offenders who were already behind bars. Researchers have compared the recidivism rates of these offenders to similar offenders who stay behind bars and serve out their sentences. If a specific deterrent effect exists, the recidivism rate of the latter offenders should be lower than that of the ones who get out early. However, research on this issue does not find this difference: The offenders who stay behind bars do not have lower recidivism rates, and sometimes they even have higher ones.

In the second kind of evidence, several states got tough on juvenile offenders during the 1980s and 1990s by transferring many of their cases to the adult criminal courts, where their punishment theoretically would be more severe. If a specific deterrent effect exists, the recidivism rate of juvenile offenders transferred to criminal court should be lower than that of similar offenders who stay in the juvenile justice system. Yet research on this issue finds that juveniles transferred to criminal court have higher recidivism rates than matched groups of juveniles who remain in the juvenile justice system. In fact, harsher treatment of juvenile offenders seems to increase recidivism, the opposite of what deterrence theory would predict.

Overall, then, the research on deterrence does not suggest that more certain and/or severe punishment can lead to significant crime reduction by deterring potential offenders in the general public or exprisoners released back to society. In drawing this conclusion, we do not mean to imply that the criminal justice system is ineffective. Its very existence certainly helps keep the crime rate lower than it would otherwise be (an effect called absolute deterrence). But the relevant issue is whether relatively small changes in the risk of criminal punishment affect crime rates to a substantial degree (an effect called marginal deterrence). Most of the many studies of this issue conclude that the threat of punishment has no effect, or at best only a small effect, on criminal behavior. If so, efforts to fight crime that rely on more certain and severe prison terms are bound to fail and, indeed, have been failing since they began in earnest three decades ago.

BIOLOGICAL AND PSYCHOLOGICAL EXPLANATIONS

Biological and psychological views obviously differ in many respects, but both assume that the roots of crime lie within the individual rather than the social environment. Simply put, these explanations say that criminals are biologically or psychologically different from noncriminals. More to the point, they say that certain biological and psychological problems make people more likely to commit crime and that criminals are more likely than noncriminals to have these problems.

Biological Views

Several contemporary biological explanations of crime exist. We do not have the space to discuss all of them here and will focus on genes, hormones, diet and nutrition, and pregnancy and childbirth complications.

Genes and Twin Studies: Predisposed from Conception?   Although no specific gene for criminal behavior has been located, many scientists think crime has at least some genetic basis. They infer this from studies of identical twins. Because such twins have identical genes, we would expect them to behave very similarly if genes affect their behavior. Many twin studies find that if one twin has a history of criminal behavior, so often does the other twin. This pattern leads many researchers to believe that some unknown genetic pattern helps produce criminal behavior, and, thus, that criminal behavior can be inherited.

But other researchers point out that identical twins are similar in many respects other than their genes. They are socialized the same way by their parents, have the same friends, and spend much time with each other. Because all these similarities may lead them to have similar behavior when it comes to crime, these researchers say a strong genetic role in crime cannot be inferred. As a middle ground, some scholars say that genes and the social environment interact in producing criminal behavior: Genes may predispose some individuals to crime, but this predisposition is not activated unless problems in the social environment exist.

Hormones: Boys Must Be Boys and Girls Go Crazy   Testosterone is often mentioned as a reason why males commit more crime than females and why some males are especially likely to commit serious crime. Several studies find a link between higher testosterone levels and greater histories of crime and aggression. Although this correlation suggests that testosterone produces aggression, it is also possible that aggression (and the dominance it involves) increases testosterone. This possibility prompts some researchers to doubt that higher levels of testosterone in humans produce higher levels of aggression.

In women, premenstrual syndrome, or PMS, has been linked in some research to greater aggression. Theoretically, some women become extremely irritable, tense, and/or depressed in their premenstrual phase and these changes increase their chances of becoming violent. Although an early study of women prisoners in Britain supposedly found this effect, other researchers soon challenged the study's methodology, and a causal link between PMS and has not been demonstrated.

Diet and Nutrition: Too Many Twinkies?   Many parents try to limit their children's sugar intake for fear that their sons and daughters may become hyperactive and even aggressive on a sugar high. Researchers have studied the possible link between aggression and diet and nutrition. One line of research involves adjusting the diet of an experimental group (of children or adolescents) and comparing the behavior of its members to those in a control group. Another line of research involves comparing the diet of actual juvenile offenders to that of nonoffenders. Some studies in both lines of research point to excessive sugar and certain chemical additives and deficiencies in certain vitamins as possible contributors to aggressive behavior. But several methodological problems limit the value of these studies, and several studies do not find a link between diet, nutrition, and aggression. A review commissioned by the National Academy of Sciences concluded that any effect of diet and nutrition on criminality is very small.

Pregnancy and Birth Complications: Trouble Right from the Start   If a fetus's neurological development does not proceed normally, a newborn can suffer lon_glasting cognitive, emotional, and behavioral impairment. Several types of problems, including poor prenatal nutrition and the use of alcohol, tobacco, and other drugs, can affect this neurological development. A difficult birth that includes a lack of oxygen can affect neurological development and lead to various impairments. These impairments have, in turn, been implicated in aggressive behavior in children and later criminality. Thus, pregnancy and birth complications may help to produce criminal behavior many years later. Fortunately, these complications are not common and at most help account for only a very small proportion of criminal behavior.

Psychological Views

Psychological views trace crime to various psychological problems arising within the individual. In general, these problems are thought to stem from childhood experiences and difficulties. We again have space only for the major explanations.

Psychoanalytic Explanations: "Id" Made Me Do It   Drawing on the work of Sigmund Freud, psychoanalytic views assume that criminal behavior stems from the failure of individuals to adjust their instinctive needs to society's needs. In particular, Freud and his followers assume that the individual personality is composed of three components: the id, the instinctive part of the personality that selfishly seeks pleasure; the ego, the rational part of the personality that recognizes the negative aspects of pure selfishness; and the superego, the part of the personality that represents society's moral code and acts as the individual's conscience. These three components need to be in harmony for an individual to be mentally healthy. When this is not so—usually because of abnormal early childhood experiences—mental disorders, including criminal behavior, can result.

While psychoanalytic explanations sound appealing, they are difficult to prove or disprove. For example, if a researcher studies someone with a history of crime, the researcher might conclude the person's superego is too weak, but, if the person's criminality is used as evidence of a weak superego, the researcher is guilty of circular reasoning. Psychoanalytic views also imply that a person with a history of criminal behavior suffers from a mental disorder, but many crime scholars dispute this point. While conceding that some criminals do have mental disorders, they argue that most criminals are not psychologically abnormal despite their history of crime.

Still, psychoanalytic explanations remain valuable because of their emphasis on early childhood experiences. Certain problems in early childhood can increase the risk of behavioral problems. Regardless of the continuing value of psychoanalytic explanations, much research today focuses on negative childhood experiences and their consequences for later behavior.

Personality Problems   One consequence of negative childhood experiences might be various personality problems, including impulsiveness, irritability, and hyperactivity, that are thought to underlie aggression and other behavior problems in children. These behavioral problems in turn contribute to juvenile delinquency and then adult criminality as the child ages. Support for the importance of early personality problems comes from research that studies children from infancy through at least young adulthood. Young children who develop various personality problems, often because of inadequate parenting, are more at risk than those who do not develop these problems for antisocial behavior during childhood and delinquency during adolescence. This does not mean that all children with personality problems are doomed to delinquency when they reach their teens, just that they are more likely than those without such problems to end up committing delinquency and crime.

Low IQ   Using standard IQ tests as measurements of intelligence, some studies find a causal relationship between low IQ and higher rates of crime and delinquency. If (and this is a big "if") this relationship exists, one reason may be that youths with low IQ are less able to understand the consequences of their actions and to be more vulnerable to the influences of friends who break the law. However, many scholars question the validity of IQ tests as measures of intelligence and say the tests are culturally biased against people who do not come from white, middle-class backgrounds. Other methodological problems, including the lack of adequate control groups, also call into question the presumed low IQ-high delinquency relationship. Because, historically, low scores on IQ tests were used to suggest that African Americans and immigrants were biologically inferior, we must be very careful in interpreting the results of the IQ research for crime and delinquency.

SOCIOLOGICAL EXPLANATIONS

Sociological explanations assume the causes of criminal behavior lie outside the individual in the social environment. The social environment does not totally determine who will commit crime, but it does have an important influence. Several sociological explanations exist.

Social Ecology: Location, Location, Location

Social ecology explanations emphasize that certain physical and social characteristics of communities increase their rates of crime and victimization. Rodney Stark calls these approaches kinds of places explanations rather than kinds of people explanations. Different types of people, he notes, can move into and out of various neighborhoods, but certain types of neighborhoods will generally have higher crime rates no matter which kinds of people live there. Among other features, these neighborhoods generally have high residential density. In neighborhoods where many people live closely together, good kids are more apt to come into contact with bad kids and thus be more likely to break the law. In neighborhoods with weak families, schools, churches, and other social institutions—generally these are neighborhoods with high rates of poverty, dilapidation, and other problems—social bonding is weak (a condition sociologists call social disorganization) and delinquency and crime thus more common.

Other ecological factors also matter. Neighborhoods with higher rates of collective efficacy, in which neighbors watch out for each other's kids, have lower crime rates. Neighborhoods with low rates of participation in voluntary organizations such as churches and community groups and higher numbers of bars and taverns have higher crime rates. Taken together, ecological explanations help us understand the reasons why some neighborhoods have higher crime rates than other neighborhoods. They suggest that criminality stems to a large extent from the social and physical characteristics of the places in which individuals live.

Poverty and Blocked Opportunity: The System's at Fault

Much criminological research emphasizes the importance of poverty (also called economic deprivation) for criminal behavior. While several poverty-based explanations exist, most assume that poverty in a society like the

United States that places such great value on economic success is especially frustrating for the poor, who feel relative deprivation as they compare themselves to other Americans who are wealthier. They face great difficulties in lifting themselves out of poverty. They have relatively little education, their children go to low-quality schools, and so forth. Their opportunity to move up the socioeconomic ladder, in short, is blocked by all sorts of factors, and such blocked opportunity is said to create angry aggression that translates into violence and other crime. As Elliott Currie has written, "[H]arsh inequality is . . . enormously destructive of human personality and of social order. Brutal conditions breed brutal behavior."

Peer Influences and Learning: They Told Me to Do It!

An important part of the social environment, according to both psychologists and sociologists, is the various people with whom we interact. The greatest influence on young children is their parents, but, as they become preteens and then adolescents, their friends become increasingly important. Accordingly, learning theories in criminology emphasize that juveniles *learn* to break the law—or, more specifically, learn attitudes and values that help them decide to break the law—from their friends and acquaintances. They may also break the law simply because they want to conform to their friends' behavior in order to fit into the crowd and avoid ridicule or even the loss of friendships.

To the extent that peer influences and learning matter, criminal behavior stems from one of the most important, normal social processes: socialization. Just as most people are socialized to become law-abiding members of society, some people are socialized to become law-breaking members. Just as adolescents' friends influence them in all sorts of ways—for example, their taste in music and clothing—so do friends influence adolescents in many aspects of their behavior, including deviant behavior. Supporting this view, many studies find that adolescents with delinquent friends are more likely than those with fewer or no such friends to be delinquent themselves. Some scholars consider the effect of delinquent peers to be a more important cause of delinquency than any other factor.

Mass Media: The TV Made Me Do It

Many people are concerned about the impact of mass media violence on youth violence, and many researchers have examined this issue. In some studies, children and college students are shown violent videos and then watched as they play (children) or given questionnaires to fill out (college students). When compared to control groups who watch a nonviolent video, the children play more violently, and the college students reflect more violent attitudes in their answers to the questionnaire. In other research, random samples of respondents are asked in self-report studies about the television shows they watch, the movies they see, the music they listen to, and the video games they play. Those who report exposure to violent media in any of these forms also report being more violent in their behavior. These studies lead many researchers to conclude that violence in mass media has a strong influence on youth violence.

However, other researchers say this conclusion is premature. The violence found among the studies involving children and college students who watch violent videos, they say, is only short-term and does not necessarily mean that mass media violence has a strong influence in the real world. They add that the correlations found in self-report studies between mass media exposure and violent behavior also do not mean a causal relationship. Perhaps individuals interested in violence are more likely to want to watch violent movies or play violent video games. Although much of the public and many scholars think the mass media share a large part of the blame for youth violence, the research evidence on this issue is far from clear-cut.

Social Bonding: Out of Control?

An important goal of any society is to control its members' behavior. A society is more stable if it has strong social bonds and can socialize its members to respect and conform to society's moral codes of behavior. Accordingly, several sociological explanations highlight the factors that keep people from becoming deviant. In this sense, they represent the flip side of the explanations discussed so far that focus on the factors that influence people to break the law. By understanding which factors help induce conformity, we better understand those that induce criminality.

Sociologists focus on two sorts of social controls: internal or personal ones such as the ability to delay gratification, and external ones found in an individual's social environment. Family and school bonds lie at the heart of *social bonding theory*, which says that strong bonds to parents and schools help keep adolescents from becoming delinquent. Teens who feel closest to their parents and like school the most presumably are more ready to accept the values of their parents and teachers and care more about what their parents and teachers think about them. For these reasons, they should be more likely to obey the law (or less likely to commit delinquency). In support of this view, many studies find that adolescents with the strongest social bonds to their parents and schools are indeed less likely to be delinquent. Citing a chicken-or-egg problem, some scholars question whether this relationship shows that these social bonds reduce delinquency or, instead, simply that delinquency weakens adolescents' bonds to their parents and schools. Still, most criminologists probably agree that social bonding does indeed reduce delinquency and thus later criminality.

Related research documents the importance of family interaction for delinquency and other antisocial behavior; harmonious families help produce well-behaved children, while conflict-ridden families are more apt to produce children with behavioral problems. Research is less clear on the importance of family structure. Some researchers find that children from single-parent households are at only slightly greater risk for delinquency, usually drinking and drug use and status offenses like skipping school. The major problem with single-parent arrangements, these researchers say, is that they are more likely to be low-income ones. Other researchers say children of single parents are indeed at much greater risk for delinquency because the one parent is less able to supervise the children and to be a good parent in other respects.

Research also documents the importance of schooling for delinquency. Here the evidence is fairly clear: Students who get good grades, who like their schools and teachers, and who are involved in school activities are much less likely to be delinquent than those who get poor grades, who dislike their schools and teachers, and who are less involved in school activities. But the chicken-or-egg question remains: Are the good students less delinquent because they are good students, or are delinquents less likely to be good students because they are delinquents?

An offshoot of social bonding theory is *self-control theory*. According to this theory, a lack of self-control is responsible for all forms of crime. People who cannot restrain themselves, who are impulsive, and who can only live for the present are much more likely to commit crime than those who have more self-control in all these respects. Low self-control stems from inadequate or poor parenting, starting in infancy and extending into adolescence and lasting well into adulthood.

Supporting the theory, much research finds that people who score low on various measures of self-control are more likely to have broken the law. Critics say a chicken-or-egg problem again exists: Does low self-control promote criminality, or does criminality lead to low self-control? They also take issue with the assumption that low self-control is more important in explaining crime than other factors such as poverty and peer influences.

Labeling

A robber who kills someone may be arrested, prosecuted, and imprisoned or even executed. A soldier who kills several people on the battlefield may well be awarded a medal. In either case, killing has occurred, yet the circumstances affect how society thinks about the killings. The sociological explanations discussed so far do not address society's reaction to crime and criminals. Other sociological views take a more critical look at crime and society. They say the definition of crime is problematic and question whether bias in the criminal justice system makes it more likely that some people and behaviors will be labeled as criminals. They also say that much crime is rooted in the very way that our society is organized.

One of these critical views is **labeling theory**. As the killing example illustrates, whether a specific behavior is considered deviant or wrong may have more to do with the circumstances surrounding the behavior than with the behavior itself. This view lies at the heart of labeling theory, which recognizes that behaviors are not automatically deviant and, in fact, become deviant if and only if society decides that they are. This labeling process, the theory adds, may be inaccurate. People may be labeled deviant when they have done nothing wrong, or they might not be labeled deviant when they have done something wrong. This inaccuracy may stem either from honest mistakes or, worse, from bias based on someone's gender, race and ethnicity, social class, age, appearance, or other factors. Thus, not only is the definition of crime and deviance problematic, so is the process by which certain individuals come to be labeled criminals.

A classic study of the "Saints" and the "Roughnecks" by sociologist William Chambliss illustrates labeling theory's view on bias. The Saints were a group of middle-class delinquents who were able to skip their high school classes without getting into trouble. They would drive to a nearby town and commit vandalism and other offenses with no one the wiser. Although some school officials and other townspeople suspected the Saints were up to no good, they considered the youths good kids from respectable, stable families and thus did not act on their suspicions. In contrast, the Roughnecks were a group of poor delinquents who often got into trouble for fighting and other offenses. School officials and police tended to watch them carefully and to sanction them when they misbehaved. Overall, the Saints were more delinquent than the Roughnecks, wrote Chambliss, but got into trouble far less often. They entered professional careers after high school and college while the Saints ended up in prison or with dead-end jobs.

Much research that tests labeling theory assesses the degree to which a suspect's gender, race and ethnicity, and social class affect the chances of the suspect being officially labeled a criminal. The evidence is mixed: some research finds considerable bias on all these dimensions, while other research finds only a relatively small amount of bias. Some research argues that much bias exists in legal processing, while other research claims that any bias is fairly minimal and that legal factors matter more than nonlegal ones like race and gender.

Labeling theory makes one more provocative point that is important for our earlier discussion of deterrence theory: people who are labeled deviant become more likely to commit deviance because they were labeled. Labeled individuals develop a deviant self-image and find themselves shunned or suspected by law-abiding individuals. Faced with these problems, they discover that it is easier to hang out with other people who have also been labeled deviant. Thus, they become more likely to commit deviant acts because they were labeled deviant. If this is true, labeling has the opposite effect from what is intended and from what deterrence theory would predict.

This argument of labeling theory is very appealing. Pretend you just got out of prison after spending five years there for armed robbery. Having paid your debt to society, you fill out application forms for several jobs, and each form asks you whether you have ever been convicted for any crimes. You write down armed robbery on all the forms. How likely is it that you will get a job? Now suppose you are in a bar or at a party trying to meet people. You start talking to someone who attracts you and who is evidently attracted by you. She or he asks you what you do for a living. You say you are looking for a job. You are then asked what you have been doing for the last few years. You respond that you just got out of prison for armed robbery. Would your companion respond enthusiastically, or would she or he make an excuse to go to the restroom? As this scenario suggests, once people are labeled criminals, they find it difficult to fit back into society. Sometimes they are even suspected of crimes they did not commit. Perhaps labeling theory, then, is correct when it says that labeling produces more deviance, not less. Ironically, harsh legal punishment may make some criminals worse, not better.

Conflict Theory: The Golden Rule or He Who Has the Gold Rules

The *conflict tradition* in sociology began with the work of Karl Marx and Friedrich Engels in the 19th century. They wrote that capitalist society was divided into two major social classes: the *bourgeoisie*, who owned the means of production such as factories and tools, and the *proletariat*, who worked for the bourgeoisie. Reflecting their class interests, the bourgeoisie aim to maintain their elite position by oppressing and exploiting the poor, while the proletariat aim to change society to make it more equal. This simple summary of Marx and Engels' views does not do it justice, but it does indicate that they saw society filled with class conflict. More to the point for our discussion, Marx and Engels thought that law played a key role in this conflict and wrote that the ruling class uses the law to maintain its power.

Conflict theory in criminology makes two major points: First, the structure of American society leads to crime by the poor. Second, the law reflects the interests of the ruling class and is used by the ruling class to reinforce its position at the top of society.

Expanding on the first point, capitalism is said to produce crime by the poor for the following reasons. As an economic system, capitalism emphasizes competition, individualism, greed, and selfish behavior. In such a system, people will be more likely to perform actions that help themselves even if they hurt others. Some of these acts are criminal acts. Although all social classes will thus commit various kinds of crime, the poor have an additional motivation, and that is economic need. Capitalism thus produces crime because it encourages selfishness, and the poor commit higher rates of street crime because of their economic need.

Expanding on the second point, much research focuses on the extent to which social class and race and ethnicity affect criminal justice outcomes. We already commented on this research in our discussion of labeling theory and reiterate that its results are mixed. Some researchers say that the strongest support for conflict theory is found in the way the legal system treats white-collar crime by corporations and wealthy individuals. Laws against such crime are weak, they say, and punishment is often minimal, even though white-collar crime can be very harmful.

Feminist Views: Battle of the Sexes

Feminism has made important contributions in many scholarly disciplines, and the study of crime is no exception. Feminist work on crime and justice deals with several issues. One is why girls and women commit crime. Most of the sociological explanations discussed just above were developed with males in mind or tested only with data on males. Their neglect of girls and women left their relevance for female crime unclear. Addressing this issue, research generally finds that the same factors that affect male criminality, including poverty and family problems, also affect female criminality. One interesting question is that, even when these problems do exist, males are still more likely than females to commit serious crime. Gender differences in socialization and opportunities to commit crime appear to explain this difference (see Chapter 3). Some researchers say that male socialization patterns greatly help to produce male criminality and call for major changes in the ways boys are socialized.

A second issue addressed by feminist work is the victimization of women. Before feminists began studying crime in the 1970s, rape, domestic violence, and other crimes in which girls and women are especially likely to be victims received little attention. In the 1970s, rape became a central concern of the women's movement as researchers emphasized that rape and sexual assault were widespread and stemmed from sexist cultural views and from gender inequality rather than from provocative behavior by women themselves. The movement soon turned its attention to domestic violence and again emphasized the high incidence of such violence and its roots in cultural views and in gender inequality. Researchers also documented how victims of rape and domestic assault were treated in the legal system where, it was said, a second victimization occurred as legal professionals doubted their word and often put them on trial. Their findings led to major reforms in the criminal justice system. Victim advocate offices now exist in many communities, and various criminal justice procedures have been changed. For example, rape shield laws preventing rape victims from being asked in court about certain aspects of their sexual history now exist throughout the country.

A third issue addressed by feminist work is the treatment of female suspects and offenders by the legal system. Are police more or less likely to arrest females than males? Once they are arrested, are females less or more likely to receive long prison terms? The evidence is again inconsistent and depends to some extent on whether adolescents or adults are being considered. Some, but not all, studies find that girls are more likely than boys to get into trouble for various delinquent offenses such as skipping school and sexual promiscuity. Most studies of adults find that the relatively few women suspected of serious crimes are treated somewhat more leniently than their male counterparts, but they also find that this gender bias in favor of women is stronger for white women than for women of color.

THEORY AND POLICY, OR HOW TO REDUCE CRIME AND HOW NOT TO REDUCE IT

We said at the beginning of this chapter that sound explanations of crime are necessary for the development of successful strategies to reduce crime. All the explanations discussed in this chapter have important implications for such strategies, but some of the explanations and strategies offer much more promise than others.

Some strategies will not work to reduce crime, even though these strategies have been at the heart of U.S. criminal justice policy for three decades. We are talking, of course, about the get-tough approach. Deterrence does not and cannot work effectively to reduce crime for the reasons discussed earlier, and the research evidence on deterrence in action (e.g., the use of harsher sentencing) yields little evidence that the get-tough approach has achieved a significant deterrent effect on crime. If our nation wanted to ensure that it will not reduce crime at all or at most will reduce crime to a very small extent at the cost of tens of billions of dollars, it would continue doing precisely what it has been doing for three decades.

The potential of crime-reduction strategies based on biological explanations depends on which explanation we have in mind. In this regard, research on pregnancy and birth complications probably holds the most promise, as it indicates that improvements in prenatal nutrition and other prenatal problems should help reduce crime rates to at least some degree. But other biological explanations might hold less promise, in part because their policy implications pose ethical and practical difficulties. For example, suppose scientists someday find specific genes that make people more likely to commit crime. How could this knowledge be used to reduce crime? Would genetic engineering be in order? Would children be tested to determine who has those genes? Once they were identified, what would their lives be like? As these questions suggest, crime-reduction policies developed from genetic research would be fraught with ethical and practical dilemmas. The same problems affect hormonal research. Suppose high testosterone is eventually proven to raise criminality. A logical step would then be to somehow identify males with high testosterone levels and to reduce these levels, perhaps by giving them a drug or even by castrating them. Again, any policy like this raises serious ethical and practical dilemmas even if we only reduced the testosterone levels of males who had already broken the law.

More-promising crime-reduction strategies are based on the emphasis in certain psychological explanations on negative childhood experiences and developmental impairments. The children at greatest risk for such problems are typically those born to young, poor, unwed mothers. Efforts to help these children should ultimately help reduce crime. These efforts include home-visitation programs in which nurses, social workers, and/or other trained professionals make regular visits to these children's homes right after birth and for many weeks thereafter. At these visits, they give the mother valuable practical advice and moral support. Other efforts include parent management training programs in which parents of children with behavioral problems are instructed on discipline and other parenting skills. While all these efforts are still fairly recent, a growing body of evidence indicates they are very effective at reducing later developmental and behavioral problems among high-risk children. For adolescent offenders, a multifaceted approach involving family therapy, parent management training, and conflict resolution counseling and programming in school and peer settings has also proven effective.

Other promising strategies stem from the various sociological explanations, although the specific measures they suggest again depend on which explanation we have in mind. For example, social ecological explanations suggest the need to focus on certain physical and social characteristics of communities. While we cannot wave a magic wand and fix everything overnight, efforts that successfully reduce residential density, promote greater involvement in neighborhood voluntary associations, and address other criminogenic community characteristics should all help reduce crime. Explanations emphasizing blocked economic opportunities also suggest the need to reduce poverty. If poverty leads to feelings of angry aggression, relative deprivation, and other social psychological states that promote street crime, then efforts to reduce poverty should help greatly to reduce crime.

Social bonding theory also has several policy implications. It directs our attention to the family and school as critical sources of attitudes that either promote criminality or inhibit it. It suggests the need to help parents, especially those who are young, poor, and unwed, to improve their parenting skills; it also suggests the need to improve our schools by having smaller classrooms and better school buildings. While little public policy has focused on school improvement, early childhood intervention programs show significant potential for crime reduction.

Finally, critical approaches to crime also have policy implications. Labeling theory implies we should be cautious in arresting and imprisoning at least some types of offenders lest they become worse as a result of their labeling. And it also suggests the need for careful attention to whether the labeling process itself is fair or biased. Conflict theory suggests a similar need, and it also highlights the possible criminogenic effects of values such as competition, individualism, and selfishness. Meanwhile, feminist research directs our attention to the need to take all possible steps to reduce the victimization of women by rape, sexual assault, and domestic violence, and it reminds us of the need to change male socialization patterns in order to reduce male crime rates.

The policy implications of all the explanations discussed in this chapter suggest the need for a multifaceted approach in reducing crime. Although the criminal justice system remains important in keeping us safe from the offenders we already have, efforts that prevent crime from arising in the first place are critical to have the safest society possible. Although all the explanations discussed in this chapter have different implications for how best to prevent crime, they hold more promise overall for crime reduction than a mere reliance on the criminal justice system via the get-tough approach.

KEY TERMS

deterrence theory: Closely related to rational choice theory, the belief that more certain and severe punishment reduces the crime rate

free will: The view that human behavior is the result of personal, independent choices and not of internal or external forces beyond a person's control

labeling theory: The view that the definition of crime and deviance depend on the circumstances surrounding a behavior rather than on the qualities of the behavior itself, that labeling produces more deviance, and that the labeling process is biased and therefore problematic

rational choice theory: The view that people act with free will and carefully weigh the potential benefits and costs of their behavior before acting

self control theory: Michael Gottfredson and Travis Hirschi's view that all crime results from the inability to restrain one's own needs, aspirations, and impulses

social ecology: In criminology, the view that certain social and physical characteristics of neighborhoods help explain their crime rates

social bonding theory: Travis Hirschi's view that social bonds to family, school, and other conventional social institutions inhibit the development of criminal behavior

SUGGESTED READINGS

Akers, Ronald L., and Christine S. Sellers. 2007. *Criminological Theories: Introduction, Evaluation, and Application*. New York: Oxford University Press.

Cullen, Francis T., John Paul Wright, and Kristie R. Blevins (eds.). 2006. *Taking Stock: The Status of Criminological Theory*. New Brunswick, NJ: Transaction Publishers.

Lilly, J. Robert, Francis T. Cullen, and Richard A. Ball. 2007. *Criminological Theory: Context and Consequences*. Thousand Oaks, CA: Sage Publications.

Muraskin, Roslyn (ed.). 2007. *It's a Crime: Women and Justice*. Upper Saddle River, NJ: Prentice Hall.

Shoemaker, Donald J. 2005. *Theories of Delinquency: An Examination of Explanations of Delinquent Behavior*. New York: Oxford University Press.

Warr, Mark. 2002. *Companions in Crime: The Social Aspects of Criminal Conduct*. New York: Cambridge University Press.

"Speed Bump" © Dave Coverly/Dist. By Creators Syndicate, Inc.

chapter nine

Taking It to the Streets: Cops on the Job

Although crime was a serious problem in early 18th century England, the idea of a standing police force was resisted for fear that such an organization could easily abuse its power. Sociologist Richard Lundman notes that "Many preferred the relative liberty and informality of community policing rather than risk the perceived threat to democracy associated with modern police." Created as a result of the Metropolitan Police Act of 1829, the London Police force was a compromise that limited police powers to the metropolitan area. It was the first urban law enforcement organization in the Western world. Under the direction of Sir Robert Peel (nicknamed Peelers or Bobbies), London was divided into 17 police districts with 165 officers assigned to each of these sectors.

Officers had to be under 35 years of age, healthy, literate and of "good character." They were clothed in a long blue-tailed coat, blue trousers, and a glazed black top hat, and patrolled the streets unarmed save for a short baton hidden under the coat. Peel introduced military-type ranks and a hierarchical command structure that is characteristic of departments around the world. In North America, the colonists typically reproduced the institutions of social control they had known in Europe. By 1800, some larger cities had adopted quasi-professional police forces with either appointed or elected officials and a regular salary.

The first modern-style police department appeared in the South primarily to control slaves. Washington, DC had a police department around 1820, while in the North, Boston (1838) and New York (1844) were the first big cities to establish such agencies. Some officers were well-trained by the standards of the day, while others were issued equipment and learned their profession via on-the-job training.

New York City police officers wore copper star badges on their Robert Peel-style uniforms, hence the word coppers and later cops, according to one version of the origin of that term. Unlike their British counterparts, American police officers were typically armed, and by 1911 some patrolled the streets in motorized vehicles.

THE ORGANIZATION OF LAW ENFORCEMENT

The defining characteristic of law enforcement in the United States is decentralization. Whereas the police in many countries are organized at the national level and administered by the federal government, in the United States they are primarily employed by local governments. There are two principal reasons for this historical emphasis on community policing. First, since the colonial era, citizens in the United States have feared allowing the federal government too much control over their daily lives. This is especially true of police who have the power to detain and arrest. Second, Americans have long believed that at least some local problems are best solved at the community level; an effective law enforcement strategy in Los Angeles is likely to be a complete failure in rural Mississippi, and vice versa.

Decentralization of the police comes at a price. To begin, no overarching agency ensures uniform standards across departments. For example, police in big cities are apt to undergo extensive training in modern police techniques, while recruits in smaller towns usually receive much less comprehensive instruction. Second, significant duplication can occur when jurisdictions of two or more agencies (state and local, for example) overlap, resulting in a waste of time, money, and effort. Finally, it is not unusual for different agencies to regard each other as rivals.

Of the approximately 19,000 publicly funded law enforcement agencies in the United States, almost 90 percent are municipal police and sheriff's departments. The largest of these organizations, the New York City Police (NYPD), has about 45,000 full-time employees, including over 39,000 sworn officers. To put this number in perspective, consider that 1 in 15 full-time police officers in this country works for the NYPD. Sheriffs' departments are organized primarily at the county level and they may be staffed by only a few officers, or they may have thousands of members, as does the Los Angeles Sheriff's Department, with 8000 uniformed officers.

POLICE: WHITE, BLACK, MALE, FEMALE, STRAIGHT, AND GAY

Historically, policing in this country has been a white, lower- and middle-class male occupation. Newly emancipated African Americans were police officers in at least one Southern city as early as 1863, although their inclusion into the profession was greeted with disdain. In July of that year a newspaper headline in Raleigh, North Carolina proclaimed: *The Mongrel Regime!! Negro Police.* By 1910, no African Americans were on Southern police forces and the nation had fewer than 600 African American officers.

After World War II, African Americans began making inroads into the policing profession, including in the South. A 1959 survey revealed that 69 of 130 southern municipalities (43.3 percent) had African American officers. However, in many of these cities, African American police patrolled in squad cars marked "Colored Police," and officers were compelled to contact white colleagues to arrest white suspects. An African American officer noted that upon appointment to the Atlanta Police Department in 1948 he took the following oath: "I do solemnly swear as a nigger policeman that I will uphold the segregation laws of the city of Atlanta." Hired in the late 1940s, Georgia's first African American police officers could not arrest white offenders without a white officer present. Black officers could not wear their uniforms to work or change to law enforcement garb at the station house forcing many to switch clothes at a local African American YMCA. Prior to 1976, African American officers in Georgia were prohibited from joining the state-sponsored police retirement fund. As a consequence, white officers who enrolled in the fund before that year now collect significantly more retirement money each year than their black counterparts of that period.

The verdict in the O.J. Simpson murder trial revealed just how deep the chasm between African American and white polices officers is in many departments. One veteran St. Louis officer stated that when the "not guilty" verdict was read by the foreperson, "The white coppers sulked, some of them looked down and shook their heads. The blacks high-fived, cheered, and applauded."

Much of the animosity between African American and white officers in large urban departments is embedded in hiring and promotion issues. Many white officers believe African Americans and women gained admittance to the force (and were later promoted ahead of white officers) because of government-mandated quota programs. Referring to the economic tension between African American and white police officers, Samuel Walker states, "There's a sense of limited job opportunities and a fight over the ones that exist."

By the mid 1990s, the number of African American police officers had increased significantly, especially in cities with substantial African American populations. Over 50 percent of police officers in Detroit and approximately 7 in 10 officers in Washington, DC are African American. While African Americans have become police chiefs in some of the nation's largest cities, overall they remain underrepresented in senior positions

Women have served as police officers (as police "matrons," initially) in the United States since 1845. However, because of widespread gender discrimination, both the scope of their activities and the number of females serving in law enforcement have been limited. Policewomen in the early 1900s were expected to be more social workers than law enforcement officers. As such, female officers dressed in street clothes, patrolled areas where children were likely to engage in deviant behavior (dance halls, skating rinks, and movie theaters), and/or come in contact with criminals. Speaking of her child-saving status, Mary Hamilton, the first female officer in New York City, noted her professional role was not unlike that of a household mother. "Just as a mother smooths out the rough places . . . looks after children . . . so the policewoman fulfills her duty."

Female officers continue to encounter sexual harassment, although this form of discrimination is not as systematic as in years past. A female officer reporting to a new district was greeted with the following remark: "Officer, I don't mean any harm, but I just want you to know that you have the biggest breasts I've ever seen on a policewoman." Another woman related the following incident:

> "My first day on the North side, the assignment officer looked up and said, "Oh shit, another fucking female." That's the way you were treated by a lot of the men. The sergeant called me in and said the training officer doesn't want to ride with you, but I've given him a direct order to work with you."

Male officers often harass female colleagues by referring to them in terms of cultural stereotypes—"ladies" and "girls"—that suggests weakness and passivity, individuals in need of protection. Female officers who challenge this gender typecasting risk being labeled butch, bitch, or lesbian. These officers often find themselves in a no-win situation, caught between stereotypical extremes in organizations wherein they have little authority.

On a more optimistic note, the overall situation for females in law enforcement may well change for the better as women could account for more than 25 percent of the nation's police officers (up from about 10 percent in 2000) in the next 15 to 25 years.

Because the majority of gay and lesbian police officers choose to conceal their sexual orientation, the number of homosexual men and women in law enforcement is unknown. Beginning in the mid 1990s, some metropolitan police forces attempted to recruit homosexuals, ending their decades-long systematic exclusion. For example, the NYPD sent 55,000 fliers and applications to gay residents via mailing lists obtained from gay and lesbian groups. A similar campaign was launched by the LAPD, an organization with dozens of officers who identify themselves as homosexual.

Anecdotal evidence indicates that homosexual male officers face more prejudicial attitudes and discriminatory behavior than do lesbian officers. In a profession dominated by heterosexual males, a homosexual male officer is a threat to the macho mystique. However, these attitudes may be changing (albeit slowly). A 2001 study concluded that the integration of openly gay and lesbian officers had improved the San Diego Police Department's organizational effectiveness. While subtle forms of discrimination remain, many straight officers take the presence of gay and lesbian officers for granted.

There is little doubt the composition of the nation's police departments is increasingly diverse, at least in major metropolitan areas. For example, in 2005, the New York City police academy graduation class was less than half white, with Asian Americans accounting for 8 percent of new officers. The December, 2006 graduating class had cadets whose racial or ethnic heritage represented 58 nations.

THE POLICE AND CRIME RATES

David Bayley, one of the foremost police authorities in the United States, notes "The police do not prevent crime. This is one of the best kept secrets of modern life. Experts know it, the police know it, but the public does not know it." Bayley presents a good deal of evidence to substantiate his claim.

First, numerous studies have failed to show any relation between the number of police officers and the crime rate in a given locale. For example, in 1995 there were 3.1 full-time law enforcement officers per 1000 population in cities of more than 250,000, and 1.9 officers per 1000 population in cities between 10,000 and 24,999 population. However, the larger urban areas had a significantly higher crime rate (especially violent crime) than the smaller cities. Bayley notes that while a police officer on every corner and on every doorstep would probably reduce crime rates, the cost of training and employing hundreds of thousands of additional officers would be prohibitive.

Research also indicates that the number officers in a given area at a particular time has virtually no impact on crime rates, victimization rates, or the public's satisfaction with police. In numerous experiments, the number of patrol officers in neighborhoods has been doubled, reduced by 50 percent, or reduced to zero without changing crime rates. While saturation patrolling or inundating a precinct with uniformed officers does depress crime rates, this strategy is "too expensive to be more than a short term expedient."

Finally, no evidence confirms that decreasing the time it takes police to respond to a crime scene will increase the chances of apprehending an offender. Unless police arrive at the locale within one minute (which is highly unlikely and an unreasonable standard) they will usually not make an arrest. Otherwise, their chances of taking a suspect into custody are probably less than 1 in 10.

Although the preceding information may appear to be a sharp indictment of police and police practices, it is not David Bayley's intent (nor ours) to chastise law enforcement entities for failing to control (much less solve) the crime problem in this country. Blaming the police for society's penchant for criminal behavior makes about as much sense as holding physicians responsible for escalating rates of diabetes. As Bayley notes, "Police shouldn't be expected to prevent crime. They are outgunned by circumstances." In other words, crime is a product of conditions well beyond the scope of police resources and authority. As we noted in Chapter 4, crime is primarily a consequence of the distribution of income and wealth, poverty and unemployment, the racial and ethnic composition of an area; family structure, age, and gender; and a host of more subtle psychological, sociocultural, and economic variables.

DOING POLICE WORK

Since World War II, television crime programs have shaped the public's perception of the police and police work. Unfortunately, even the most supposedly realistic shows fall short of the reality mark. The following section provides an overview of patrol officers, detectives, and traffic enforcement officers and what they do.

On Patrol

On virtually every police force, the majority of departmental personnel are patrol officers. Sometimes called peacekeepers, these officers are summoned when the relative harmony of everyday life has been disrupted. Patrol officers routinely deal with rowdy teenagers, neighbors arguing or fighting over a petty incident, homeless people panhandling too aggressively, a house party blasting earsplitting music, or a mentally impaired man threatening passersby with a weapon. No more than 30 percent of all calls for service are about crime, and even that figure may be high, as responding officers often find that the alleged law-violating incident was not a criminal offense.

Family disputes are some of the more difficult situations patrol officers deal with on a regular basis. On arriving at a domestic altercation, police must determine whether a crime has been committed, who is the offender, and who is the victim. The answers to these questions are not always apparent, with participants in the dispute offering police a version of the incident that will cast them in the best light. A major difficulty police have in their peacekeeper role, as David Bayley notes, is that people routinely and blatantly lie to them. This is a major reason why patrol officers "become cynical and hard to convince" and increasingly jaded over time.

Criminologist John Van Maanen suggests that police tend to view their "occupational world" (those people they come into contact with on a regular basis) in terms of three groups of citizens: (1) Suspicious persons are individuals whom the police believe have committed an offense. Because of their possible criminal involvement, these suspects are treated in a "brisk" yet "thoroughly professional manner." (2) Assholes are treated harshly because they are (or are perceived to be) disrespectful, difficult, and/or confrontational. Assholes are prime candidates for "street justice," physical attacks aimed at teaching them a lesson for an insulting attitude, remark, or behavior. (3) Know nothings are average citizens whose only encounters with the police are requests for assistance. They are so labeled because as civilians they cannot know (according to officers) what the police and policing is about. Using Van Maanen's typology, one important aspect of a patrol officer's peacekeeping/order maintenance role is to make sure persons in category two (assholes) are held in check. As one officer stated:

I guess what our job really boils down to is not letting the assholes take over the city. Now I'm not talking about your regular crooks . . . they're bound to wind up in the joint anyway. What I'm talking about are those shitheads out to prove they can push everybody around. Those are the assholes we gotta deal with and take care of on patrol . . . They're the ones that make it tough on the decent people out there. You take the majority of what we do and it's nothing more than asshole control.

Police officers are society's frontline troops when it comes to maintaining and restoring order and are summoned when people believe no one else can help them. These calls for help range from the silly and absurd to the poignant and tragic. A man stops a patrol car and asks officers if they have needle-nosed pliers to fix the zipper on his pants. A woman brings her parakeet to the station house to be weighed because she thinks the bird is ill. An invalid falls out of bed and calls for assistance. A man wants the police to notify his sister (who does not have a phone) that their brother has died.

Many police officers think it is impossible for the average citizen to understand and appreciate what they do. Although this perception may be exaggerated, it is certainly true that few non-police individuals are called on to perform what patrol officers routinely experience (especially in big cities) during their years on the street. Consider what a young patrol officer told Van Maanen:

"I'll tell ya, as long as we're the only sonofabitches that have to handle ripe bodies that have been dead for nine days in a 90-degree room, or handle skid row drunks who've been crapping in their pants for 24 hours, or try to stop some prick from jump'en off the Liberty Bridge or have to grease some social misfit who's trying to blow your goddam head off, then we'll never be like anyone else . . . as far as I can see, no one is ever gonna want to do that shit. But somebody's got to do it and I guess it'll always be the police. But hell, this is the only profession where ya gotta wash your hands before you take a piss!"

Criminal Investigation

When a crime has occurred, plainclothes detectives may be summoned to analyze the situation. These officers talk to victims, witnesses, and suspects to ascertain what happened and decide if an investigation is warranted. The decision to extend their inquiry is contingent on two primary factors, whether a credible perpetrator has been fairly clearly identified, and if the crime is serious enough to attract public attention.

The key ingredient in solving crimes is whether information passed on to police by victims and witnesses will help identify a suspect. Because they are typically overwhelmed with cases, detectives (especially in big cities) are not inclined to investigate an offense if a suspect has not been identified. This means that relatively few burglaries and only a limited number of robberies will be pursued seriously by police. David Bayley notes that a citizen's request (or demand) that officers dust a crime scene for a burglar's fingerprints is likely to be met with controlled anger or muted laughter depending on how busy officers are at the time.

A particularly inaccurate perception the public has about detective work involves the role of physical evidence in solving crimes. Mystery stories are replete with examples of criminal investigators painstakingly searching a crime scene and eventually finding clues that lead to the perpetrator's identity and arrest. However, police seldom use physical evidence gathered at a real crime scene in the apprehension of a suspect. Rather, this evidence is used as confirmation once police identify an offender. Bayley notes that detectives "begin with an identification, then collect evidence; they rarely collect evidence and then make an identification."

Although programs that focus on forensic evidence make for interesting television drama, a significant number of cases (some would argue a majority of cases) are solved by nothing more than chance. In the aftermath of the Washington-area serial sniper killings in October, 2002—one of the most intensive manhunts in U.S. history—Montgomery County Maryland Police Chief Charles A. Moose stated: "As a police chief I know you need a lot of luck. I'd like to say it's all skill, moxie, and brains. But it's mostly luck."

When a suspect is questioned, detectives may resort to various forms of deception. The authors of a well-known police interrogation manual argue that criminal suspects are reluctant to confess because they are ashamed of what they have done and/or are fearful of punishment. The manual states that a certain degree of "pressure, deception, persuasion, and manipulation" is necessary for the "truth" to be revealed. Fred E. Inbau and his colleagues, authors of *Criminal Interrogations and Confessions*, state that "We do approve . . . of psychological tactics and techniques that may involve trickery and deceit; they are not only helpful but frequently indispensable in order to secure incriminating information from the guilty." The most common form of deception or trickery involves presenting the suspect with false evidence of guilt during questioning. For example, the police may state (falsely) that a codefendant has confessed and implicated the suspect being questioned. Other forms of deception include exaggerating the strength of the evidence against the suspect, falsely claiming that the police possess conclusive forensic evidence of the suspect's guilt, stating that an eyewitness saw the suspect commit the crime, and/or lying about the results of a polygraph test.

Police typically defend the practice of deceiving suspects by way of the following: (1) In the end justifies the means argument, if detectives cannot bend the truth on occasion, many guilty people will not confess; (2) because of liberal court rulings such as the *Miranda* decision, police already have one hand tied behind their backs, and lying to suspects during an interrogation is one way to even the score; it gives the police a fighting chance of bringing the guilty to justice; and (3) using the "It's all part of the game" interpretation as one detective put it; "Yes, you an lie to a person in an interrogation—they lie to you."

It is important to note that with few exceptions, police officers do not knowingly try to obtain confessions from individuals they believe are factually innocent. Rather, in their zeal to solve a crime (especially a high-profile offense), detectives may unwittingly coerce innocent suspects into making false confessions.

In southern California a 12-year-old girl was stabbed to death in her bedroom during the night. Because her 14-year-old brother appeared to be unaffected by the tragedy, and, according to the boy's story, because he walked by his sister's room two hours before the body was found without noticing anything, the youth became a primary suspect. The following excerpts are from the police interrogation in this case:

> **Detective:** We found blood in your room already.
> **Suspect:** God, where did you find it?
> **Detective:** I'm sure you know. It's easy to make mistakes in the dark. [No blood was found in the suspect's room. The detective would say later that he thought he saw blood.]

At another point in the interrogation the detective told the boy, "We can prove that nobody came into the house. So we know that the person who did this was inside the house."

The detective told the suspect that all the doors and windows were locked, and that the house showed no signs of entry. That was a lie. The sliding glass door from the master bedroom to the backyard was unlocked. The victim's blood was eventually found on the shirt of a local transient (evidence that was missed when this individual was questioned and his clothing examined shortly after the murder), and the case against the boy was dismissed.

Although deception can be a very effective strategy for obtaining confessions, this policy may well come at the expense of some public trust and support of the police. After details of the murder case just discussed were published in the newspaper, one woman stated: "Yes, find the murderer and bring him/her to justice—but not like this. Reading those articles has shaken my faith in both justice and the police, and I for one, will never believe anything they say again."

Traffic Enforcement

Metropolitan police departments typically assign about 7 percent of their personnel to traffic units, with these officers responsible for controlling the flow of traffic, investigating motor vehicle accidents, and regulating traffic at sporting events, rock concerts, and political rallies. While many of these officers believe what they are doing is important (although not appreciated by the public), other officers think that traffic regulation is "chicken shit work." Jerome Skolnick notes that many traffic officers dislike their assignment for two reasons. First, stopping vehicles for moving violations is one of the most dangerous components of police work; of the officers injured and killed annually, a significant number are assaulted after stopping armed, dangerous offenders. Second, traffic stops represent the most frequent point of contact police have with the public, resulting in low-level but persistent tension between these groups. Many people resent being stopped, detained, and lectured by traffic officers. Police are in a no-win situation as they issue traffic tickets to the very group (mainstream America) they count on for political and economic support.

The public especially resents receiving citations perceived to be driven by quotas. Although police chiefs typically deny their officers must issue a specific number of tickets within a designated time period, the evidence clearly indicates that ticket quotas are part and parcel of departments across the country. Criminologist Lawrence Sherman notes that "quotas have been widely used in regulatory work," although this policy is "rarely admitted." New York City police officers picketed traffic courts in their protest of ticket writing quotas. A police union secretary and 30-year department veteran stated, "These cops are hustled and harangued to give out more and more tickets. Time off and holidays are rewarded by the number of tickets you write."

As of 2004, police officers in Falls Church, VA were required to write three tickets or make three arrests every 12-hour shift as well as accumulate 400 citations and arrests per year. Failure to realize this quota results in a 90-day probationary period without a pay raise and the possibility of demotion or dismissal if the number of arrests and citations issued does not reach acceptable levels. Two officers from a small-town New Jersey police department went to court, arguing their quota of writing 20 tickets a month in a low-traffic, rural locale was unattainable. The judge disagreed, noting that the department's policy did not violate any public policy or constitutional principles.

POLICE DISCRETION

The criminal justice system can be thought of as an unfolding series of situations and decisions, with police officers making one of the first decisions (whether to arrest someone) in a chain of events. The police not only decide which laws will be enforced, but also when, where, and under what circumstances legal statutes are implemented. Criminologist Samuel Walker considers police officers gatekeepers of the system, as their arrest decisions determine in large measure the workload of prosecutors, probation officers, judges, and parole officers as well as prison guards and administrators.

Because laws are written in generalized language and must be interpreted in the context of a given situation, discretion—the freedom or power to act on one's own judgement without the direct supervision of others—is an inescapable part of a police officer's job. In addition, no law, statute, or departmental regulation can possibly cover each and every setting to which it might be applied. Consider officers responding to a call for assistance at a local nightclub, arriving just as a barroom brawl has ended. Should some or all of the combatants be taken into custody for simple assault (a misdemeanor) or aggravated assault (a felony)? Summoned to a restaurant, police find a waiter physically restraining a shabbily dressed elderly woman. The waiter states that his captive ordered and consumed a meal then stated she could not pay for it. Upon learning the woman is homeless and possibly mentally impaired, officers must decide if the elderly suspect should be arrested, escorted to a homeless shelter, taken to the county mental hospital, lectured and released, or simply released.

Discretion is also an inescapable part of police work, because officers are, to a certain extent, future oriented. Regarding the woman who ate without paying, on a relatively slow Sunday afternoon officers might decide to take her to a nearby homeless shelter, while on a busy Friday evening, they may opt to release her. Police make decisions, in part, based on their perception of the best course of action for all concerned. James Q. Wilson, who has written extensively on crime and criminal justice, argues that police practice a kind of distributive justice, exercising discretion based on their evaluation of the moral character of victims and suspects. Police perceptions, therefore, of good guys and bad guys will be a factor in what course of action they pursue.

Attempts to reduce or eliminate police discretion can result in undesirable and highly embarrassing moments for police. As a consequence of a departmental policy that permitted no exceptions, officers in Highland Park, TX, stopped and arrested a 97-year-old-woman for having an expired registration sticker. Dolly Kelton was handcuffed, taken to jail, fingerprinted, and photographed. After her release, the mayor sent Mrs. Kelton a letter stating that he was saddened by the incident.

Although police discretion is inevitable and, to a certain extent, desirable, uncontrolled discretion, as Walker notes, is problematic. This behavior can result in the unequal treatment of citizens, as the power to exercise discretion is also the power to engage in discriminatory behavior. Racial profiling is a perfect example of the undesirable application of discretion.

THE POLICE AND PROFILING

A young man from the West African nation of Liberia attending college in the United States was pulled over by police in Maryland for not wearing a seat belt. The driver and his passengers were detained for two hours while officers searched the car for illegal drugs, weapons, and other contraband. After their initial search turned up nothing out of the ordinary, the officers began dismantling the car by removing part of the door panels, the seat panels, and a portion of the sunroof. When this more comprehensive search also failed to yield any illegal material, one of the officers handed the young African a screwdriver and said, "You're going to need this." The police then drove away.

Profiling is the widespread police practice of viewing certain characteristics of individuals and situations as indicative of criminal behavior. For example, U.S. customs agents are alert for individuals who appear particularly nervous when they enter this country, and these people are subject to routine questions and possibly searches. Anxiety is likely to be interpreted as a sign the individual is attempting to smuggle illegal drugs or other contraband into the United States. Racial profiling is the police practice of stopping, questioning, and searching a disproportionate number of racial and ethnic minorities (especially African Americans and Hispanics). This practice is predicated on the belief that these individuals are involved in a significant amount of crime, particularly the use and distribution of illegal drugs.

Some have argued that if minorities are disproportionately involved in the possession and distribution of illegal drugs, racial profiling is not only justifiable, but a useful strategy in combating the drug trade. However, the government's own statistics undermine this perspective. Approximately 80 percent of cocaine users in the United States are white, and the typical user is a middle-class white suburbanite.

Although individuals and civil rights groups have accused police departments of racial profiling for decades, law enforcement officials have steadfastly denied condoning this practice. However, in the past few years, both anecdotal and social science research has supported the contention that racial profiling is a fundamental component of policing in much of the nation. For example, after denying that New Jersey troopers engaged in profiling for more than five years, the state attorney general admitted that troopers routinely used this strategy. In a sworn statement, one trooper stated that he was trained to target African Americans and Hispanics for traffic stops. An Alabama state trooper testified that traffic stop indicators included Texas license plates and Mexicans. Testifying in a drug interdiction case, a police officer in Kentucky reported that race and out-of-state plates were indicators used in his department. A study by the General Accounting Office found that African American women were subject to intrusive searches by U.S. Customs agents more than other groups, and were nine times more likely than white women to be x-ray searched after a pat-down.

The American Civil Liberties Union (ACLU) conducted a rolling survey to identify both the race and the ethnicity of drivers, and the number of individuals violating the law (mostly speeding) on a stretch of Interstate 95 in Maryland over a 42-hour period. Of the 5741 drivers observed, 5354 (93.3 percent) were violators and, therefore, eligible to be stopped by the police. Slightly less than 75 percent of the violators were white and 21.8 percent were minorities. Maryland State Police records revealed that, between January 1995 and September 1996, 823 motorists on Interstate 95 were stopped. Of those drivers pulled over 661, or 80.3 percent were African Americans, Hispanics, or other minorities. Slightly less than 20 percent were white.

Racial profiling can send law enforcement personnel off in the wrong direction, wasting time and limited resources and thereby jeopardizing public safety. In the aftermath of the Oklahoma City bombing in 1995, the two white male assailants fled the scene while police reportedly worked on the assumption that the perpetrators were Arab terrorists. During the sniper attacks in the Washington DC area in 2002, police conducted their investigation based on the standard serial killer profile of a lone white male. As a consequence of this racial assessment, the two black shooters passed through police roadblocks multiple times before they were finally arrested.

In June 2003, President Bush issued a set of guidelines banning racial profiling by federal law enforcement agents. The ACLU noted that while this is a positive first step, these guidelines lack enforcement mechanisms and do not mandate data collection or any provisions for remedies. In addition, the ban does not apply to local and state enforcement officials where most racial profiling occurs.

MILITARIZING THE POLICE

Peter Kraska and Victor Kappeler have investigated a relatively little-known aspect of law enforcement—the militarization of American police. This militarization exists primarily in the form of police paramilitary units (PPUs). The two criminologists surveyed all 690 state and local law enforcement agencies in the United States that serve jurisdictions of 50,000 or more residents and employ at least 100 officers. Of the 548 departments that responded, almost 90 percent (490) had PPUs, and over 20 percent of agencies without such units were planning on establishing one in the next few years.

PPU members often refer to themselves in military jargon ("heavy weapons units") and have a variety of high-powered weapons and high-technology devices unavailable to other units. Mission-ready PPU members look more like combat soldiers than police officers. Their attire typically consists of battle dress uniforms (BDUs), combat boots, full body armor, Kevlar battle helmets, and in some instances hoods. These officers see themselves as members of an elite unit, a view that is promoted by police administrators.

This elite self-perception and elevated status in the law enforcement community is in part a consequence of the training PPU members receive. When queried by Kraska and Kappler about the sources of training that were influential during the start-up period of their PPUs, departments reported they gained expertise from "police officers with special operations experience in the military" as well as active-duty military special operations personnel, including Army Rangers and Navy Seals. Speaking of this training, one respondent stated:

> We've had special forces folks who have come right out of the jungles of Central and South America. These guys are into real shit. All

branches of the military are involved in providing training to law enforcement. We've had teams of Navy Seals and Army Rangers come here and teach us everything. We just have to use our judgement and exclude information like: "at this point we bring in the mortars and blow the place up."

Not only have the number of PPUs increased significantly over the past 25 years, but their utilization has risen markedly, with most deployments drug related. Carl Klockars argues that the perception of PPUs as a "domestic army engaged in a 'war' on crime" is grounded in three themes. First, by associating the police with military heroes and victories as opposed to the often-sordid side of local politics, militarization confers honor and respect on the police profession. Second, the notion of a "war" on crime creates a sense of urgency and heightened importance. Denying police the necessary resources to battle criminals is like "siding with the enemy and metaphorically tantamount to treason." Third, just as generals make battlefield decisions in combat, militarizing the police is a way for police administrators to gain more control of their departments (at the expense of local politicians) in terms of hiring, firing, assignment, and disciplinary matters that are necessary to manage these quasimilitary organizations.

While many police administrators and officers are enthusiastic about PPUs, arguing that they are necessary to preserve order and control law-violating behavior in a society characterized by high crime rates, other members of the law enforcement community adamantly oppose these units. Some officers think that PPUs pose a danger to team members and civilians. One police captain who had a negative view of paramilitary drug raids recounted the following incident:

We did a crack-raid and got in a massive shoot-out in an apartment building. Shots were fired and we riddled a wall with bullets. An MP5 round will go through walls. When we went into the next apartment where bullets were penetrating, we found a baby crib full of holes; thank god those people weren't home.

Cato Institute researcher Radley Balko discovered that SWAT teams made mistakes in over 200 raids over the past 10 years, resulting in the deaths of 24 people not guilty of any criminal activity. Some died when they picked up weapons to defend themselves from the unknown invaders. A Mexican immigrant in Denver was killed when police raided the wrong home. In Modesto, CA, an 11-year-old boy was shot dead, and in Harlem a 57-year-old woman died of a heart attack after a flash grenade exploded during a raid based on faulty information. Balko notes that "while courts have been extremely deferential to police who fire on innocent civilians, they've been far less forgiving of citizens—even completely innocent civilians—who fire at police who have mistakenly raided their homes. Victims who have used force to defend themselves from improper raids have been prosecuted for criminal recklessness, manslaughter, and murder and have received sentences ranging from probation, to life in prison, to the death penalty."

Joseph McNamara, former chief of police in Kansas City and San Jose, California, considers PPUs a "disaster," resulting in now-routine levels of force that would have been unacceptable in the past. Ron Hampton, executive director of the National Black Police Association believes that the militarization of the police is dangerous, and that departments so organized will begin attracting applicants who are not oriented toward community service, but motivated by the "thrill of chasing criminals and the excitement of the weapons." While opponents of PPUs are often dismissed as little more than misguided liberals, some conservatives have also been highly critical of these paramilitary groups. Diane Weber of the conservative Cato Institute stated that the mindset of a soldier is not appropriate for civilian law enforcement personnel. "Police officers confront not an 'enemy' but individuals who are protected by the Bill of Rights. These units are going into public housing projects, into black neighborhoods, as if they're in hostile territory."

While PPUs are almost universally considered necessary in certain hostage-related tasks and antiterrorist situations, this activity comprises only a small fraction of their work. Research indicates no correlation between overall paramilitary policing and crime rates. Peter Kraska states that the number of PPUs is increasing not as a result of rising crime or any evidence that these units can reduce law violating behavior, but simply because police departments want them.

POLICE DEVIANCE: WHEN THE GOOD GUYS ARE BAD GUYS

Most Americans have a positive image of the police, an image that goes back to childhood when they were taught that police officers were their friends. Although the vision of the hard-working, scrupulously honest cop on the beat is certainly true in the majority of cases, there is another dimension to policing in this country. As Victor Kappler and his colleagues note, "To study

the history of police is to study police deviance, corruption, and misconduct." In 1895, the Lexow Commission, one of the first formal commissions
to study police wrongdoing in the United States, concluded that in New York
City "almost every conceivable crime against the elective franchise [citizens]
was either committed or permitted by the police."

Over the past 20 years, a significant amount of both individual and
organized police corruption has been drug related. A report by the Government Accounting Office cites examples of drug-related police corruption in
Atlanta, Chicago, Cleveland, Detroit, Los Angeles, Miami, New Orleans,
New York, Philadelphia, Savannah, and Washington, DC. Of the 508 police
convictions for criminal activity between 1994 and 1997, almost 45 percent
were drug related. Forty-two police officers and correction officers from five
Cleveland-area agencies were arrested in an FBI sting operation and charged
with conspiracy to distribute cocaine. Police cars were used to transport
cocaine, and one officer routinely arrived late for narcotics transactions
because he was giving antidrug lectures at local schools. In 2004, a former
New York City detective said that, along with fellow officers, he stole narcotics and money from drug dealers.

Drug-related corruption reached new highs (or lows) in New York City
in the 1980s. Officers sold confidential information (for example, who was
being investigated and targets of forthcoming drug busts) to drug dealers,
provided protection for the transportation of illegal substances and drug
money, became active drug entrepreneurs, and harassed competing traffickers. Some officers also robbed drug dealers and burglarized known drug
locations. The stolen drugs were sold to other dealers, other police officers,
and back to the dealers from whom they were taken.

The NYPD was rocked again by a scandal as two former detectives were
convicted in June, 2006, of drug dealing, kidnaping, and murder. Dubbed
the "Mafia Cops," the officers received $4000 a month from a local crime
family in exchange for information on law enforcement investigations. On
one occasion they were paid $65,000 for a contract killing. A federal prose-
cutor described the case as "the bloodiest, most violent betrayal of the badge
this city has ever seen."

Drug corruption among police raises an important question. How can
individuals sworn to uphold the law flagrantly violate these statutes? A
related issue involves how deviant officers maintain a positive self-image. In
other words, if drug dealers are bad guys, in what sense are police officers
now engaged in this very activity any different from narcotics traffickers?

Consider the following rationalization by an officer who stole money from drug dealers rather than arrest them.

> Even though I was a bad guy, I had the feeling, "Hey I'm bad on one side, but on this side I'm making up for it." . . . We weren't taking money from honest workers. I know it doesn't matter whether it's an honest worker or a skell [criminal], it's still wrong. I know that. But we were taking money that was illegal to begin with. Drug money. It's weird but I never thought I was robbing people. I was robbing low life. A drug dealer.

Joseph McNamara, former police chief of San Jose and Kansas City, believes that the impossibility of controlling drugs in a society wherein tens of millions of people use these substances is a telling factor in understanding why so few police officers experience guilt when breaking the law. "The sheer hopelessness of the task has led many officers to rationalize their own corruption. They say 'Why should the enemy get to keep all the profits?' Guys with modest salaries are suddenly looking at $10,000 or more, and they go for it."

Just as every organization has some members who at least occasionally act in an inappropriate manner, all police departments experience some corruption. Lawrence Sherman has offered a three-part typology of police corruption.

Type I: Rotten Apples and Rotten Pockets

This is the most common and least serious form of police wrongdoing. For example, a small number of uniformed patrol or traffic officers acting independently—the "rotten apples"—engage in low-level corrupt acts such as taking bribes from motorists for not issuing tickets. Regardless of the seriousness of the corruption problem, police administrators and city officials are likely to invoke the rotten apples explanation should police deviance become public. This interpretation both minimizes the scope of the wrongdoing and absolves high-ranking officers from any direct responsibility for the problem.

Rotten pockets, or small work groups, is a form of police deviance that exists in departments where corruption occurs more frequently and is an accepted practice. Whereas rotten apples are apt to be patrol officers, rotten pockets are plainclothes detectives and vice officers. These latter individuals typically are corrupted by people with whom they come in contact, such as prostitutes, pimps, and bookies.

Type II: Pervasive Unorganized Corruption

The defining characteristic of this level of police corruption is that even though a majority of personnel are corrupt, individuals are acting primarily on their own. Each corrupt officer accepts his or her bribery money when the opportunity presents itself; no formal organization of corruption exists.

Type III: Pervasive Organized Corruption

These departments are thoroughly corrupt, with illegal behavior organized from the top down. Type III departments are typically embedded in a network of corruption that includes judges and officials at the highest level of local government. Corrupt officers work together in a variety of illegal enterprises, including gambling, prostitution, and drug trafficking, often in connection with organized crime.

Criminologist Samuel Walker argues that police corruption has a significant impact on the law enforcement community, the courts, and society. Besides being illegal, this behavior undermines the integrity of police not only in the department where it occurs, but also, to some extent, in all police agencies. When a community loses respect for police and distrusts them because of corruption, it becomes more difficult for officers to do their job. Police professionalism is undermined when crooked cops lie to protect one another. Internal discipline is also thwarted as corrupt supervisors (often known to patrol officers and detectives) are always at risk of being exposed by disgruntled subordinates.

SEXUAL PREDATORS IN BLUE

A troubling minority of police officers engage in duty-related sexual misconduct. Criminologist Allen Sapp notes that in addition to institutionalized authority and the power to detain and arrest, police officers have ongoing contact with citizens, much of which occurs in relative isolation. As such, law enforcement personnel occupy a unique position in society, a position that affords them numerous opportunities to engage in sexual misconduct.

Speaking of police sexual misconduct, Norm Stamper, former chief of the Seattle Police Department stated that over the course of his 34-year career, "I would see it all: cops fingering and fondling prisoners, making bogus traffic stops of attractive women, trading freedom for a blow job with a hooker, making 'love' with a fourteen-year-old police explorer scout."

Allen Sapp interviewed police officers and their supervisors in major metropolitan police departments across seven states. He found that various forms of police sexual misconduct occurred, including:

- *Sexually motivated contacts:* This misconduct occurs when officers stop females without legal basis or probable cause to get a closer look at the driver and/or obtain information about her. A traffic patrolman stated: "Sure I see a good-looking fox driving around by herself or even a couple of foxes together. I pull them over and check them out. You can always claim a stoplight isn't working . . . Lots of times it doesn't lead anywhere, but you'd be surprised how much action I get . . .
- *Contacts with crime victims:* Emotionally upset female crime victims are particularly vulnerable to sexual harassment by police. Unnecessary call-backs to these women is one of the most common forms of police sexual misconduct. A burglary detective told Sapp: "When you drop in a few times to check on the victim or to tell her that you are still working on the case, you kind of establish a connection with her. She will offer a cup of coffee or a drink . . . You can't do this with everyone, but if you pick them carefully it gets you a lot of action."
- *Contacts with offenders:* Criminal offenders are particularly vulnerable to an officer's sexual advances. A theft squad detective stated: "You bet I get [sex] once in a while by some broad who I arrest. Lots of times you can just hint that if you are taken care of you could forget about what they did." Speaking of female shoplifters the officer noted: "If it's a decent looking woman, sometimes I'll offer to take her home and make my pitch . . . I never mess around with any of the kids, but I know a couple of guys who made out with a couple of high school girls they caught on a B and E [breaking and entering]."
- *Sexual shakedowns:* Demanding sex from prostitutes, homosexuals, and drug users is one of the more serious forms of police sexual misconduct. A vice squad sergeant told Sapp: "I know several dozen guys who have worked vice . . . I believe every one of them has gone beyond the rules on sex with prostitutes . . . Sometimes the officer goes ahead and has sex and then makes the arrest and files the report saying he followed procedures. If the whore claims otherwise, no one believes her anyway since they think she is just trying to get her case tossed out."

Samuel Walker and Dawn Irlbeck found that targeting female drivers for sexual favors occurs throughout the country, with abuse ranging from "harassment to sexual assault and even murder." Their review of the national print media from 1990 to 2001 revealed hundreds of allegations of driving while female (DWF), with over a dozen substantiated cases recorded each year.

Walker and Irlbeck argue that documented allegations of law enforcement sexual improprieties are the tip of the iceberg, as many victims do not come forward because of humiliation and fear of police reprisals. In addition, some police departments fail to acknowledge this category of victim complaint.

Norm Stamper believes "about 5 percent of America's cops are on the prowl for women." Assuming the 1 in 20 estimate is correct, over 35,000 law enforcement personnel nationwide may engage in sexual misconduct, some on a regular basis. If Stamper significantly overestimated the number of law enforcement sexual predators, and the true figure is closer to 1 in 40, that still amounts to almost 18,000 officers likely to commit sexual improprieties in the line of duty.

POLICE AND THE USE OF FORCE

Few people question the need for police to use necessary force in the performance of their duty, as many individuals they encounter will resist arrest, run from the officers, attack them, and on occasion, attempt to kill them. The question, therefore, is not whether force should be used, but how much force and under what circumstances. More specifically, there are two fundamental issues regarding the use of force by police. First, was the use of force justified? Was it necessary, or could the situation have been resolved in a nonphysical manner? Was the force administered within the confines of the law and/or departmental policy? Second, if the use of force was needed, was it administered at the appropriate level, or was more force used when a lesser application of force would have achieved the desired result?

If legitimate force is a necessary and inescapable part of police work, then police brutality is "the use of excessive physical force or any force than is more than reasonably necessary to accomplish a lawful police procedure." The difficulty with this and similar definitions is determining what constitutes reasonable force. This point was duly noted by Sergeant Anthony Miranda of the NYPD: ". . . but what's reasonable force in a moment when a cop who thinks he's in trouble, and what's reasonable in a courtroom can often be open to interpretation."

That well-meaning, intelligent people can disagree on what constitutes excessive force was demonstrated more than 40 years ago when the President's Commission on Law Enforcement and the Administration of Justice documented 37 cases of what they believed involved the improper use of force. When a panel of experts reviewed these same incidents, they concluded that only 20 of 37 situations fit the definition of unreasonable force. In other words, one group of observers determined that the cases included almost twice as many examples of excessive force as the second group agreed on. Whose interpretations were correct?

Tasers

One of the more controversial weapons in the law enforcement arsenal—especially as it relates to the excessive use of force—tasers were first employed by the Los Angeles Police Department in 1974. These pistol-shaped weapons use compressed nitrogen gas to fire twin darts up to 21 feet that can penetrate two inches of clothing. Electricity is then conducted through the wires that connect the darts to the gun. The newest generation of tasers can deliver up to 50,000 volts of electricity, significantly more than the original weapons used in the 1970s and 1980s. The newest generation of tasers are designed ". . . to incapacitate dangerous, combative or high risk subjects that may be impervious to other less-lethal means, regardless of pain tolerance, drug use, or body size."

Amnesty International (AI) reports that from 2001 through the end of September 2007, 276 people in this country have died after being struck by tasers. "While medical examiners had usually attributed cause of death to other factors, such as drug intoxication, more research was needed." In at least 20 autopsy reports gathered by Amnesty "coroners have cited the taser as a causal or contributory factor in the deaths, sometimes combined with other factors." In a 2006 report that examined 152 taser related deaths, AI concluded that:

- Most of those who died in custody were unarmed and were not posing a serious threat to police officers, members of the public or themselves
- Those who died were generally subjected to repeated or prolonged shocks
- Taser use was often accompanied by the use of restraints and/or chemical incapacitating sprays
- Many of those who died had underlying health problems such as heart conditions or mental illness, or were under the influence of drugs

- Most of the individuals who died went into cardiac or respiratory arrest at the scene
- Most of who died were shocked more than once and 92 were subject to between 3 and 21 shocks

Rather than being implemented in limited situations to avoid lethal force (firearms), many police departments are using tasers as "routine force options." In a 2004 report AI concluded that most of those who died via tasers were unarmed men. While they often displayed bizarre or disruptive behavior, these individuals "did not appear to present a serious threat to the lives or safety of others." Tasers have been used against unruly school children, unarmed mentally disturbed individuals, intoxicated citizens, suspects fleeing a minor crime scene, and people who fail to comply immediately to a police command.

In October 2007, researchers at Wake Forest University Baptist Medical Center released a study of almost 1,000 law enforcement tasings and concluded that 99.7 percent resulted in minor injuries—scrapes and bruises. Only three subjects were admitted to hospitals. The report noted that although two of these individuals died, their deaths were not taser related. Lead investigator Dr. William Bozeman stated that "These results support the safety of the devices."

The Wake Forest research has been roundly criticized. Aram James of the Coalition for Justice and Accountability noted that the study concluded—with no supporting data—that tasers reduce injuries to both police and the individuals tasered. Richard Konda of the Asian Law Alliance argued that the study failed to mention the effect of tasers on vulnerable individuals such as pregnant women, the elderly, and mentally ill, people who "are far more likely to suffer serious injury and even death as a result of being tasered." As of 2007, approximately 7,000 of the nation's 18,000 law enforcement agencies had tasers.

Police Violence: Why Does It Happen?

While many cases of alleged "police brutality" are open to interpretation, others leave little room for doubt. The officers who beat Rodney King in 1991, bragged and joked about the number of times they hit the suspect, often using baseball analogies. "We played a little ball tonight, didn't we Rodney . . . You know, we played a little hardball tonight, we hit quite a few home runs . . . Yes, we played a little ball and you lost and we won." In

August 1997, Abner Louima, a Haitian immigrant, was arrested in New York City and dragged into a restroom. A police officer jammed a broom handle up his rectum, then waved the feces-covered stick under Louima's nose threatening to kill him if he ever told anyone what happened. Louima suffered extensive internal injuries. The officer who perpetrated the attack eventually confessed and was convicted of crimes carrying a minimum 30-year prison sentence.

There are a number of explanations why some police officers engage in excessive force. One or more of these interpretations may be a factor in any given incident of police brutality.

Excessive Force Is Tolerated   The Christopher Commission, the investigatory body that critically examined the Los Angeles Police Department in the wake of the Rodney King incident, concluded that excessive force was treated leniently at the command level because this behavior did not violate the department's internal moral code. One officer told the commission that "some thumping" was permissible as a matter of course. A veteran New York City police officer noted that "Cops develop the sense that they can exercise power without too great a risk of being called . . . strictly into account for its use."

Excessive Force as a Rite of Passage   The Commission to Investigate Allegations of Police Corruption and Anti-Corruption Procedures of the New York City Police Department found that a willingness to abuse people who challenge police authority on the street or in the station house is a way for officers to demonstrate they are tough cops who can be trusted by fellow officers. "Brutality, like other forms of misconduct, thus sometimes serves as a rite of initiation into aspects of police culture . . ."

The "Dirty Harry" Strategy   In the Dirty Harry films popularized by Clint Eastwood, Inspector (Dirty) Harry Callahan of the San Francisco Police Department has no compunction about using excessive force to obtain information he deems necessary to solve crimes and save lives. In the first movie, Harry stomps on the bullet-mangled leg of a kidnapper in order to find the location of a kidnaped girl. Like Harry, some officers regard procedural laws (laws of search and seizure, for example) as little more than legal obstacles in the apprehension and conviction of criminals. Hence the laws are disregarded as officers resort to means such as intimidation and torture in their roles as professional crime fighters. James Fyfe, a 16-year New York City

police veteran, notes that these officers work on the presumption of a suspect's guilt.

Police Violence and Race   The relationship among the police, racial minorities,
and police violence is complex and exists on several levels. In the United
States, most police officers are white, and the people suspected of committing crimes they deal with on a regular basis (street crimes as opposed to
white-collar and corporate crimes) are disproportionately African American
and Hispanic. Speaking of policing minority neighborhoods, sociologist
James Marquart argues that "white officers don't understand a lot of things
that go on in these areas. One way to deal with that is to use force. It goes
across all cultural boundaries."

One Los Angeles police officer told the Christopher Commission that
the prone-out technique in which suspects are made to lie face down on the
pavement was "pretty routine" in minority neighborhoods because the police
had been taught that "aggression and force are the only things these people
understand." Stereotyping in a police-citizen encounter can work in both
directions. An officer who stops a "typical black gang member" believed to
be involved in the drug trade is perceived by the youth as the "typical white
ugly cop" out to hassle African Americans. The youth's hostile reaction confirms and reinforces the officer's stereotype and leads to an arrest decision.
The move to arrest confirms the young man's worst expectations of the officer, leading him to resist being taken into custody and thereby inviting the
use of force or excessive force.

Anxiety and Fear   One researcher found that police officers in stressful, sometimes dangerous situations who could not admit to themselves that they were
afraid tended to get angry beyond the point of control. These officers were
most likely to use excessive force. A former police chief believes some of his
officers were fearful of African American suspects. The manifestation of this
fear (in the form of nervousness, for example) might provoke resistance on
the part of some African Americans, which in turn, could trigger a forceful
response by officers. In addition, fearful officers may use force unnecessarily
to mask their anxiety, action that could elicit resistance on the part of suspects.

"Contempt of Cop"   While investigating the Los Angeles Sheriff's Department
in the aftermath of the Rodney King incident, the Kolts Commission concluded that police brutality (especially against racial and ethnic minorities)
was often directed against individuals who criticized them. "This is the worst

aspect of police culture, where the worst crime of all is 'contempt of cop.' The deputy cannot let pass the slightest challenge or failure to immediately comply." Some officers consider a verbal affront no different then a physical assault, inasmuch as verbal defiance can lead to loss of control, which in turn increases the risk of danger. To maintain (or regain) control of a situation, some officers may resort to force.

The Code of Silence   Officers who witness the use of excessive force by a colleague often have a difficult decision to make. If the action is ignored, the use of force has been condoned tacitly, increasing the likelihood that it will happen again. In addition, officers who remain silent are trampling the rights of individuals they are duty bound to protect and serve. However, if they are forthcoming to superiors regarding police deviance, these officers may encounter a variety of informal sanctions at the hands of colleagues. As one Los Angeles police officer noted: "What do you do if you see your partner do something wrong? If you can't stop him from doing it you're supposed to tell your commander. But if you squeal, no one will work with you." Former New York City police officer Frank Serpico, who was forced to leave the department in the early 1970s after he helped expose internal corruption, stated: "Why does the blue wall exist? Because there's a code that says, 'Keep your mouth shut and you will be taken care of.'" Both the Christopher and Mollen Commissions found that officers who break the code of silence are not only ostracized, they also become the object of complaints and physical threats from colleagues. Perhaps the greatest fear an officer contemplating breaking the code has is that he or she will be left alone in a dangerous situation.

Some observers believe the blue wall of reluctance is eroding and that "Good officers are anxious to weed out the bad ones . . ." The officer who provided key evidence in the prosecution of Abner Luima's police assailant was motivated by the seriousness of the crime. Nonetheless, as a result of threats, he and his family were forced to live in protective custody. Overcoming the wall of silence is arguably the most critical factor in reducing police deviance.

POLICING THE POLICE

Historically, police departments in the United States have not been strictly accountable for their policies and activity, although individual officers answered to superiors within the department. In other words, to the extent that accountability existed, the police were policing themselves. David Bayley argues that internal mechanisms of accountability can adequately control police deviance. For example, because internal affairs officers are intimately familiar with their agencies, in-house investigations can be much more thorough than external inquiries, and as such, are likely to focus on a whole range of police activities and not only the visible incidents in question.

John Crew, head of the Police Practice Project of the ACLU, disagrees with Bayley, contending that internal mechanisms for dealing with police misconduct are doomed to fail. Crew believes that politicians, and police administrators in particular, lack "the will to take steps necessary to institute real [police] oversight, real accountability, real scrutiny. It's a pattern that stretches from the president to the Department of Justice, to statehouses, down to the local city council and police chief."

Local communities have been moving toward independent citizen review boards to investigate police misconduct, especially the use of excessive force. Police unions and associations have been highly critical of civilian oversight, insisting that only law enforcement personnel understand the pressures and demands of police work, and have vigorously opposed such groups. Mary Powers of the National Council of Police Accountability believes that civilian review boards are a necessity if police misconduct is to be curtailed. "Without watchdog organizations that have teeth, police know they can hide behind civil-service protections until the latest scandal passes by. Then they come out again when the coast is clear." This does not mean that mutual trust and respect cannot exist between the police and civilian oversight groups. In San Francisco and Pittsburgh, this relationship appears to work. A member of the Pittsburgh Citizen's Police Review Board stated: "If you are going on a witch hunt or cop chase just for the chase, it won't work. The key is being fair and balanced."

Police unions and organizations have mounted legal challenges against citizen review agencies. These groups deny allegations against officers, "even those they know are brutal; encourage noncooperation with investigators, and the 'code of silence' when allegations arise . . ." In approximately 25 states, police organizations have obtained special bills of rights for law enforcement personnel that make it difficult to discipline or dismiss officers who engage in human rights violations. One notable exception to police stonewalling formal action against officer misconduct is Black Cops Against Police Brutality or B-CAP. This organization "was developed for the purpose of assuring that the rights of all citizens are not abridged by the police, especially in urban America." According to B-CAP, the victims of police brutality are often African American and Hispanic males who are victimized by African American and white officers. Research by Ronald Hunter indicates that not all police are hostile to civilian oversight. Hunter surveyed 65 police officers ranging in experience from academy cadets to 30-year veterans and found that 41.5 percent agreed or strongly agreed with the statement, "A good means of regulating police conduct is using citizen review boards." The exact same percentage of respondents disagreed or strongly disagreed with the statement.

In an important 2006 legal decision, *Copley Press v. County of San Diego*, the California Supreme Court ruled that the public no longer had the right to information regarding citizen complaints against police. "According to the Court's interpretation of California law, an officer found guilty of misconduct has a 'right to privacy' that outweighs the public's right to know." As a consequence of this decision, California's public disclosure of police misconduct is one of the most restrictive in the nation. Legislation introduced in the state assembly to overturn this decision has stalled in the face of strong opposition by police unions. According to one interpretation, union representatives have made it clear to law makers considering supporting the bill to overturn the court's ruling that "they would be branded as anti-law enforcement."

KEY TERMS

contempt of cop: Defiant attitudes of individuals (often those suspected of crimes) for the authority of a police officer; examples include profanity and fleeing after being stopped for questioning

Dirty Harry strategy: The disregard of procedural laws (laws of search and seizure, for example) and the use of means such as intimidation and torture by police officers in their role as professional crime fighters; named for the Dirty Harry movies starring Clint Eastwood

discretion: In criminal justice, the power of officials to decide whether to take a particular course of action

police accountability: The philosophy that individual officers and the department as a whole should be accountable for police behavior

police corruption: A catchall term that covers a wide variety of police misconduct, including accepting money, illegal drugs, or articles of value for not arresting law violators, committing burglaries or robberies, selling illegal drugs confiscated in drug busts, and so on

profiling: The informal practice of some officers and the formal policy of some law enforcement departments that the characteristics of certain individuals and groups (especially racial and ethnic) are indicative of criminal behavior

rotten apples explanation: A common interpretation of police deviance stating that, unbeknownst to senior police officials and administrators, a small number of officers are engaging in corrupt behavior; this explanation minimizes the scope of the problem and insulates police officials from any blame regarding the deviant behavior of subordinates

sexual shakedown: A form of sexual misconduct on the part of police wherein an officer will forgo arresting a female suspect (often a prostitute) in exchange for sexual favors

SUGGESTED READINGS

Aplert, G.P., R.G. Dunham, and M. S. Stroshine. 2006. *Policing: Continuity and Change.* Long Grove, IL: Waveland Press.

Chevigny, Paul. 1995. *The Edge of the Knife—Police Violence in America,* New York, NY: The New Press.

Crank, John. 2004. *Understanding Police Culture.* Cincinnati, OH: Anderson Publishing Company.

Dunham, Roger and Geoffrey Alpert. 1997. *Critical Issues in Policing: Contemporary Readings.* Prospect Heights, IL: Waveland Press, Inc.

Gaines, Larry and Victor Kappler. 2005. *Policing in America.* Anderson Publishing, Company,: Cincinnati, OH.

Kappler, Victor. 1999. *The Police and Society.* Prospect Hills, IL: Waveland Press, Inc.

Kappler, Victor, Richard Sluder, and Geoffrey Alpert. 1998. *Force of Deviance.* Prospect Heights, IL: Waveland Press, Inc.

Manning, Peter. 1997. *Police Work: The Social Organization of Police.* Prospect Heights, IL: Waveland Press, Inc.

Sampler, Norm. 2005. *Breaking Rank: A Top Cop's Expose of the Dark Side of Policing.* New York, NY: Nation Books.

Walker, Samuel. 2005. *The New World of Police Accountability.* Thousand Oaks, CA: Sage Publications.

Walker, Samuel and Charles M. Katz. 2004. *Police In America: An Introduction.* Boston, MA: McGraw Hill.

Wells, Sandra K. and Betty L. Alt. 2005. *Police Women: Life with the Badge.* Westport, CN: Praeger.

"Speed Bump" © Dave Coverly/Dist. By Creators Syndicate, Inc.

ten

chapter ten

Pretrial Procedures and Plea Bargaining: From Arrest to "Let's Make a Deal"

In this chapter we begin our examination of the criminal courts from the moment the accused enters the system by way of arrest. Keep in mind that the administration of criminal justice is not a streamlined process with individuals moving with assembly-line precision through a machine that stamps out "guilty" or "innocent" as its final product. Rather, the criminal courts should be viewed as a series of critical situations, in which practitioners with discretionary authority make processing decisions.

For example, prosecutors determine how many charges will be brought against a defendant and at what level (misdemeanor or felony). District attorneys also determine whether a case is adjudicated by a jury trial or plea bargaining. As the vast majority of guilty convictions are obtained via plea negotiation rather than a jury trial, we will look at this crucial aspect of the American justice system via the vantage points of the major courtroom players: prosecutors, judges, and public defenders.

PRETRIAL PROCEDURES

Arrest

For a person to move through the criminal court system, he or she must be arrested. With the exception of those small number of crimes in progress that police witness, someone has to be aware that something that might be a crime has occurred and then decide to call for help. On arrival, the police must determine whether what has been described and what they see constitute evidence of a criminal event. If the officers decide the event was a crime, the culprit must be identified and located. Arrests are also a function of departmental priorities. Because police departments have limited budgets, garden-variety burglaries and other property crimes are low-priority offenses. The police are not likely to pursue the perpetrators actively, because the police typically do not know who committed the crime.

The two basic forms of arrest in the United States are those with and those without warrants. An **arrest warrant** is issued after one person (the complainant) files a complaint against another individual, and, on reviewing the complaint, a judge decides that probable cause for arrest exists. Arrests without warrants occur when police officers witness criminal events or when law enforcement personnel have probable cause to believe that someone has committed (or is about to commit) a crime. About 95 percent of arrests are made without a warrant.

Initial Appearance

Defendants make their **initial appearance** before a magistrate typically within a few hours of arrest. The magistrate informs the defendant of the charges, reaffirms the defendant's right to remain silent, have a lawyer assigned to him or her, and, in felony cases, have a preliminary hearing. In crowded urban courts, defendants may be advised of their rights in groups. Magistrates have no authority to accept a plea in felony cases. Judges may also set bail at this time. Most defendants in court on misdemeanor offenses plead guilty and are sentenced.

Charging

The judge informs the defendant of the charges at the initial appearance. David Neubauer notes that there are four types of charging documents (documents that describe the date and time of the offense): complaint, information, arrest warrant, and indictment. A complaint must be supported by an oath on the part of the witness or arresting officer. An information is the same as a complaint, except that it is a document filed by a prosecutor stating that sufficient evidence against the individual exists to justify an arrest, and he or she directs police to do so. A magistrate, typically a lower-court judge, issues an **arrest warrant**. Indictments are discussed in the section on grand juries.

Preliminary Hearing

States that do not have a grand jury system (about 50 percent) use a preliminary hearing or preliminary trial to determine whether the accused is to be subjected to a criminal trial. The central purpose of the preliminary hearing is to "prevent hasty, malicious, improvident, and oppressive prosecutions, to protect the person charged from open and public accusations of a crime, to avoid both for the defendant and the public the expense of a public prosecution, and to discover whether or not there are substantial grounds upon which a prosecution may be based." Legal scholars Ellen Hochstedler Steury and Nancy Frank note that although the prosecution must present evidence of the crime, the district attorney does not have to present all the evidence in his or her possession. Rather, just enough evidence must be revealed to convince the judge of probable cause. In many jurisdictions evidence that will be inadmissible at a trial is allowed in a preliminary hearing. That is, in most instances preliminary hearings are held prior to a judges ruling on defense motions to suppress evidence as inadmissible.

Defendants have a right to be present at preliminary hearings, to hear accusations made against them, as well view the prosecutions evidence in support of these allegations. The defendant also has the right to be represented by counsel during these proceedings. Defense attorneys use preliminary hearings to gauge the strength of the government's case against their clients. If defense attorneys come to believe (correctly or incorrectly) that prosecutors have an especially strong case, defense attorneys may be more willing to plea bargain.

Grand Juries

Grand and *petit* are French for large and small, respectively, with the **grand jury** so named because it consists of between 12 and 23 members (16 and 23 for federal grand juries); a petit jury usually has 12 members; both are integral parts of the criminal court system. The petit or common jury is the trier of facts and determines the guilt or innocence of a defendant through a criminal trial. The purpose of the grand jury is to determine whether a crime was committed and whether the person accused of this offense is the likely culprit and should face criminal prosecution. If the grand jury determines by a majority vote that the state's charges against the accused are well-founded then the jury presents an **indictment** or true bill. When jury members are not satisfied with the weight of the evidence against the defendant, they do not render an indictment or bill. Grand jury service usually lasts from one month to one year, and jurors can hear up to 1000 cases during this period.

The grand jury is supposed to be an institutional safeguard that prevents prosecutors and the government from using the criminal courts to harass citizens for personal and political reasons. However, some people argue that the prosecutor—whose power that body is supposed to check—is the jury's legal advisor to court proceedings and thus has little difficulty obtaining a true bill. The district attorney calls witnesses and determines which individuals should be issued subpoenas and compelled to testify. The defendant and his or her legal counsel are excluded from grand jury proceedings.

Critics who believe grand juries do little more than rubber stamp the district attorney's agenda note that, in a recent year, federal grand juries returned more than 23,000 indictments and only 123 no-bills. In other words, federal grand juries agreed with the prosecutor almost 99.5 percent of the time. However, this does not necessarily mean that grand juries are the tools of overzealous district attorneys. What critics do not understand is that prosecutors rarely seek indictments against defendants if they do not have the evidence to substantiate their assertions. Saltzburg and Capra argue that the high percentage of indictments is a result of district attorneys screening out weak cases before they reach a grand jury.

Grand juries also have investigative powers and can summon anyone to testify before them. Any aspect of a person's life that may have relevance for criminal investigation falls under the scope of a grand jury inquiry. Regarding federal grand juries, Susan Brenner and Lori Shaw note that these bodies offer prosecutors numerous advantages, especially in "high-profile, factually complicated" investigations. To begin, because grand juries operate in secret, prosecutors have the ability to shield evidence. Second, unless a witness decides to reveal that he or she testified before a federal grand jury, no one will ever be aware of this occurrence, as grand jury transcripts are sealed. Third, a federal grand jury can subpoena documents, computers, blood samples, and virtually any type of physical evidence without having to demonstrate probable cause.

Finally, grand juries can hear evidence—hearsay evidence, for example—that would not be admissible in court. Few people realize how powerful grand juries in this country are, especially at the federal level.

Arraignment

After the information or indictment has been presented to the court, the defendant is **arraigned**. During this formal proceeding, the accused stands before a judge in open court and hears a clerk or district attorney read the charges against him or her. Once again, the individual is informed of his or her constitutional rights, including the right to legal counsel provided at state expense. At this juncture of the criminal proceedings, the defendant is asked to enter a plea of guilty, not guilty, or in many jurisdictions, *nolo contendere* (no contest). Although, strictly speaking, nolo contendere is not a plea, it means that the defendant will not formally contest or refute the charges. By way of a *nolo contendere* plea, the defendant can later claim that technically no guilty verdict was reached (even though he or she was punished for the offense); therefore, the defendant may be spared the civil penalties that result from a guilty plea or conviction. In other words, *nolo contendere* is an implied confession only to the offense charged and cannot be used against the defendant in another criminal or civil case. These pleas are often entered in tax fraud cases where a civil suit is likely to follow. Most defendants plead not guilty and the proceedings move toward a criminal trial. David Neubauer observes that the significance of the arraignment is the signal it sends to the courtroom working group (discussed below) that the defendant is probably guilty and will most likely be convicted via trial or a negotiated plea of guilt.

Pretrial Release and Detention

The primary reason for detaining defendants prior to trial is to ensure that they appear in court and accept the lawful disposition of their cases. A U.S. Department of Justice survey of the nation's 75 largest counties from 1990 to 2004 reported that between 62 and 64 percent of felony defendants were released prior to their trial (ranging from a low of 19 percent for those charged with murder to a high of 82 percent for individuals accused of fraud offenses). Regarding minor infractions, the police issue the defendant a summons or ticket, informing the individual when he or she is to appear in court. This saves law enforcement personnel much time and energy, as they do not have to transport defendants to the station for booking and the courthouse for arraignment. Judges may disallow pretrial release if they believe a defendant is likely to commit additional crimes, attempt to intimidate witnesses or the victim, and/or destroy evidence upon discharge.

Bail

A judge may decide that releasing a defendant on his or her own recognizance is too risky and opt for a financial bond that is forfeited should the individual fail to appear in court. The next issue is the monetary amount at which the bail should be set. Almost all felony court judges use three primary criteria in determining the amount of bail a defendant will have to post: (1) the seriousness of the offense (despite the lack of evidence linking the gravity of the crime to the defendant's likelihood of appearing at trial); (2) the strength of the case against the accused, information that is usually passed on to judges by prosecutors. As the strength of the state's case increases, so to does the likelihood of the defendant not appearing in court, hence the greater bail amount for strong prosecutorial cases; and (3) the defendant's prior criminal record. Individuals with numerous criminal convictions are more likely to have a higher bail than people with no or limited criminal backgrounds. Judges also consider character references and an individual's ties to the community. For example, people employed full time or full-time students are considered less likely to flee. Therefore, lower bail amounts are set for these individuals.

Bail Bondsmen

Just under 20 percent of defendants secure their freedom through bail bondsmen. Bondsmen can be thought of as insurance agents whose companies guarantee the courts that on a specified date the defendant will appear, or the bondsmen will be responsible for the full amount of the bond. Thus, if a judge sets bail at $10,000, the defendant or family and friends pays the bondsmen a non-refundable sum equal to 10 percent of bail, in this case $1000. If the defendant appears in court at the required time and date, the bond company has profited by the latter amount. To prevent themselves from being stuck with paying the courts $9000 if the defendant fails to appear, bondsmen may purchase a surety bond from a large insurance company for 30 percent (in this example, $300) of the bondsmen's fee.

Once a contact has been made with a potential client (often by way of defense attorneys), the bondsmen must decide if the company will do business with this individual. Keep in mind that bondsmen are not part of the criminal justice system. Rather, they are private business people, and as such, can choose whom to accept as customers. Professional criminals and individuals with extensive crime histories are good risks because they know the ins and outs of the system; they have been this route before and can be expected to appear in court when required. Narcotics users are acceptable, as they have to stay close to their supply of drugs. First-time offenders charged with serious crimes are considered poor risks, as they may panic and flee. Violent offenders pose a threat to the bondsmen and are less likely to secure financial backing. Profit-oriented bail bondsmen have long been associated with corruption, as some individuals in this highly competitive business pay off police officers, jailers, and attorneys to steer clients in their direction.

The overwhelming majority of people who secure their freedom via a bail bondsman appear in court as required, meaning that forfeitures (the money bondsmen must pay the courts when an individual does not appear) are low. Most states have a grace period to locate a truant defendant before the bondsman must make good on the money now owed the state. A bail or bond jumper can be certain that one of the estimated 2500 to 5000 skip tracers (bounty hunters) will be looking for him or her intently in a matter of days. Depending on their experience and state laws, bounty hunters receive between 5 and 20 percent of the bail bond amount when they apprehend a fugitive. These individuals are credited with making approximately 30,000 arrests annually (many of dangerous criminals), saving taxpayers money and police time.

However, because the failure to live up to the terms of a bail agreement is considered a civil as opposed to criminal matter, bounty hunters are not limited by the rules of due process and extradition (if the defendant flees to another state or country) that law enforcement personnel must follow. For example, bounty hunters may enter, without an arrest warrant, the residence of a suspect or third party believed to be sheltering the suspect, they may enter or burst into a dwelling without the knock-and-announce requirement by which law enforcement personnel must abide, and bounty hunters may imprison a suspect until he or she can be returned to the state where the bond was secured. Bounty hunters have sweeping powers because defendants have consented to broad search and arrest rights by entering into a bail contract.

The negative consequences of these far-reaching legal rights becomes apparent when bounty hunters target the wrong individual, a not-uncommon occurrence. In Kansas City, bounty hunters broke into the house of an innocent man and shot him three times. The victim was African American; the defendant they were looking for was Hispanic. In Arizona, five heavily armed bounty hunters invaded a home after they broke down a door with a sledgehammer. A woman, her four children, and an adult male were rounded up and held at gunpoint. The intruders then headed for the locked bedroom of 23-year-old Chris Foote and his 20-year-old girlfriend Spring Wright. As the men entered the room Foote grabbed a handgun and began firing, wounding two of the bail bondsmen. The other men returned fire, killing both Foote and Wright. Another case of mistaken identity as the fugitive the bounty hunters were after had no connection to anyone in the house. Because bounty hunters are hired as independent contractors by bond agents, these agents have little if any liability if they are sued for the actions of skip tracers. Few bounty hunters have the financial assets to make civil suit worthwhile. This leaves the innocent victims of bounty hunter mistakes with no meaningful civil or criminal court recourse.

Discovery and Pretrial Motions

Discovery is the pretrial procedure by which the prosecution and defense gather information by means of the disclosure of deeds, facts, documents, and other material that may be necessary to present an adequate defense or aggressive prosecution. The discovery process eliminates surprise witnesses or pieces of evidence and mandates that the prosecution disclose to defense counsel any exculpatory evidence. **Exculpatory evidence** "refers to evidence and/or statements which tend to clear, justify, or excuse a defendant from alleged fault of guilt." The federal government and most states now have statutory pretrial discovery stipulations. Legal writers Steven Emanuel and Steven Knowles have outlined disclosure requirements that the prosecution must satisfy:

1. On request, the prosecution must provide the defense with copies of any statements made by the defendant; many states also require statements made by codefendants.
2. The defense is entitled to copies of medical and physical examinations and scientific tests made for the prosecution.

3. Most states give defense counsel access to documents and tangible objects that will be presented by the district attorney during the trial.

4. Many states require the prosecution to present the opposing attorney(s) with a list of witnesses who will be called to testify as well as any prior written or recorded statements made by these individuals.

Although the prosecution has no constitutional right to the defendant's materials, the federal government and most states provide the prosecution with discovery rights that tend to be less broad in their scope than the rights given to the defendant's counsel. From Emanuel and Knowles again:

1. The defense counsel must inform the district attorney if the defendant's attorney intends to present an alibi defense (a defense contending that the defendant was elsewhere when the crime was committed).

2. Defense counsel must inform the prosecution if an insanity defense will be introduced or if an expert witness will be called to testify that the defendant lacked the mental capacity necessary to commit the offense.

3. Some states require defense counsel to provide the prosecution with the names and addresses of witnesses it plans to call as well as any previously recorded statements made by these individuals.

4. In some states prosecutorial discovery is contingent on defense disclosure; that is, if the defense counsel does not seek discovery, then the prosecution has no right to information in possession of the defendant's attorney.

Discovery does not provide access to the opposition's **work product**, i.e., the attorney's trial strategy. Work product also includes reports of private investigators as well as a list of witnesses in the order they will be called. Because work product is not evidence, it is not subject to discovery as outlined in state and federal statutes.

In the period between the arraignment and the beginning of the trial, the prosecutor and/or defense counsel can file **pretrial motions**—petitions to the court requesting orders or rulings favorable to the applicant's case. Judges make their rulings on these motions after reading written briefs, after hearing the oral arguments of the competing attorneys, and/or after a fact-finding hearing. The most important pretrial motions filed by defense attorneys are those that attempt to exclude a confession and/or suppress evidence on the grounds that it was obtained illegally, for example, by way of an illegal search and seizure. According to the **exclusionary rule**, evidence obtained in violation of a defendant's rights may not be used by the prosecution for the purpose of establishing an individual's guilt. Obviously, the judge's ruling on whether evidence is admissible can be of utmost importance to the outcome of a trial.

THE COURTROOM AS A CLUBHOUSE

The major players in the courtroom workgroup are judges, prosecutors, and defense attorneys (most often public defenders) all of whom share one goal: to move cases through the criminal courts rapidly. Although they often disagree on how to achieve this goal, these courtroom players work together in interrelated and interdependent informal groups. In one way or another, workgroup members are rewarded for expediting cases and penalized for letting a backlog of cases mount. For example, prosecutors who do not engage in delaying tactics (from the perspective of the judge) may be rewarded accordingly. As one prosecutor stated: "You might have 10 or 15 preliminary hearings any given morning and you need a little time to talk to a witness you didn't have a chance to talk to, or you need time, a break say, to talk to a defense counsel. If you have a good rapport with that judge, you get that time."

Courtrooms are affected by the number and type of offenses they process. During the course of a year, urban judges, prosecutors, and defense attorneys deal with thousands of burglaries, street robberies (muggings), and assaults. These offenses become normal crimes, in that the circumstances of these events (offender and victim background, degree of injury suffered by the victim) are much the same. The sameness of these crimes provides a common frame of reference for workgroup members. Because the burglary case being negotiated today falls in the normal-crime category, the work group members can dispose of it in the same manner as yesterday's case and tomorrow's case. A high volume of similar cases, therefore, greatly facilitates the speed in which they can move through the criminal court system.

In a classic article entitled "The Practice of Law as a Confidence Game," sociologist Abraham Blumberg views the defense lawyer as a double agent whose goals are more in line with the courtroom workgroup than the welfare of his or her client. Blumberg found that, of 724 individuals convicted by way of a negotiated plea (plea bargaining), in 411 cases (56.7 percent) the defendant's attorney had the greatest influence on his final decision to plead. In the 16.5 percent of the cases wherein the defendant's wife influenced the plea bargaining decision, this suggestion on her part was usually "sparked and initiated by defense counsel."

The working relation between prosecutors and defense attorneys is clearly evident in one southern California locale. The Orange County Attorney's Association is a union that comprises prosecuting attorneys and public defenders. When the county board of supervisors recommended cutting the salary for public defenders (which is the same as that of prosecutors), district attorneys threatened to strike if the pay cuts were implemented. The county backed off and made no pay reductions. Joint prosecutor/public defender unions also exist in two Minnesota counties.

External forces such as the police, the prisons, and the media also influence courtroom workgroups. Prosecutors cannot obtain convictions without the investigative work of police, who provide the necessary evidence and witnesses to build a case. Therefore, district attorneys are likely to give high priority to cases that are important to law enforcement personnel and will often check with officers before plea bargaining with some defendants. District attorneys prosecute assaults on police officers vigorously, and defense attorneys know that the rules of the game are different in these situations.

PLEA BARGAINING

A **plea bargain** is an agreement between the state (represented by the prosecutor) and the defendant (typically represented by his or her attorney) wherein each side gives up something and receives something in return. For example, the accused pleads guilty to a less serious charge (or fewer charges) than the state would have brought had the case gone to trial. Pleading to a lesser offense results in a lighter sentence. However, the state gains a measure of control over the now-convicted defendant (a jail or prison sentence or probation). If the case had gone to trial and the jury voted to acquit, the state achieved nothing. Plea bargaining allows the state to win a conviction while expending little time and effort compared to a trial and, in the case of a capital offense, a potentially lengthy trial.

The Supreme Court both affirmed and advocated plea bargaining in a 1971 decision (*Sontobello v. New York*). Chief Justice Warren Burger wrote that "plea bargaining is an essential component of the administration of justice. Properly administered, it is to be encouraged. If every charge were subjected to a full-scale trial, the States and the Federal Government would need to multiply by many times the number of judges and court facilities."

Plea bargaining has been maligned by the general public in large measure because people do not fully understand the concept. Many individuals are confused about the term "bargain," which is supposed to convey a sense of compromise, but which often has another connotation, as in "I got a real bargain on a car I bought yesterday." They have the impression that criminals are getting a good deal at the expense of the victim and the criminal justice system.

There are **explicit** and **implicit plea bargains**. In the majority of situations defendants negotiate for explicit concessions from the state through their attorneys. For example:

- Reduction in the number and/or type of charges, e.g., a reduction from five to two counts of felony burglary, from armed robbery to unarmed robbery, or from rape to sexual assault. In the latter two instances, although all of the crimes are felonies, unarmed robbery and sexual assault are not punished as harshly as armed robbery and rape.
- A specific sentence recommendation (typically, a reduced sentence) in terms of months or years to be incarcerated. In exchange for a guilty plea, the district attorney recommends that the defendant serve two years of a possible seven-year sentence.
- A promise by the district attorney that he or she will not oppose the defense attorney's request for probation.
- Some other specific considerations. For example, the offender will be incarcerated at a prison near his family.

Implicit bargaining is more difficult to detect. It occurs when the defendant, through his or her attorney, comes to believe that, if he or she does not plead guilty and is convicted at a trial, the sentence will be harsher than would have resulted from a plea. Researchers have discovered that in many jurisdictions defendants who pled guilty consistently received lighter sentences than those individuals who took their cases to trial when relevant variables such as age, gender, race/ethnicity, and previous criminal record were held constant.

Since its inception, the primary criticism of plea bargaining is that offenders are not punished severely enough, are released prematurely from prison, and often continue their criminal careers. A vocal opponent of plea bargaining, John Langbein argues that at least three evils are associated with this non-trial procedure. To begin, because plea bargaining negates both the jury and the trial, it concentrates the power to punish completely in the hands of the state. Inasmuch as government officials are now the judge, jury, and executioner, citizens are removed from the everyday allocation of justice. In addition, high-visibility trials are replaced with behind-the-scenes decision making, and the central figure in these negotiations (the prosecutor) is not directly accountable to the public for her or his actions.

Second, plea bargaining is based on coercion, as the defendant is under significant pressure to bear witness against himself. For Langbein, "A legal system that comes to depend upon coercing people to waive their supposed rights is by definition a failed system." As the disparity grows between the sentence offered a defendant via plea bargaining and the punishment threatened by way of a jury trial, the pressure to confess intensifies. Because prosecutors routinely overcharge (that is, charge offenders with crimes they know they could not prove at a trial), defendants often end up pleading guilty to a crime of which they may have been acquitted had the case gone to trial. Langbein argues that this is plea pressure rather than plea bargaining.

Finally, negotiated justice makes a mockery of criminal statistics. A person arrested and charged with multiple rapes may plead to one or more counts of sexual assault. As approximately 90 percent of all criminal convictions in both state and federal courts are the result of plea bargaining (in some jurisdictions it is as high as 99 percent), there is virtually no correspondence between what individuals are arrested for and the offenses for which they have been convicted.

A staunch defender of negotiated pleas, New York State Supreme Court Justice Carolyn Demarest argues that plea bargaining offers significant benefits to the community, including many crime victims. She notes that testifying and undergoing rigorous cross examination can be very difficult for the young, the old, the infirm, as well as victims of sexual abuse and rape who may be reluctant to take the witness stand because of shame and/or the fear of public humiliation. Plea bargaining may also be the best prosecutorial strategy when witnesses are truthful but not credible, for example, when a known prostitute informs authorities that she has been raped.

Prosecutors and Plea Bargaining

In the police chapter we saw that officers have significant discretionary power and continually make decisions regarding what specific course of action they will follow. Similarly, after a decision has been made to prosecute, district attorneys must determine if they will take the case to trial or plea bargain. In their *Plea Bargaining in the United States* study, Herbert S. Miller and his colleagues offer a list of questions that, when answered, will influence a prosecutor's decision:

- If I go to trial, will I have credible witnesses? Do any of them have criminal records? Are they believable? Will they crumble under rigorous cross-examination?
- If I go to trial, what are my chances of gaining a conviction?
- Is the opposing attorney an efficient and capable advocate who will provide his or her client with a first-rate defense?
- Will he or she get continuances and prolong the proceedings until my case is worn out?
- Is the defense attorney capable of swaying the jury to his or her point of view?
- Will this trial be so long and arduous that it takes time and energy away from other cases?
- Would the judge and police chief want the case concluded without going to trial?
- What would the public want me to do?
- What should I do: take a plea of guilty (the sure thing) and a lesser term than this case may deserve, or take a chance, gamble for a longer sentence by going to trial, and risk a not-guilty verdict and no punishment?

Of these considerations, the most important appear to be the prosecutor's perception of the strengths and weaknesses of the state's case. Other major factors include the seriousness of the offense, the suspect's past record, and the reputation of the defense attorney.

Judges and Plea Bargaining

In his study of plea bargaining, Milton Heumann noted that judges are typically forceful advocates of plea bargaining for at least four reasons. First, bargaining makes their jobs easier, as they have no need to prepare for a trial, rule on legal issues that come up during a case, or write jury instructions. Second, upon striking a deal, one of the most difficult decisions criminal court judges face—sentencing offenders—has been removed. Third, plea bargaining expedites moving the business, a euphemism for processing cases through the system. Judges state that they are under pressure from colleagues and court administrators to dispose of cases quickly. Finally, plea bargaining significantly reduces the chances that judicial decisions will be reversed by an appellate court.

Like district attorneys, criminal court judges work under the assumption that the overwhelming number of defendants are guilty. Judges are not opposed to these guilty individuals taking their cases to trial if they have a realistic, contestable issue justifying this course of action. Judges object to legal maneuvering that they view as frivolous and that results in an unnecessary trial that serves no purpose. As one judge noted:

> The circumstances are these: If a fellow tries a case where he is obviously guilty, where it is sheer folly to try the case, and yet the accused is insisting on trying the case, he deserves to be penalized for the trial because he takes up court time, there is expense involved, and so on. On the other hand, in a situation where he had a good defense, where it is a close question, and he happens to be found guilty, I would not penalize him for trying a case like that. He's got a right to try a case where he has a reasonable position, and it happens not to prevail.

Judges are also responsible for insuring that the rights of the defendant have been protected and that factually innocent people are not pleading guilty to crimes they did not commit. This is ostensibly accomplished by way of a public ritual (sometimes referred to as the litany) where the judge formally accepts the guilty plea. Michael McConville and Chester Mirsky offer an example of this ritual:

> **Judge:** Have you had an opportunity to consult with your lawyer, Mr._____, and to discuss the matter with him/her before choosing to plead guilty?
> **Defendant:** Yes
> **Judge:** Do you understand that by pleading guilty you have given up your right to confront and cross-examine witnesses against you, to testify, and to call witnesses on your own behalf?
> **Defendant:** Yes
> **Judge:** Do you understand that you have given up your right to remain silent and your privilege against self-incrimination?
> **Defendant:** Yes
> **Judge:** Do you understand that at a trial you are presumed innocent and that the prosecution had to prove you're guilty beyond a reasonable doubt?
> **Defendant:** Yes
> **Judge:** Has anybody threatened or coerced you?
> **Defendant:** No

Judge: Is your plea voluntary and of your own free will?

Defendant: Yes

Judge: Do you understand that in pleading guilty you have given up all these rights and that the conviction entered is the same as a conviction after a trial?

Defendant: Yes

Critics contend this procedure is a sham that masks the assembly line justice (or injustice) meted out. The guilty plea is hardly given voluntarily, in that the defendant has learned (either explicitly or implicitly) that, if he requests a jury trial and is convicted, he will be punished more severely. In some cases the judge is an active player in this coercive process.

Public Defenders and Plea Bargaining

A study utilizing in-depth interviews with 20 public defenders in New York, North Carolina, and South Carolina commented on judicial pressure put on these attorneys to plea bargain cases. Almost all of the attorneys interviewed said that judges routinely hand out stiffer punishments (often the maximum sentence) to defendants who go to trial and lose. A New York public defender commented that judges often say: "How could you be trying this stupid case, how dare you." A public defender in Carolina noted what happens when attorneys fail to convince their clients to accept a deal. "We go back and say, I've done everything I can, judge. He screams at you."

Respondents believed that most prosecutors did not take advantage of the fact that many judges excessively punished defendants who insisted on going to trial. However, they were hostile to those who did capitalize on this situation. A Carolina public defender stated that "some [prosecutors] are not merely unreasonable. They are astronomically unreasonable." A New York defender summed up the feelings of many defense attorneys who are forced to deal with prosecutors in jurisdictions where a jury trial is not a viable option: "You want to go to the district attorney's office and say, You self-righteous son-of-a-bitch, who the hell do you think you are?"

In a recent year, public defenders handled 82 percent of the 4.2 million criminal cases in the nation's 100 largest counties. With heavy caseloads, these attorneys have little choice but to plead as many cases as possible. Public defenders in Kentucky began fiscal year 2004 with an average of 484 cases.

Since the mid-1980s, the use of a **contract system** for defending indigents has increased dramatically. Under this system, a local government agency engages in a contractual relationship with a law firm or individual attorneys to provide necessary legal services. The major criticism of contract programs is that they emphasize cost over quality, as local governments accept the lowest bid from law firms regardless of the caliber of representation provided. A contract lawyer in Georgia who represented approximately 300 indigent clients never took a case to trial. Rather, he encouraged all of his clients to plead guilty. In one case, a client pleaded guilty to injuring someone while driving drunk and received a 15-year sentence. During the 13 months this individual spent in jail before entering a plea, he never conferred with his attorney, nor had his attorney interviewed a single witness. Another Georgia contract lawyer stated it was a grave error for a defense attorney to assume his clients were innocent. His contract with the state was later revoked.

A study of federal prosecutions from 1997 to 2001 found that defendants with court-appointed lawyers who bill by the hour receive substantially longer sentences than defendants represented by full-time public defenders. As 95 percent of federal prosecutions are adjudicated via plea bargaining, the sentence differences are explained by knowledge of which cases to plead out and which cases to take to trial, as well as how skillful and experienced defense attorneys are in making deals with prosecutors. Radha Inyengar, one of the study's authors, noted, "The court-appointed lawyers tend to be quite young, tend to be from small practices and also tend to be from lower-ranked law schools . . . They have a smaller client base and fewer interactions with prosecutors."

Let's Make a Deal

Plea bargaining is best understood as a process involving a number of decisions that occur over a period of a few days to a few weeks rather than as a single decision. In an observational study of plea bargaining in southern California, Debra Emmelman observed 15 attorneys employed by a private non-profit organization that had a contract with Smith County (a fictitious name) to defend indigent persons. The organization had a reputation for providing high-quality defense service. In Smith County, plea negotiations include the judge, the prosecuting attorney, the defense attorney, the defendant, and, on rare occasions, the victim. Judges in this locale were encouraged to promote plea bargaining and minimize jury trials. Emmelman found that pleas could be negotiated almost anywhere (the judge's chambers, in hallways outside the courtroom, and holding tanks where defendants awaited their court appearance) and at almost any time or juncture of an ongoing criminal case.

In some cases, pleas were bargained successfully in one session, while in other instances the negotiators met several times. A decision on the part of defense counsel to reach a settlement immediately or continue the process involved assessing the prosecutor's plea bargain offer, negotiating the terms of the deal, and conferring with the defendant and deciding what should be done. These steps do not necessarily occur in order, as when a public defender discusses with his or her client possible terms to be offered by the prosecutor prior to meeting with the district attorney.

Assessing a prosecutor's offer as acceptable or unacceptable is in large measure a function of the defense attorney's understanding of the value of the case. This value is determined by several factors, beginning with the seriousness of the charge and the circumstances of the case. Prosecutors are likely to consider beating a helpless senior citizen as a more serious offense than beating a rival gang member, although both crimes are technically the same offense (aggravated assault). Case value is also contingent on the strength of the state's evidence, as interpreted by the defender and his or her perception of how successfully this evidence can be refuted. Finally, the defendant's background is assessed, including prior criminal convictions.

Even when district attorneys make a reasonable offer in exchange for a guilty plea, defenders may hold out for a better deal in the belief that subsequent offers will be less punitive. This is often the case, as prosecutors lose witnesses over time, or evidence appears that is favorable to the defense. However, in some instances, prolonging the culmination of a plea bargain may work against the accused. As one defender noted in the Emmelman study: "I have a case now, I think the client was offered local time [in municipal court]. They just came up with two prison priors on this guy. He swears they aren't his. I don't believe him. I think they're his . . . In which case the deal's gonna get worse." An often-overlooked aspect of plea bargaining is that both the prosecutor and defender are playing an information game. Who knows (or believes) what at a given stage of the proceedings will, in part, determine if a case is negotiated and what the settlement is likely to be.

Is Plea Bargaining Here to Stay?

Over the past 30 years, a number of criminal court jurisdictions and a few states have attempted to limit or eliminate plea bargaining with various degrees of success. Typically, these efforts have done little more than shift negotiations from a visible process to a more covert or hidden stage. If legislators or a district attorney bars plea bargaining, prosecutors simply consult with defense counsel before charges are filed officially and make a deal at that time. As George Fisher notes: "The form of the plea becomes, I will seek only these charges if you will promise that your client will plead guilty."

The most frequently given explanation for the existence and persistence of plea bargaining is the sheer number of criminal defendants that the courts have to process. However, Milton Heumann has presented compelling evidence that "guilty pleas characterize high and low volume courts" and have done so since the late 1800s. The Rutgers political scientist notes that "First and foremost, plea bargaining is not a function of case pressure. This is a huge myth that continues to bedevil analysis. Even if courts had unlimited resources 70 to 80 percent of all criminal cases would be negotiated." If plea bargaining is not driven by necessity, then why is it the most frequently used method of case disposition in the country? Based on his research, Heumann argues that, through the course of their work, judges, prosecutors, and defense attorneys learn that, because most defendants are both legally and factually guilty, plea bargaining is the most realistic and beneficial method of dispensing justice. Judges want to move cases along and avoid trials with complex legal issues. Prosecutors learn that adequate punishments relative to the seriousness of the crime can be achieved via bargaining. District attorneys also want high conviction rates—prosecutorial batting averages—much more easily obtained via plea bargaining than jury trials. Defense attorneys learn their guilty clients will get a better deal if they bargain than if they go to trial and are convicted. Overburdened public defenders want relief from staggering case loads. "As the actors spend more time in the system, these assumptions become so ingrained that they no longer think in terms of a trial."

KEY TERMS

arraignment: The proceedings by which an individual is formally charged and required to enter a plea

arrest warrant: A document issued by a judge or court having the authority to do so that directs or sanctions the arrest of one or more persons

contract system: A system wherein a government agency engages in a contractual agreement with a law firm or individual attorneys to provide legal services for indigent defendants

exclusionary rule: A rule of law stating that an otherwise admissible piece of evidence cannot be used in a criminal trial against a defendant because it was the product or result of illegal police conduct; an illegal search, for example

exculpatory evidence: Evidence or statements that tend to clear, justify, or excuse a defendant from alleged guilt

explicit plea bargain: Specific considerations, such as charge or sentence reduction from the state, that defendants negotiate through their attorneys

discovery: A pretrial procedure by which one party (defense counsel, for example) gains important information about the case form the other party (the district attorney)

grand jury: A jury of between 12 and 23 individuals who determine whether an individual should be indicted; grand juries also have the power to investigate crimes in some jurisdictions

implicit plea bargain: Defendants learn through their attorneys that if they do not plead guilty and are convicted of a crime at a jury trial, they will by punished more severely than if they would have pled to an agreed-on offense(s)

indictment: A formal accusation of a felony that is delivered by the grand jury after carefully considering the evidence that is presented by the prosecutor

initial appearance: The first appearance before a judicial officer of a person who has been arrested; at this appearance, the individual is informed of his or her constitutional rights, the terms of bail or other forms of release are set, and counsel is appointed if the individual is indigent

nolo contendere: A statement by the defendant that he or she will not contend or contest the charge(s) made by the government

plea bargain: An agreement between the state and the defendant wherein each side gives up something and receives something in return; for example, a charge and/or sentence reduction in exchange for a guilty plea

pretrial motions: Petitions to the court requesting orders or rulings favorable to the applicants case

work product: The strategy that an attorney will utilize as part of his or her trial

SUGGESTED READINGS

Champion, Dean J. and Gary A. Rabe. 2006. *Criminal Courts: Structure Process and Issues.* Upper Saddle River, NJ: Prentice Hall.

Cole, David. 2000. *No Equal Justice.* New York, NY: The New Press.

Emanuel, Steven L. and Steven Knowles. 2005. *Criminal Procedure.* Larchmont, NY: Emanuel Publishing Company.

Fisher, George. 2003. *Plea Bargaining's Triumph: A History of Plea Bargaining in the United States.* Stanford, CA: Stanford University Press.

Israel, Jerold H. and Wayne R. LaFave. 2001. *Criminal Procedure in a Nutshell.* St. Paul, MN: West Publishing.

Vogel, Mary, E. 2007. *Coercion to Compromise: Plea Bargaining, the Courts, and the Making of Political Authority.* New York: Oxford University Press.

Walker, Samuel. 1993. *Taming the System: The Control of Discretion in Criminal Justice 1950–1990.* New York: Oxford University Press.

"Speed Bump" © Dave Coverly/Dist. By Creators Syndicate, Inc.

chapter eleven

Criminal Trials and Courtroom Issues: Convicting the Innocent, Exonerating the Guilty

The single most distinguishing feature of the system of justice in the United States is the jury trial, guaranteed by the sixth amendment of the constitution. "In all criminal prosecutions, the accused shall enjoy the right to a speedy public trial, by an impartial jury of the State and district wherein the crime shall have been committed." The Sixth Amendment also states that the defendant has the right to be informed of the charges prior to the trial so that he or she can prepare a proper defense and confront those who are making accusations. Harry Kalven and Hans Zeisel characterize the jury trial as an "exciting experiment in the conduct of serious human affairs," one that has been "the subject of deep controversy, attracting at once the most extravagant praise and the most harsh criticism." We will examine criminal jury trials from jury selection and composition through the appeals process.

We include a discussion of **jury nullification**, the ability or power of juries to return verdicts of "not guilty" despite belief on the part of jury members that defendants are guilty as charged. In another form of jury nullification the jury nullifies a law it considers either immoral or wrongfully applied to a particular defendant. While it is abundantly clear that juries have the power to nullify, the question is whether they have the right to do so. Legal scholars have argued both sides of this issue, positions that will be explored in this chapter.

We conclude with the one of the most serious problems facing the American criminal justice system, the wrongful conviction of innocent people. Since 1990, hundreds of individuals, many of whom served lengthy prison sentences and some of whom were on death row, have been exonerated of crimes they did not commit. How were these innocent people wrongfully convicted? How did the criminal courts fail them? And of utmost importance, how can these travesties of justice be prevented?

THE CRIMINAL TRIAL

Annually, state courts conduct about 150,000 jury trials in this country, of which almost half are felony criminal trials. In 2006 California had the largest number of trials (16,000) while Vermont and Wyoming conducted the fewest (126). Federal courts conduct approximately 6,000 jury trials every year. Since 1990, the number of jury trials has declined as non-trial dispositions such as plea bargaining, settlements, and summary judgements have increased.

Jury Selection, Composition, and Voir Dire

While the probability of being impaneled on a jury in any one year is small, over the course of a lifetime more than one third of Americans will serve as jurors. The first step in selecting a jury is compiling a roll from which individuals can be selected. In an effort to remove discretion from local authorities, who could include and/or exclude people (or categories of people) from serving at trials, states use voter registration lists to draw the names of prospective jurors. These lists typically contain the names of between 60 to 80 percent of the population that is over 18. To construct a more inclusive jury list, a number of states have supplemented registered voters with names from city directories, drivers' licenses, motor vehicle registrations, and telephone books, as well as lists of taxpayers, welfare recipients, recent high school graduates, naturalized citizens, hunting licenses, utility customers, and dog licenses. Multiple lists are used to select names from a representative cross section of the population. The final list from which names are selected is called a master jury list or a jury wheel. The use of some highly inclusive lists such as census data, federal income tax, returns and social security numbers is prohibited by law. The federal court system develops a plan for the random selection of jurors on a district-by-district basis. The national average for jury compensation is $22 a day.

Jury Composition

A 1970 Supreme Court decision (*Williams v. Florida*) ruled that, with regard to the representativeness of the body, the difference between a 12- and 6-person deliberating body "seems likely to be negligible." Examining this decision from a mathematical perspective, Michael J. Saks argues that the high court was dead wrong in its ruling: "If we draw juries at random from a population consisting of 90 percent one kind of a person and 10 percent another kind of person (categorized by politics, race, religion, social class, wealth, or whatever), 72 percent of juries of size 12 will contain at least one member of a minority group, compared to only 47 percent of juries of size 6."

People between the ages of 18 and 64 are usually employed or attending school, making jury service of more than a few days burdensome. Some surveys indicate that juries serving on trials that exceed 20 days tend to have fewer college graduates and more unemployed and retired individuals as well as a disproportionate number of females and single people. Trials of this duration often involve murder charges, meaning that defendants in capital punishment cases may have their fates decided by a less-educated segment of the population.

Almost all states have a literacy and language requirement stating that jurors must speak and understand the English language. Twenty-six states stipulate that exemptions from jury duty are not possible. The remaining states excuse a variety of state officials and employees, with Nevada adding to that list physicians, dentists, optometrists, attorneys, state legislators (when the legislature is in session), and locomotive engineers.

Jury yield is the number of citizens found to be qualified and available for jury service as a percentage of qualification questionnaires or summonses mailed to prospective jurors. The national average jury yield is 46 percent, with urban courts having lower jury yields (38 percent) than rural courts (50 percent).

Voir Dire

Voir dire (literally "to see, to tell" also translated as "to speak the truth") is the mechanism by which opposing attorneys and judges gain information about prospective jurors. As such, it is a crucial part of the trial process. In

42 percent of state courts, attorneys conduct voir dire with limited input from judges, while attorneys and judges have equal roles in 19 percent of courts. Judges conduct void dire with limited attorney participation in 18 percent of state jury selection procedures. Judges have much greater control over the voir dire procedure in federal courts. There are three types of challenges that, if successful, will prevent an individual from being selected as a juror.

To the array   If an attorney or judge believes the master list is less then representative of the community or has been selected improperly, he or she can make a challenge to the array. A challenge to the array of a seated jury can occur if either the prosecution or defense believes that the jury has been stacked on the basis of gender, race, or some other characteristic. In the trial of professional football player Rae Carruth, the defense argued that jury selection had been biased against blacks, especially black males. The judge asked the prosecutor to give reasons other than race or gender for dismissing 8 of 11 potential black jurors. The prosecution provided answers (age too close to that of defendant, which could elicit sympathy for the defendant, for example) to the judge's satisfaction.

For Cause   An attorney may make a challenge for cause when a juror is believed to be unfit for service because of his or her past experiences, occupation, bias, or relationship with the defendant or victim. Challenges for cause (which are unlimited in number) may be overruled by the judge, and the prospective member remains seated.

Peremptory   Attorneys also have **peremptory challenges**, which they do not have to justify or explain, and which were often based on "hunch, insight, whim, prejudice, or pseudoscience." However, in *Batson v. Kentucky*, the Supreme Court ruled that the equal protection clause forbids prosecutors to peremptorily challenging potential jurors solely on account of their race or to make such challenges on the assumption that African Americans as a whole are unable to impartially consider the state's case against an African American defendant.

The number of peremptory challenges varies by seriousness of the offense and state. For example, in misdemeanor cases, municipal courts in Connecticut permit only 2 challenges each for the prosecution and the defense, while New Jersey allows 10. In death penalty cases, Virginia law stipulates 4 challenges; at the other extreme, Connecticut permits 25 for each side. A few states give the defense more challenges than the prosecution, especially in capital cases. In these trials, Georgia grants the prosecution 12 challenges and the defense 20.

Both the prosecution and the defense use voir dire to select jurors they believe will be sympathetic to their arguments. In other words, they are seeking individuals who are predisposed to a greater or lesser degree to their positions, and hostile to the arguments that will be presented by their counterpart.

During the past 30 to 35 years, social scientists and others who believe that they can identify salient personality characteristics in prospective jurors (and determine if they are more likely to convict or acquit) have become increasingly influential in both civil and criminal American courts. The practical application of social science to the justice system has resulted in jury consultants. In 1982, there were 35 jury consultants in the United States; in 1994, there were 250. As of 2008, the American Society of Trial Consultants claimed more than 500 members.

Both attorneys use peremptory challenges to exclude individuals they believe their opponents have designated as highly desirable. Because the ability to mold a jury to one's liking is limited, consultants are most effective in advising lawyers how they should proceed with a case once jurors have been chosen. In other words, instead of trying to match jurors to the facts of the case, attorneys may shape the presentation of the case to the composition of the jury. Consultants may tell lawyers which evidence should be stressed or the most effective way to make a point. Some consultants may use a shadow jury, that is, individuals hired to follow the trial closely and inform the consultants what they are feeling and thinking as the case unfolds.

How effective are jury consultants in helping to bring about a verdict that is favorable to their clients? Matthew Hutson notes that because real-world success rates are impossible to measure, a controlled study with two parallel juries, one selected at random and one selected via jury consultant techniques would have to be compared. In their book, *Scientific Jury Selection*, Joel Leiberman and Bruce Sales state that knowledge of a juror's demographic profile (age, gender, race) and personality indicators, improve the ability to predict his or her vote (innocent or guilty) by only 10 to 15 percent. Psychologist and law professor Shari Diamond notes that jurors do not make decisions alone. Rather, they are part of a decision making dynamic with 11 other people. "How can you possibly make sure there's a certain type of outcome given the complexity of the information on one side, the complexity of the information on the other side, and then the human dynamics involved in the jury?"

Despite these jury selection limitations, one jury consultant claims a 90 percent success rate in the over 100 cases on which she has worked. However, would some of the jurors have returned the same verdict had she not provided her services? Most social scientists believe that jury consultants may well make a difference in close cases where the evidence is ambiguous. However, in trials where one side of the other has a demonstrably stronger case, jury selection is most likely irrelevant.

Opening Statement

When the voir dire process has been completed and the jury is sworn in, the trial begins. Each side has the opportunity (although is not required) to make an opening statement, beginning with the prosecution. The opening statement is the first opportunity for the attorneys to present their case to the jury. Some judges permit the attorneys to speak at length, while others place time constraints on opening statements. Attorneys consider this initial contact with the jury an important part of their overall trial strategy as "repeated studies and experience have demonstrated that the lawyer who 'wins' the opening statement, in the great majority of cases, will ultimately receive a favorable verdict." Attorneys typically present a case summary, an overview of the relevant people, and the chronology. During the case summary, the attorneys tell the jury what happened, and why they should convict or acquit the defendant. The attorneys also tell the jury about the major players—the cast of characters—and then provide an overview of the time sequence of the significant facts and events. At this juncture of the trial, the attorneys are not permitted to argue the facts of the case, nor can they try to convince the jury to return a verdict of guilty or innocent.

Presenting the Evidence

The presentation of evidence by opposing attorneys constitutes the bulk of the trial. In its attempt to demonstrate that the defendant is guilty beyond a reasonable doubt, the prosecutor presents the state's case and the evidence first. It is important to understand exactly what reasonable doubt entails in a legal sense. It means that the evidence "must be so conclusive and complete that all reasonable doubts of the fact are removed from the mind." However,

"It does not require that the proof should be so clear that no possibility of error can exist." If we used the latter—and much more stringent—criterion for reasonable doubt, juries would return far fewer guilty verdicts. David Neubauer notes that four types of evidence are presented during a jury trial:

- *Testimony:* any statement made by a witness under oath.
- *Direct evidence:* evidence that is derived by way of the physical senses (sight, hearing, taste, touch, and smell). For example, testimony that the defendant was seen pulling a knife from his jacket pocket is direct evidence.
- *Real evidence:* tangible objects, for example, the knife with the defendant's fingerprints on it and the receipt for the purchase of this object.
- *Circumstantial or presumptive evidence:* evidence that is inconclusive because it is not based on a witness's direct experience. Testimony that the defendant was seen walking out of the park sweating profusely with a tennis racket under his arm is circumstantial evidence that he had been playing tennis. This form of evidence is treated as true until it is refuted by other evidence.

When a district attorney calls witnesses who will testify on the state's behalf, he or she is engaging in direct examination. The purpose of this form of questioning is for witnesses to tell their stories, and, in so doing, to make an emotional connection with the jury. According to Small, direct examination has at least three primary objectives: first, to introduce and validate the witness, that is, to convince the jury by way of the individual's testimony that he or she is a credible (believable) person; second, to present a "dramatic and persuasive" picture of the principal facts in the case; and finally, by way of testimony, to introduce real, tangible evidence. In planning their trial strategy, attorneys must determine which witnesses have what information and in what order they should testify. Because attorneys want "their" witnesses to speak to jurors in a persuasive, unrestrained manner, they ask them open-ended questions. For example, "Mr. Jones, on the Saturday night in question, what did you see as you approached in your car?" or "Mr. Smith, what did you and the defendant talk about on that occasion?" This type of questioning is predicated on the assumption that, although attorneys shape and direct a case, it is ultimately the testimony of the witnesses, and, through them, the introduction of real evidence, that will determine the outcome of a trial.

When a defense attorney questions a prosecution witness, or vice versa, he or she is conducting a cross-examination. One of the goals of cross-examination is to undermine the credibility of a witness. If an attorney can show that the witness has a prior criminal record or has engaged in previous acts of dishonesty, jurors may conclude that the witness is untrustworthy. Small argues that cross-examination is also a vehicle for attacking the testimony of a witness directly by way of demonstrating that he or she has had a longstanding hatred of the defendant, that there are inconsistencies in his or her testimony, and/or that the ability of the witness to hear or see things accurately is impaired for some reason. Simply stated, the overall purpose of cross-examination is to undermine the witnesses, the evidence, and, therefore, the case of opposing counsel.

John Wigmore states that "cross-examination is the greatest engine for determining the truth." Defense lawyer Roy Black maintains that more defendants end up in prison because of inept cross examination than any other legal mistake. During this type of questioning, lawyers attempt to get witnesses to say only what attorneys believe will support their view of the case. Towards this end, attorneys want to control the testimony of the witness as much as possible. This is accomplished by asking short questions that can only be answered by a "yes" or "no" response.

For Black, the "formula for successful cross-examination is simply stated: Use plain declarative sentences, add only one new fact per question . . . Think of cross-examination as a series of statements by the lawyer only occasionally interrupted by a yes from the witness." In the following example offered by Black, a skillful attorney cross-examines a police detective:

Attorney: And your interview with Officer Alvarez took place at 6:35?

Detective: Yes, sir.

Attorney: By then the crowd was even larger?

Detective: Yes, sir.

Attorney: Still a steady stream of rocks and bottles?

Detective: Yes.

Attorney: Did people begin pounding on the windows of the arcade?

Detective: Yes.

Attorney: Were the officers handling the crowd outside forced back inside the arcade?

Detective: Yes, sir.

Attorney: Things were too hot for them standing outside the front doors?

Detective: Yes, sir.

After the prosecution has presented its case, the defense may enter a motion for a directed verdict (sometimes called judgment of acquittal) stating that the prosecution has not presented a case strong enough to warrant a continuation of the trial and that the defendant should be acquitted of all charges. These motions are rarely granted. The defense will call witnesses whose testimony challenges aspects of the prosecution's evidence, for example, that the defendant was at a location other than the crime scene when the criminal event occurred. During cross-examination, prosecutors may attack the credibility of these witnesses and/or search for inconsistencies in their story. By way of an affirmative defense, the defendant's attorney does not deny the charge, but rather introduces new evidence that will avoid a judgment (guilty verdict) against his or her client. For example, entrapment and self-defense are affirmative defenses. Note that in all affirmative defenses the burden of proof rests with the defendant; that is, the defendant must demonstrate why he or she should be legally excused from conviction and punishment.

When the defense rests, the prosecution may choose to call rebuttal witnesses. The purpose of rebuttal evidence is to undermine or directly attack the just-concluded defense case, and to buttress any weaknesses in the prosecution's case that defense counsel has exposed. For example, if the defendant's attorney presented a highly credible witness whose testimony was not damaged in cross-examination, this testimony can be contested by rebuttal witnesses. The expert testimony of a defense witness is likely to be called into question by the introduction of another expert witness who contradicts that testimony. Rebuttal witnesses are subject to cross-examination.

In a recent survey, the Center for Jury Studies found that in 15.1 percent of state and 11 percent of federal trials, judges allowed the highly controversial practice of jurors submitting written questions to witnesses. As criminal trials can be long and complicated, with jurors hearing the testimony of dozens of witnesses, 69 and 71 percent of jurors respectively at state and federal trials are allowed to take notes during the course of the proceedings.

Closing Arguments

After the prosecution and defense have presented their cases, each side has the opportunity to address the jury in closing arguments: first, the prosecution, then the defense, with the prosecution having a final opportunity (because it has the burden of proof) to refute the assertions of defense counsel. Typically the most dramatic part of the trial process, Steven Goldberg refers to the closing argument as pure persuasion, in which each side ties elements of the case together and attempts to convince the jury of the truth of its position. This is the last chance for opposing attorneys to refute each other's position as well as shore up weaknesses in their own cases. The prosecution will argue that the state has met its obligation as the evidence clearly demonstrates that the defendant is guilty beyond a reasonable doubt. Conversely, defense counsel will state emphatically that the state clearly failed to overcome the reasonable doubt criterion.

In their closing arguments, Daniel Small advises attorneys to "use powerful, descriptive, catchwords, and phrases . . . use all the tools at your disposal . . . use silence, and contrast, use movement and body language." Jurors are often reminded that they are part of an honorable judicial position and that the outcome of the trial now rests with them and them alone.

Instructing the Jury and Jury Deliberations

After the attorneys have addressed the jury in their closing arguments, the judge charges (or instructs) the jurors about the issues of the case and the applicable points of law that must be observed during their deliberation. Educating jurors at the conclusion of the proceedings is based on the assumption that people are more likely to remember and utilize what they have just heard. However, instructing the jury after as opposed to before the trial begins has been harshly criticized by many practitioners in both criminal and civil courts. One judge noted that this practice is comparable to "telling jurors to watch a baseball game and decide who won without telling them what the rules are until the end of the game." A team of researchers concluded that when jurors received instructions both before and after the trial, they were better able to ascertain the relevant evidence as well as remember more accurately what they heard. Yet another alternative is to present the jurors with instructions during the course of the proceedings as the need arises. Regardless of when the jury is addressed, the instructions are not likely to provide guidance if they are not presented in plain English. Consider the following paragraph excerpted from 81 pages of jury instructions in a civil case:

> The outer boundaries of a product are determined by the reasonable interchangeability of use or the cross-elasticity of supply and demand between the product itself and substitutes for it . . . The average variable cost test is a double inference test because if you find that _____ & _____ priced below its reasonably anticipated average variable cost . . .

Attorney Stephen Adler notes that this was not a case of jurors unable to understand some of the instructions; rather, they did not comprehend anything they were told. Little wonder that many jurors are incapable of making an informed decision. Although some attempt has been made to render instructions in plain English, what is straightforward and understandable to attorneys may remain unclear to jurors.

For example, the video used in Florida to provide standard jury instructions stresses that a defendant is presumed to be innocent until he or she is proven guilty by the presentation of evidence beyond a reasonable doubt. In a startling and worrisome discovery, researchers found that after viewing this film, only 50 percent of the jurors fully understood the concept of reasonable doubt. Ten percent were confused as to what the presumption of innocence meant, and 2 percent believed that the burden of proof of innocence rested with the defendant.

Social scientists have examined the important question of how individuals arrive at a determination of guilt or innocence. A juror's vote can be viewed as the result of a two-fold decision making process. The first step is the individual juror's evaluation of the defendant's guilt or innocence at the end of the trial but before deliberations commence. The second part of this process is an examination of group dynamics, that is, the deliberation process of the jury as a whole and how this process affects the vote of constituent members.

In their investigation of jurors' evaluations before deliberations began, Barbara Reskin and Christy Visher interviewed 331 of 456 jurors who served in 38 sexual assault trials and observed the ongoing trial proceedings. The researchers concluded that jurors' predeliberation verdict was a consequence of at least three factors: 1) the jurors used trial evidence in determining a defendant's guilt or innocence; 2) jurors interpreted evidence selectively; they tended to ignore available eyewitness testimony and focused on evidence of the force and seriousness of the assault; and 3) when the prosecution presented substantial evidence such as a recovered weapon or other physical evidence, eyewitness testimony, or physical injury to the victim, jurors were more likely to believe the defendant was guilty without considering extralegal evidence such as the jurors' perception of the defendant's physical and social attractiveness or any reference to whether he was employed as well as negative comments about the victim's moral character and the jurors' perception of the victim's carefulness or carelessness when she was assaulted.

When the prosecution presented a weak case and the guilt of the defendant was in doubt, jurors reached a predeliberation decision by incorporating their personal values (attitudes about crime) and the extralegal values concerning the defendant and victim. The weaker the state's case against the defendant, the more likely a juror's sentiments would affect his or her decision. Other researchers have found that the congruence of the evidence with jurors' beliefs, values, and experiences frees them from looking at the case solely in terms of how it is presented by the prosecution and defense. Under these circumstances, jurors may interject their personal sentiments into the decision-making process, a phenomenon known as the liberation hypothesis.

Because a person's life experiences are likely to affect the way he or she interprets evidence, and one's experiences are in large measure a function of gender, race, age, social class, occupation, and personality traits, these factors have been examined as they relate to jurors' decisions in criminal trials. However, after reviewing dozens of studies, Marilyn Chandler Ford found that "the influence of social and demographic factors on juror behavior is unclear." In other words, we cannot say with any certainty that African Americans, whites, men, women, or college-educated jurors are likely to vote one way or another in certain cases.

The Verdict

After the jury has reached a verdict, the jurors return to the courtroom and deliver the decision. Although uncommon, either attorney may request that the court poll the jury, with the judge asking each juror if the verdict reflects his or her decision. Approximately two-thirds of all criminal trial juries convict the accused. There is significant variation among states, ranging from 59 percent in Florida to 83 percent in California. This number reflects the notion that district attorneys are loathe to invest time and human resources in trials they do not believe (with a high degree of certainty) will result in a guilty verdict. In other words, if prosecutors believe the outcome of a trial is doubtful—the credibility of a key witness (a drug dealer, for example) is likely to be successfully attacked by defense counsel—they are likely to plea bargain (offering a reduced sentence in exchange for a guilty plea) and gain a conviction and control over the defendant. However, if prosecutors are of the opinion they have a "slam dunk," can't lose case, they will go to trial. There is no reason to "bargain" with a defendant if a jury trial conviction, and the maximum penalty it will bring, is a near certainty.

In about five percent of cases, even after prolonged deliberation, juries cannot reach a unanimous verdict or the designated majority vote (in some states) for a conviction or acquittal. This is a **hung jury**, with the judge declaring the proceedings a mistrial. Defense attorneys routinely ask the judge to order the acquittal of the defendant in the aftermath of a hung jury. If the judge denies this request, the prosecutor must decide whether to dismiss the charges or try the case again. In a high-profile case where the district attorney feels that the reputation of the department is at stake and/or has a perception that the public wants the case tried again a second trial is likely. In a non-jury or **bench trial**, the judge is both the trier of the law and the fact-finder. As such, in addition to reaching a decision to convict or acquit, the judge must write a findings of fact and conclusion of law to support that decision.

Appeals

After a guilty verdict, a defendant can appeal points of law that resulted in his or her conviction. Although there is no constitutional right to appeal, the prerogative to do so is granted by state and federal statutes. All criminal court jurisdictions provide the defendants with at least one direct appeal of a conviction. The appeal is sent to an appellate court by way of a written brief summarizing the legal error that was allegedly made during the trial. If the court finds that a brief has merit, the attorneys in the contested case (that is, the prosecutor and defense counsel) argue the issue raised in that document. (At the federal level, overburdened appellate courts are deciding cases without oral arguments in about one in three appeals.) Upon listening to both presentations and discussing the disputed point of law, the appellate judges have several options. They can vote to affirm or uphold the conviction, reverse the conviction with no further action needed, or reverse and remand meaning that the decision has been overturned, and the case is sent back to a lower court for further proceedings. These additional proceedings may include holding a new trial or changing the original judgment. The appellate judges may also remand a case to a lower court with specific instructions for additional proceedings.

An appellate court may conclude that, although an error occurred during the trial, it did not affect the outcome or the fairness of the hearing. Courts typically only overturn convictions when the lower court proceedings violate a fundamental right of the defendant, or the error affected the outcome of the case. Appeals are much more the exception than the rule, and only a small percentage of them are successful.

If the conviction is upheld, the defendant may appeal to the court of last resort overseeing that jurisdiction, which is usually the state supreme court. However, this judicial body decides which appeals it will hear and this second appeal is not a right and far from automatic. Should the defendant be unsuccessful at the state level, the final court of appeal is the U.S. Supreme Court.

JURY NULLIFICATION

There are two primary types of jury nullification. In its original form, the jury decides that, even though an act has been prohibited by law, it is not a crime and, therefore, should not be punished. This form of nullification dates back to the colonial era, when juries refused to convict defendants charged with crimes against the Crown (England). In the 19th century, juries in northern states often acquitted defendants charged with harboring runaway slaves, an act explicitly prohibited by the Fugitive Slave Act of 1850. During Prohibition, some juries failed to return guilty verdicts in cases where individuals were charged with making and distributing whiskey. Vietnam War protestors, motorcycle helmet law opponents, and individuals who use marijuana for medical purposes have benefitted from juries who believed the laws defendants violated were unjust. In recent years juries are returning not guilty verdicts in some drug possession cases that would result in life in prison for conviction of a third felony. Jurors' values also have clashed with the law in assisted suicide prosecutions. Jurors sympathetic to medical practitioners or family members who help terminally ill people in severe pain end their lives have voted to acquit. In San Diego, just as Operation Rescue defendants accused of trespassing and other offenses near a medical clinic were about to be tried, a three-quarter-page ad appeared in a local newspaper:

Attention Jurors and Future Jurors

You Can Legally Acquit Anti-Abortion

"Trespassers" Even If They're "Guilty"

The second form of nullification occurs when a jury bases its decision on how a law is being applied to a specific category of defendants. In the aftermath of the Civil War, all-white juries (especially in the South) routinely convicted blacks accused of crimes against whites regardless of the strength or weakness of the evidence. Conversely, whites on trial for physically assaulting, raping, and killing blacks were exonerated.

In Albany, New York, 11 white jurors concluded that an African American man tried for distributing cocaine was guilty. The 12th juror, a black man, refused to convict arguing that he was sympathetic with African Americans who struggle to make a living. He chastised his fellow jurors for not understanding the economic difficulties that racial and ethnic minorities face on a daily basis.

In another case, after an all-black jury acquitted an African American man accused of murder, an anonymous juror sent a letter to the court stating that, even though the majority of individuals deciding the defendant's fate believed he was guilty, they changed their verdicts to accommodate the hold-outs who "didn't want to send anymore Young Black Men to Jail." Paul Butler believes "African American jurors are doing a cost benefit analysis" and concluding that "defendants are better off out of jail, even though they're clearly guilty." Butler argues that jury nullification is not a matter of race per se: "It's life experiences. Blacks are more likely to have been jacked by the police, and less likely to view police testimony with quite the same pristine validity as a white male from the suburbs."

The number of hung juries remained relatively constant for decades, at about 5 percent. However, since the late 1970s that number has doubled, even quadrupled in some jurisdictions. Joan Biskupic notes that much if not most of this increase can be attributed to race-based decisions on the part of jurors. Michael Markowitz and Doloers D. Jones-Brown examined jury-acquittal rates and trial rates in the five counties that comprise metropolitan New York City. In Bronx County where three of four residents are African American or Hispanic, 38.4 percent of juries returned not guilty verdicts. In 80 percent white Richmond County (Staten Island) the acquittal rate is 21.3 percent.

The researchers note, "Our data suggest that it is, indeed, the presence of Blacks and Hispanic jurors that spells the difference between conviction rates from county to county. The Bronx is the epitome of the 'inner city,' and the greater inclination of Bronx juries to give the benefit of the doubt to defendants reflects points of view common to that milieu ... Staten Island's 'whiteness' appears to account for the greater tendency of its jurors to convict."

While these findings point to race as the key explanatory conviction rate variable, Markowtiz and Jones-Brown are skeptical of the jury nullification interpretation. They argue that juries are reluctant to acquit individuals whose guilt is indisputable noting "There is no reason to believe that blacks and Hispanics are any more willing to defy judicial authority than anyone else." Former jurors who acquit defendants in racially "tinged" cases are adamant in their denial that race—or any other nonlegal factor—was a criterion in their decision making.

Advocates of jury nullification contend this process is one way imperfections of the law can be corrected. In other words, jury nullification serves as a check by citizens on the power of legislatures to create laws that may be contrary to the will of the people. Some of the founding fathers articulated this position. Speaking of a juror in 1771, John Adams, who would become the nation's second president 26 years later, stated that "it is not only his right, but his duty . . . to find the verdict according to his best understanding, judgement, and conscience, though it be in direct opposition of the court." Alexander Hamilton, Secretary of the Treasury under George Washington, said that jurors should disobey a judge's instructions and find for the defendant "if exercising their judgement with discretion and honesty they have a clear conviction that the charge of the court is wrong."

Opponents of jury nullification contend this practice undermines the notion of the law of the land. Jurors voting to convict or acquit based on their own values of right and wrong or the race of the defendant reduce the law to what a particular group of people believes at a given time as it applies to a specific defendant. From this perspective, justice would have little if anything to do with the strength of the evidence against the accused or the skill of the attorneys presenting or refuting the evidence. Former District Judge and Deputy Attorney General Eric Holder Jr. believes that for practitioners who are regular members of the criminal court system, "there is a real cynicism that grows out of nullification." A much more hostile view of nullification was given by former Federal Judge Simon Rifkind who noted that informing juries of their power to nullify would lead to "a society without laws, without regulations. That is a monstrosity."

A 1997 ruling by the U.S. Court of Appeals for the 2nd district stated that by definition nullification is a "violation of a juror's oath to apply the law as instructed by the courts . . ." In 2001, the California Supreme Court ruled that jurors must follow the law and not their consciences even if they believe following the law will produce an unjust verdict. New trial guidelines resulting from the Court's decision require jurors to notify the judge if a fellow juror is following his or her personal values and not the law in question. Law professor Gerald Uelman questions the effectiveness of this directive noting that "jury nullification is not explicit . . . It is almost subliminal. The jury applies a higher standard of reasonable doubt because they don't like the law."

Currently, only Maryland and Indiana instruct jurors that they are triers of the law as well as the facts of the case although they are cautioned against making arbitrary decisions. "Members of the jury, under the Constitution of Maryland the jury in a criminal case is the judge of the law as well as the facts. Therefore, anything which I may say about the law, including any instructions which I give you, is merely advisory and you are not bound by it. You may feel free to reject any advice on the law and to arrive at your own independent conclusion . . ." In his book, *We the Jury*, Jeffrey Abramson reports that judges in Maryland and Indiana have not detected any significant increase in jury nullification not-guilty verdicts as a consequence of these instructions.

Abramson relates an interesting study conducted by Alan Scheflin and Jon Van Dyke suggesting that jury nullification may be a function of the issue involved. Mock juries given nullification instructions were not more likely to acquit a college student on trial for killing a pedestrian while driving drunk than juries given standard instructions. To the contrary, the nullification instructed juries were more likely to convict. However, mock juries given nullification instructions were more like to acquit a nurse charged with the mercy killing of a terminally ill cancer patient.

WRONGFUL CONVICTIONS: FACTUALLY INNOCENT BUT LEGALLY GUILTY

How many factually innocent people are wrongfully convicted in the United States each year? In an effort to answer this question, C. Ronald Huff, Arye Rattner, and Edward Sagarin surveyed 353 criminal court judges, prosecutors and public defenders around the country and concluded that during a 10-year period almost 100,000 individuals were punished for crimes they did not commit. This figure only includes wrongful convictions for index offenses, i.e., murder, robbery, rape, aggravated assault, burglary, larceny, and motor vehicle theft. The investigators determined that erroneous conviction is primarily the result of the following nine factors.

Eyewitness Errors

Four of five respondents to the Huff questionnaire stated that good faith misidentification of suspects by witnesses was a major factor responsible for improper convictions. "Aside from the prejudices of police, prosecutors and jurors, [a person's] being of a race different than that of witnesses may increase the possibility of misidentification." Former public defender David Feige notes that it is difficult to overestimate the impact eyewitness testimony has in criminal cases. "In thousands of cases every year, testimony of a single witness, uncorroborated by forensic or any other evidence, is used to sustain serious felony charges, including robbery and murder." A study of 200 innocent men who served an average of 12 years in prison discovered that the leading cost of wrongful convictions was erroneous identification, a factor in 79 percent of the cases examined.

A significant amount of research supports the cross-race effect. In general, African Americans and whites recognize faces of individuals of their own group more accurately than they do the faces of the other race. Because juries give so much credibility to eyewitness testimony, it is reasonable to assume that a significant number of innocent people have been imprisoned on the basis of eyewitness identification error. Compounding the problem is the fact that courts have traditionally prohibited expert testimony on the cross-race effect.

Prosecutorial and Police Misconduct and Errors

A 20-year-old immigrant from the Dominican Republic was convicted of killing a New York City man in 1990. After serving five years in prison, a state appeals court overturned the conviction, stating that prosecutors knowingly allowed their chief witness to perjure herself and failed to disclose a second witness whose testimony might have helped the defense. This was one of 15 Bronx cases "in which serous misconduct or error by prosecutors has led to wrongful convictions." In only one of these cases was a prosecutor disciplined. The Center for Public Integrity reported that since 1970 more than 2000 cases of prosecutorial misconduct resulting in dismissed charges or reversed convictions have occurred in the United States. (During this same period only 44 prosecutors faced disciplinary hearing and 2 were disbarred). Some observers believe that the actual number is significantly higher. As the attorney for the falsely accused Dominican man stated, "Most of the time when prosecutors withhold evidence, no one finds out about it."

In arguably the worst scandal to hit the Los Angeles Police Department, Officer Rafael Perez (who was caught removing 8 pounds of cocaine that he intended to sell from an evidence room) told investigators there was widespread perjury in LAPD. According to Perez, "90 percent of officers who work CRASH [Community Resources Against Street Hoodlums] and not just Rampart CRASH, falsify a lot of information. They put cases on people . . . it hurts me to say this, but there's a lot of crooked stuff going on in the LAPD." Perez identified 57 cases involving 90 defendants wherein he and his partner perjured themselves and fabricated evidence, mostly false drugs and weapons charges. Some Rampart officers specialized in particular kinds of misconduct; one planted guns on people, while another concealed crack cocaine on unsuspecting individuals. Within months of this investigation, more than 40 criminal convictions were overturned and as many as 17,000 cases involving the testimony of 71 officers had to be reviewed.

In 2007, a federal judge in Boston awarded over $100 million to four defendants framed by the FBI for a 1965 murder of a local gangster. A government attorney argued the FBI was not duty bound to release evidence that could have cleared the men because it was a "state" as opposed to federal case. Judge Nancy Gertner stated that, "The FBI's misconduct was clearly the sole cause of this conviction. The government's position is, in a word, absurd." Richard Moran states that the Boston case was hardly an isolated incident of government misconduct:

> "My recently completed study of 124 exonerations of death row inmates in America from 1973 to 2007 indicated that 80, or about two-thirds, of their so-called wrongful convictions resulted not from good faith mistakes or errors but from intentional, willful, malicious prosecutions by criminal justice personnel . . . If a death sentence is overestimated because of malicious behavior, we should call it for what it is: an unlawful conviction, not a wrongful one."

False Confessions

In Alabama, three "poor, black, and retarded" defendants confessed in 1999 to killing a newborn infant and were sentenced to prison. However, there was no evidence the baby ever existed, and a fertility expert testified that because one of the defendants had undergone a tubal ligation years prior to the alleged birth, she could not have become pregnant. When asked what evidence he had that a baby had been born and killed, the prosecuting attorney stated, "Well, they told us that." The defendants' lawyer believes police interrogators "planted that idea [the killings] in the minds of these mentally retarded people." The Alabama Court of Appeals agreed ruling that "a manifest injustice" occurred in the case.

Steven Drizin has studied false confessions and notes that people with developmental disabilities, individuals with mental illness, and juveniles are most likely to make such admissions of guilt. Sociologist Richard Ofshe notes that, "Mentally retarded people get through life by being accommodating whenever there is a disagreement. They've learned that they are often wrong; for them, agreeing is a way of surviving." A Florida attorney who trains police on interrogation techniques throughout the country stated that she and her colleagues are aware of approximately 100 cases of possible false confessions by mentally impaired defendants.

According to Gisli Gudjonsson, of London's King College, one of the foremost experts on this phenomenon, there are three categories of false confessions:

1. Voluntary false confessions are made by people without any external pressure from law enforcement officials because of a desire for notoriety, an inability to distinguish fact from fiction, or a desire to protect the guilty party. Those who desire notoriety have a pathological need to become known, even if fame comes at the cost of punishment, including imprisonment, and, in extreme cases, execution. Individuals with various forms of mental illness suffer from a breakdown in reality monitoring and are unable to differentiate between real and imaginary events. Individuals may try to protect the real criminal with whom they have some relationship. Godjonsson believes this is an important reason why suspects volunteer false confessions. Although these confessions are often made in relation to minor offenses, they can sometimes occur in serious cases, including murder.

2. Coerced-compliant false confessions result from police pressure during the interrogation process. The suspect does not confess voluntarily, rather, he or she caves in to the demands of interrogators for some "immediate instrumental gain." These gains include:
   • being allowed to go home until the next phase of the justice process, in the case of minor offenses;
   • bringing an extended stressful and "intolerable situation" (the interrogation) to an end (i.e., suspects may believe that even if they make a false confession, the truth—which will include their innocence—will come out in the end); or
   • avoiding incarceration (e.g., drug addicts may choose to confess and be released and take their chances in court rather than remain in police custody cut off from their source of drugs).

3. Coerced-internalized false confessions occur when suspects come to believe during the course of an interrogation that they committed the offense in question, even though they have no recollection of the event. There are two types of memory distrust syndrome. (MDS). In one type, as a consequence of amnesia, alcohol, and/or drug induced memory problems, suspects have no recollection of what they were doing at the time

the offense was committed. By way of the interrogation process, these individuals are led to conclude that they committed the offense. In the second type of MDS, suspects who have a clear recollection of not having committed the crime at the onset of the interrogation, begin to distrust their memory as a result of "subtle manipulative influences" by the police.

One possible solution to the problem of false confessions is to have the entire interrogation videotaped from beginning to the end, a procedure that is currently mandated in Alaska, Minnesota, and Illinois as well as some 500 other jurisdictions nationwide. A 2004 study conducted in Illinois concluded that police departments in that state embrace the idea of videotaped interrogations. However, resistance to taping interrogations is likely to occur for the indefinite future. For example, as of October 2007, the California State Sheriff's Association was combating two bills that would mandate the electronic recording of interrogations and corroboration of informant testimony.

While an increasing number of law enforcement agencies record confessions, these admissions often come at the end of an interrogation session. One observer noted that this procedure is all but useless. "Asking a jury to judge the credibility of a confession without seeing the interrogation is like a medical examiner conducting an autopsy without the body."

Plea Bargaining

An unknown number of innocent defendants are convicted after pleading guilty to one or more offenses. Through the influence of prosecutors and/or defense attorneys, individuals conclude that the state has an airtight case against them, and that if the case goes to trial, they surely will be convicted and punished more harshly than if they accepted a negotiated plea. Because many plea bargains result in immediate freedom, a suspended sentence, or probation, innocent defendants may decide to reduce their losses rather than sit in jail until they are tried and possibly convicted. Some individuals may plead to crimes they did not commit in order to save themselves and their families legal fees that can be substantial if the case goes to trial.

Community Pressure for Conviction

The public may pressure criminal justice officials to increase their efforts and put more offenders behind bars. Special interest groups, such as women's organizations (regarding domestic violence and rapes), racial/ethnic groups, and homosexual advocates may demand that authorities be especially strident in their prosecution of hate crimes. Public pressure can be intensified when these groups receive media coverage.

Inadequacy of Counsel

In 1987, 18-year-old Jimmy Ray Bromgard was sentenced to three 40-year terms of imprisonment (to be served concurrently) for raping an 8-year-old girl. His attorney performed no investigation, did not file pretrial motions, declined to give an opening statement, did not prepare for closing arguments, failed to file an appeal, and did not provide an expert witness to refute the fraudulent testimony of the state's forensic witness. When the victim was asked at trial to state her confidence that Bromgard was the assailant, the young girl stated, "I am not too sure." Even with this degree of uncertainly the court allowed the victim to identify the defendant as her assailant and Bromgard's attorney did not object to this testimony. Bromgard was released from prison in 2002 after his innocence was established via **DNA** testing.

Individuals can be convicted of crimes they did not commit when their lawyers are inexperienced, have heavy caseloads, have little interest in their clients, lack the resources to carry out in-depth interviews, or are grossly incompetent. In some of the most egregious instances of attorney incompetence, defense lawyers have been disbarred shortly after concluding a death penalty case.

Law professor and attorney Barry Sheck (a member of O.J. Simpson's defense team) stated that "nothing guarantees the conviction of an innocent person faster than a lawyer who is incompetent or lacks resources. It's terribly risky to be poor, or even middle-class, and unable to afford a good attorney." Supreme Court Justice Ruth Bader Ginzburg stated that she has "yet to see a death case, among the dozens coming to the Supreme Court on the eve on execution petitions, in which the defendant was well-represented at trial."

Accusations Against the Innocent by the Guilty

Incarcerated criminals trade their cooperation with the district attorney for sentencing considerations or immunity from prosecution, and have testified against and helped convict other wrongdoers as well as innocent people. After spending the night with a woman in a coed dormitory, a young man was arrested the following evening and charged with rape based on her accusation. The defendant had a past criminal record while the young woman "appeared to be the epitome of everything righteous and proper." He was convicted and sent to prison. Later, the alleged victim was arrested for arson and claimed her criminal behavior was related to rape trauma. A curious prosecutor learned that she had a history of mental illness and was being treated by a therapist. The therapist said that he knew the woman had not been raped but was prohibited from informing the police because of doctor-patient confidentiality.

Previous Convictions

Having a criminal record significantly increases one's chances of being convicted wrongfully. Prior arrests and convictions means that a person's photo will be in police files, where it can be mistakenly singled out by crime victims. Because they open their previous criminal history to scrutiny during cross examination, innocent defendants are less likely to take the stand and testify on their own behalf. If they do not testify, these individuals are depriving jurors of information that could refute the prosecution's assertions. Even though judges inform jurors that a defendant's failure to testify cannot be construed as an admission of guilt, this directive may be ignored. Jurors may also believe that, even if a defendant with a felonious history is not guilty of this particular crime, he is a criminal, and it is a good idea to put these individuals behind bars when one has a chance to do so.

Race as a Factor

Although many people found guilty of crimes they did not commit are white, a disproportionate number are African American and Hispanic, with some of these latter individuals convicted by all-white juries. In their study of wrongful convictions, Samuel R. Gross and Barbara O'Brien found that "black men accused of raping white women face a greater risk of false convictions than other rape defendants . . ."

In his study of 200 cases in which innocent defendants served an average of 12 years in prison, law professor Brandon L. Garrett, found that exonerated convicts were more likely to be minority group members than was the prison population generally. Seventy-three percent of convicts cleared of rape charges were African American or Hispanic compared to 37 percent of all rape convicts who were African American or Hispanic. Thirty-one of the wrongfully convicted prisoners appealed to the Supreme Court with the justices refusing to hear 30 of the cases. In the one case they did review, the justices ruled against the (innocent) inmate. Adam Liptak notes that appeal courts typically focus on asserted procedural errors as opposed to a jury's factual findings. Appellate courts, therefore, cannot be counted on to overturn wrongful convictions.

DNA Testing

On July 30th 1981, a woman was abducted from an Atlanta parking lot by an armed man who forced his way into her car and threatened to kill her. The victim was bound, beaten, and raped three times. Days later she poured through books of photographs provided by the Atlanta Police Department and pointed to one that resembled her assailant. On a subsequent date she was shown a photographic array and stated that Robert Clark looked very much like the perpetrator. Clark was not the man she initially identified. Clark was later convicted of the crime with the victim testifying there was no doubt in her mind that he was the man who beat and raped her. He was sentenced to life plus 20 years in prison. Twenty-three years later (December, 2005) Clark was released after DNA evidence proved his innocence and implicated another man for the crime as well as rapes in 1993 and 1996.

In January, 2008, 47-year-old Charles Chatman was released on his own recognizance after serving almost 27 years of a 99-year sentence on the basis of new DNA evidence. Chatman was the 15th inmate freed from Dallas County—the most in the nation—since 2001. While DNA testing is routine in television crime shows, it is only available in approximately 10 percent of violent crime cases. In addition, this sophisticated evidence testing procedure is often opposed by prosecutors.

Since 1973, over 120 convicted murderers in 23 states have been released from death row via evidence of their innocence. In a number of cases, this evidence was a result of DNA testing. DNA is deoxyribonucleic acid, a substance containing genetic material that is unique to the individual except in the cases of identical twins. The Innocence Project, a non-profit legal clinic housed in the Benjamin N. Cardozo School of Law in New York City, has helped secure the release of more than 200 people since 1992, including 15 who were sentenced to death. The organization "only handles cases where postconviction DNA testing of evidence can yield conclusive proof of innocence." Since the FBI began DNA analysis in 1989, approximately 25 percent of primary suspects have been cleared. This raises the disturbing question of how many innocent people have been imprisoned and executed prior to sophisticated biological testing. Forty-two states now give inmates varying degrees of access to DNA evidence that might not have been available to their attorneys when they were convicted.

FINGERPRINTS: PROOF POSITIVE?

Although Huff and his associates did not examine fingerprinting errors, recent findings indicate that erroneous evidence of this type may result in numerous wrongful convictions. When Rick Jackson was arrested for the brutal murder of a friend, police showed him photos of bloody fingerprints they said placed him at the murder scene. Maintaining his innocence, Jackson felt relieved to learn about the prints, knowing they were not his. "Ah, somebody made a mistake," he thought, a mistake that would soon be rectified. However, after three local police fingerprint "experts" stated unequivocally that the prints were Jackson's, the family obtained the services of FBI field examiner, George Wynn. Along with fellow examiner Vernon McCloud, the two FBI experts (with a combined 75 years of fingerprint experience) reached same conclusion. The alleged match between Jackson's fingerprints and the fingerprints found at the crime scene was "not even a close call." In court, the trio of police examiners testified with 100 percent certainty that the crime scene fingerprints were the defendant's. Rick Jackson was convicted of first degree murder and began serving a life sentence with no chance of parole.

So how is it that five experts reached completely divergent conclusions about the two sets of fingerprints? To begin, the computer does no more than narrow the search. At this point the human eye goes back and forth between one set of prints and another looking for similarities, a process that can take hours, sometimes days.

And herein lies the problem, argues Robert Espstein. "There's complete disagreement amongst fingerprint examiners themselves as to what they need in order to declare a match." Consider the following: To prove a match between sets of prints in Sweden, examiners must show seven points of similarity (or points of comparison), in Australia 12, and in Brazil 30. In the United States, most examiners, including the FBI, do not even use a point system. Epstein argues that the reliability of fingerprint matching has never been proven. Ralph Haber agrees, stating that "There isn't a single experiment that's ever been done, literally."

The training and testing of fingerprint examiners further complicates the problem. About half of current fingerprints examiners in the United States have failed the International Association of Identification's (IAI) rigorous certification test, yet continue to practice their profession. Ken Smith, the IAI's certification chairman, stated: "There are very few employers who will terminate an employee for not passing the test." In fact, the majority of examiners have never taken the test. A proficiency test of 156 examiners administered by the IAI concluded that 20 percent of these individuals made at least one false positive identification. Based on his research, Simon Cole argues that as many as 1000 incorrect matches are made annually in spite of efforts to prevent such mistakes.

In the aftermath of the Madrid terrorist attacks that killed 191 people, Brandon Mayfield, a 37-year-old former military officer and convert to Islam, was arrested by the FBI when the Bureau linked him to the incident by way of his fingerprints on a bag of detonators recovered in Spain. In an FBI affidavit used to obtain a material witness warrant, a Bureau agent noted that the FBI was 100 percent positive about the fingerprint identification. As Seton Hall Law School Professor David Feige noted, the FBI turned out to be 100 percent wrong in its identification of Mayfield who was released after spending two weeks in jail.

Although they have been unsuccessful to date, defense attorneys are attempting to get prosecutorial fingerprint testimony barred from courtrooms under standards set by the Supreme Court. A forensic scientist who has worked with both prosecutors and defense attorneys is of the opinion that challenges to the legitimacy and reliability of fingerprint testimony will eventually bring about changes in the courtroom. Edward Imwinkelried notes there is a "very good possibility" judges will instruct juries that a fingerprint analyst is not a scientist stating an exact conclusion, but an expert giving an opinion. After serving two years in a state prison, Rick Jackson was released when the IAI stated the crime scene fingerprints were not his. One of the prosecution's fingerprint examiners was decertified and lost his job, the other two were not sanctioned.

"IS THAT ALL THE EVIDENCE YOU HAVE?" THE CSI EFFECT

The world of forensic crime scene investigation as shown on television is having an impact on the criminal justice system. District attorneys note that some jurors now expect prosecutors to exhibit forensic evidence before they are willing to return a guilty verdict. As juror expectations increase, district attorneys are having to change the way they conduct trials. To the extent the threshold for high-tech evidence has increased, the "beyond a reasonable doubt" criterion for conviction may also be changing in favor of defense attorneys. An Illinois prosecutor told a jury that "this is your CSI moment," stating DNA analysis matched the saliva on the victim's breast to the defendant. This was not enough to sway jurors, who were of the opinion that police should have tested (and matched) the debris found on the victim to the soil where the attack occurred. After the not guilty verdict was announced, the prosecutor stated: "They [the jury] said they know from CSI that police could do that sort of thing." Some prosecutors are using "negative evidence witnesses" in an effort to convince jurors that DNA and fingerprints are not found at all crime scenes.

Andrew P. Thomas, chief prosecutor of Maricopa County, AZ surveyed 102 of the 300 attorneys in his office. His findings indicate that the CSI effect is "no myth." Thirty-eight percent of experienced district attorneys were of the opinion that at least one trial had "resulted in either an acquittal or hung jury because forensic evidence was not available, even though prosecutors believed the existing testimony was sufficient by itself to sustain conviction." In some of these cases prosecutors reported questions from jurors about evidence such as "mitochondrial DNA," "trace evidence," or "ballistics" even though these topics were not brought up during the course of the trial. Perhaps the most disturbing finding of Thomas' survey is the potential impact CSI watchers have on fellow jurors. "In 72 percent of cases, prosecutors suspect that jurors who watch shows like CSI claim a level of expertise during jury deliberations that sways other jurors who do not watch these shows."

Many of the gadgets seen on television do not exist, or do so in a more limited capacity. The computer that scans millions of fingerprints in a matter of seconds and produces an exact match is a perfect example. On television, unintelligible phone calls are stripped of background noise until the conversation is crystal clear. Not so easy in real life. As Andrew Thomas notes, although some jurisdictions have access to high-tech laboratory equipment, these resources are employed toward solving only the most serious crimes. Also, the completion of DNA and other tests can take weeks or months rather than minutes or hours. In reality, the results of many laboratory tests are subject to interpretation, hence conflicting expert witnesses mean jurors are left to ponder competing explanations of scientific findings. Although DNA analysis is a powerful tool that can make or break a case for either the prosecution or defense, blood is only found at five percent of crime scenes.

Finally, many crime laboratories are understaffed, underfunded, and/or manned by poorly trained personnel. In March, 2002, the public learned that between 180,000 and 500,000 rape kits (which contain DNA from crime scenes) across the country had not been analyzed because of inadequate funding. Cutting-edge technology is of little value if monies needed to apply that technology are lacking. Some crime laboratories are not accredited, and the accreditation process of a number of organizations has been called into question.

KEY TERMS

acquittal: A decision by a judge (bench trial) or a jury that a criminal defendant is not guilty of a crime(s)

bench trial: A trial before a judge as opposed to a jury trial

DNA: Deoxyribonucleic acid, a substance containing genetic material that is unique to an individual except in the case of identical twins

hung jury: A jury unable to reach a final determination. In a criminal trial if a mistrial is declared the prosecutor decides to try the case again, offer a plea bargain, or dismiss the charges against the defendant.

jury nullification: A decision by a jury to acquit a defendant who has violated the law because the law is considered unjust, or how the law is applied to a specific category of defendants

peremptory challenge: During jury selection a decision by an attorney to excuse a potential juror from the case without having to offer a valid reason for the dismissal

voir dire: From the French to speak the truth, an examination of prospective jurors to determine their qualification for jury duty

SUGGESTED READINGS

Black, Roy. 2000. *Black's Law: A Criminal Lawyer Reveals His Defense Strategies in Cliffhanger Cases.* New York: Simon & Schuster.

Conrad, Clay S. 1998. *Jury Nullification: The Evolution of a Doctrine.* Durham, NC: Carolina Academic Press.

Culbertson, Robert G. and Ralph A. Weisheit. 2001. *Order Under Law: Reading in Criminal Justice.* Prospect Heights, IL: Waveland Press, Inc.

Friedman, Lawrence, M. 1993. *Crime and Punishment in American History.* New York: Basic Books.

Lowenthal, Gary T. 2003. *Down and Dirty Justice: A Chilling Journey into the Dark World of Crime and Criminal Justice.* Far Hills, NJ: New Horizon Press.

Neubauer, David W. 2007. *America's Courts and the Criminal Justice System.* Belmont, CA: Wadsworth.

"Speed Bump" © Dave Coverly/Dist. By Creators Syndicate, Inc.

chapter twelve

Prisons and Jails: Punishment at Any Cost?

Ernie Preate, Jr. was once an advocate without peer for being tough on crime. As district attorney in Pennsylvania, he won one guilty verdict after another and 19 murder cases in a row. He put 5 murder defendants on death row and passionately supported capital punishment. He also urged mandatory sentences for drug offenses. When he became the attorney general of Pennsylvania in 1988, Preate continued to press his tough approach to fighting crime and drugs.

Several years later, Preate found himself on the other end of the law when he pleaded guilty to mail fraud in the wake of a campaign financing scandal, resigned as attorney general, and entered a federal prison in Duluth, MN. There he saw that many of the inmates were African Americans serving long terms for minor drug offenses. He recalls, "I'll never forget it. In January of 1996, I walked into the mess hall the first night I was up in Duluth, and I turned around and I said, 'Oh, my God, what have we created?' It was a sea of black and brown faces." Preate was also shocked by the conditions he found in his prison. When he was released after spending 14 months behind bars, he emerged as a prison reformer and urged an end to capital punishment, a reconsideration of the nation's harsh sentencing policy for drug offenders, and a renewal of rehabilitation efforts for inmates. He explained, "Hey, look, this is common sense. We can't keep building prisons, or pretty soon we're going to wind up with more people in prisons than we have kids in school."

Preate's story reminds us of the continuing controversy surrounding prisons, a key component of the nation's get-tough approach to fighting crime. As we noted in the first few pages of this book, our number of prison and jail inmates quadrupled between 1990 and 2008 and now exceeds 2.2 million inmates, yielding an incarceration rate that is more than three times higher than any other Western nation. Our corrections system costs more than $60 billion annually and forces states to cut their budgets elsewhere. What are we getting for our money? Has the surge in imprisonment lowered the crime rate appreciably? Has it made our society safer or has it set the stage for further problems down the road? What is life behind bars like anyway? We examine all these questions in the pages that follow.

SUCH AN ORDEAL: PRISONS YESTERDAY AND TODAY

A Brief History of Prisons

Before the 18th century, jails and prisons were not used to punish criminals in Europe or what was to become the United States. Instead, a common form of punishment was corporal punishment, which took on many forms: beating, whipping or flogging, branding, mutilation, and the use of devices such as the rack and pillory. Thanks to criminal justice reform movements, such punishment finally fell out of favor during the 19th century. It did not disappear altogether, however: flogging remained a possible form of legal punishment in Great Britain for at least some offenses until 1967 and occurred in the United States in Delaware as late as 1952.

Confinement (or **incarceration**) was used before the 18th century, but not as a form of punishment. Instead, people were confined while they awaited trial or until they could receive some other form of punishment such as flogging or banishment. In one exception to this general rule, debtors in medieval Europe were confined until they could pay the money they owed. Social reform movements in Europe then gave rise to the idea that imprisonment should replace corporal punishment as the dominant mode of punishment. Thus, during the 1700s, more and more Europeans were punished by being put behind bars. Although that was an improvement over corporal punishment, the prisons in which they were put were filthy and overcrowded. Men, women, and children were confined together, as were violent and minor offenders. Rape and other abuses were common. The crowded and dirty conditions made communicable diseases rampant, and these diseases routinely killed inmates, guards, and even, sometimes, lawyers and judges. Criminal justice reformers continued to call attention to all these decrepit conditions and said prisons should become places where offenders could learn the error of their ways and repent the offenses they had committed. Britain began to build **penitentiaries** to accomplish this goal, ushering in the age of the modern prison.

The development of corrections in the United States mirrored the European experience. During the colonial period, jails were used mainly to hold either debtors until they paid their debts or defendants for trial or other kinds of punishment. Like their British counterparts, colonial jails were filthy and overcrowded and the repositories of communicable disease. Punishment in the colonies took four primary forms: fines, shaming, flogging, and execution. Fines were used for minor offenses while shaming was used both for minor and more serious offenses. Common forms of shaming involved the use of the pillory or stocks, which were wooden frameworks with holes in them for the offender's arms, legs, and head. The offender would be placed in these devices for several days at a time in the town center and subjected to taunting and ridicule by passersby. Sometimes they would also be whipped or the target of stones, eggs, or other thrown objects. Ducking stools were also used; here an offender would be tied to a chair at the end of a long plank which would then be dunked into a pond or river.

After the Revolution with Britain, the new United States ushered in an era of criminal justice reform. Many of the grievances outlined in the Declaration of Independence centered on British abuses of the criminal justice system, and the Constitution and Bill of Rights included many protections for people suspected of crimes. Inspired by the European criminal justice reformers, American reformers devised a system of punishment that would limit executions and avoid public humiliation and flogging and other corporal measures.

In the post-Revolutionary period, a major event in the development of the modern prison was the opening in 1790 of the Walnut Street Jail in Philadelphia. Devised by that city's Quakers, the Walnut Street Jail was meant to be a penitentiary in which inmates lived in solitary confinement and had no contact with other inmates, even during the day. This solitude was intended to allow inmates time for reflection, Bible reading, prayer, and other activities that would help reform them. Male and female inmates were kept apart, as were minor and serious offenders, and guards were prohibited from abusing inmates. However, overcrowding and other problems plagued the Walnut Street Jail and prevented inmates from being isolated from each other. To the extent it did occur, solitary confinement, however well intended, did not prove an effective instrument of rehabilitation. If anything, it impaired prisoners' mental health and led to other problems.

Despite these problems, the Walnut Street Jail served as a model for new prisons elsewhere. During the 1820s, Pennsylvania built two additional, large prisons in Philadelphia and Pittsburgh, and other states also followed suit. All these prisons involved solitary confinement, little or no contact among prisoners, and considerable time for reflection and prayer. The Pennsylvania model, as it was called, involving solitary confinement, soon gave way to the Auburn or New York model, named after a prison opened in 1817 in Auburn, New York. The key difference in the two models was that the Auburn prison did not practice continuous solitary confinement. Inmates were housed separately at night but worked and ate together during the day, although they still were not allowed to talk with each other.

Despite this key difference between the Auburn and Pennsylvania models, both assumed that crime was the result of problems in the external social environment rather than a sign of personal moral problems. Both thus also assumed that prisoners could be rehabilitated through reflection and hard work that would inmates overcome the negative influences of their social environments.

Prison conditions during the nineteenth century alarmed many observers. Prisons suffered from overcrowding, filth, lack of sanitation, and abuse by guards. About 15 to 20 percent of many states' prisoners died each year from these problems. A prison reform movement that began after the Civil War led to a new strategy at a prison in Elmira, NY, that became known as the Elmira Reformatory. This strategy involved awarding inmates points for good behavior and for achievement in work and educational programs. When they earned enough points, they would get certain privileges, including more comfortable clothing and a better mattress, and could eventually win early release from prison on parole. It was thought that this incentive would motivate inmates to behave well and to become fully rehabilitated. Despite this innovation, living conditions in the Elmira prison were no better than those at other prisons. Guards routinely whipped or beat prisoners for almost any misbehavior, and rape was common.

Still, the Elmira model became popular, and by the 1920s nearly every state had adopted it. The goal of rehabilitation guided corrections throughout the nation. During the 1960s and 1970s, the U.S. Supreme Court and other federal courts rendered several important decisions about prison conditions and prisoners' rights. In one of the first decisions, a federal court in 1970 declared the entire Arkansas prison system unconstitutional because of appalling conditions in the state's prisons; six years later, another court reached the same conclusion for an Alabama prison. During this same period, however, the goal of rehabilitation lost favor as concern grew over rising crime rates, and the get-tough approach to crime control began in earnest.

Jails Versus Prisons

Prisons house offenders who have been found guilty of felonies and sentenced to at least one year of incarceration. They are run by the 50 states, by the federal government, and by private companies contracting out to the states. Each state has at least a few prisons, and some big states, such as California and Texas, have many prisons. The United States has about 1700 prisons of all types, most of which are filled beyond capacity.

The word prison probably brings to mind a massive structure with high towers, heavily armed guards and search lights, and long, bleak hallways with cell after cell. This is one kind of prison, but other kinds of prisons exist. The one just described fits the portrait of a **maximum-security prison**, more commonly called the "Big House." These grim places hold inmates who are thought to be the most violent and dangerous to the outside community. Several maximum-security prisons in use today were built a century or more ago and include such famous names as San Quentin in California (just north of San Francisco) and Leavenworth in Kansas. Today San Quentin serves as death row for all California men who have been sentenced to death, and, 150 years after it was built, still holds some 6000 inmates behind its forbidding walls. Several other Big Houses built years ago have closed, but some, such as Alcatraz, remain notorious in popular lore.

Medium-security prisons hold property offenders and other inmates who are considered less dangerous than those in maximum-security institutions. Instead of high towers and walls with armed guards and searchlights, they typically are enclosed by barbed-wire fences. In many of these prisons, inmates live in large rooms, similar to a military barracks or summer camp cabin, rather than in small cells. They are less expensive to build and maintain than maximum-security prisons and usually have more educational and vocational programs and drug and alcohol counseling.

Minimum-security prisons hold inmates, typically property and drug offenders, who are considered the least dangerous and who are serving the shortest sentences. They typically lack even fences and instead rely on locked doors and electronic equipment to detect escape attempts. Inmates are allowed much more freedom of movement than their counterparts in the other types of prisons and are offered a greater number of programs and services.

Jails are run by towns, cities, and counties and hold several kinds of offenders and other individuals: (1) people who have been arrested and are awaiting arraignment, trial, or some other procedure (amounting to about half of all jail inmates); (2) people convicted of a misdemeanor and sentenced to less than one year behind bars; (3) people convicted of a felony who are waiting to be sentenced; (4) people convicted of a felony and sentenced to more than one year behind bars but for whom no room in a state prison is available; (5) inmates who have completed their sentence and are awaiting release to the community; (6) individuals held for contempt of court, for military legal action, or for protective custody; (7) mentally ill persons awaiting movement to a health facility; and (8) inmates awaiting transfer to other correctional facilities. About 3400 jails exist throughout the United States, and more than 12 million Americans (including repeat admissions) enter a jail each year for at least one of these reasons.

Jails often have worse conditions than prisons in many ways. They generally are more overcrowded and dilapidated. Because jails are intended for short-term confinement, they also lack the vocational, educational, and drug treatment programs found in prisons. A special problem of jails is that they confine people awaiting arraignment or trial (and thus in pretrial detention) with those already convicted of crimes, some of them quite serious. Although the pretrial detention inmates have not yet been convicted of any crime and some may well be innocent of any wrongdoing, their confinement with convicted criminals subjects them to the possibility of violent assaults and negative peer influences. Even if they emerge from jail unharmed physically, they may emerge very much harmed emotionally and perhaps more apt to commit crime than before they were placed in jail.

Is It Worth It? Prisons and the Crime Rate

With more than 2.2 million people behind bars at any one time and more than $60 billion spent on corrections every year, it is important to ask, "Is it worth it?" With any expensive effort, it is important to determine whether dollars are being spent wisely. This determination is no less important for corrections. What is our society getting for the tens of billions of dollars it spends each year on corrections? For most of us, the most important issue in deciding whether the surge in imprisonment has been worth it involves whether it is keeping society safer by reducing crime to a considerable degree. If any crime decline is substantial, then we have grounds for concluding that the money spent on incarceration has indeed been worth it. If the crime decline is small, then we may decide our dollars have not been spent wisely, especially if other strategies might achieve more cost-efficient declines in the crime rate.

For these reasons, it is important to determine whether the surge in imprisonment has in fact reduced the crime rate, and, if so, the extent of this effect. In Chapter 8 we argued that the deterrent effect of the get-tough approach has been small or nil. When we discuss here whether incarceration reduces the crime rate, we are talking about the **incapacitation** effect: the reduction in crime caused by keeping offenders behind bars where they cannot commit crime against the public. The rise in incarceration accompanying the get-tough approach has prompted many scholars to determine the size of this incapacitation effect.

Much of their research has focused on the 1990s, when incarceration increased by 67 percent and the crime rate dropped by 22 percent. This correlation might mean that the increased incarceration caused the crime drop, but other evidence casts doubt on this conclusion. The states with the largest increases in incarceration during the 1990s actually had smaller crime rate declines than the states with the lowest increases in incarceration. Moreover, although the incarceration rate in the United States has increased steadily since the 1970s, the crime rate has fluctuated. It declined and then rose during the 1980s, for example, even though incarceration soared steadily during that time. Although the crime rate finally fell during the 1990s as incarceration continued to rise, other factors were probably at work, say many criminologists, including a strong economy, stabilizing crack cocaine trafficking in major cities, and the aging of the large baby boom generation, whose sheer size had helped fuel the crime rate increase during the 1960s. Assessing such evidence, The Sentencing Project, a national non-profit criminal justice research and advocacy organization, found "little support for the contention that massive prison construction is the most effective way to reduce crime."

Supporting this conclusion, research sponsored by the National Consortium on Violence Research and the Harry Frank Guggenheim Foundation concluded that the increased imprisonment during the 1990s accounted for only one-fourth of the crime drop during that decade. The study's author concluded that imprisonment is "an incredibly inefficient means of reducing crime." Another study sponsored by the two groups found that the increased imprisonment accounted for about 25 percent (or 100 homicides per year) of the drop in homicides during the first half of the 1990s. This study's author called this finding "good news" but added that each of the homicides averted by the increased incarceration cost $13 million per year in prison costs. Thus, the 100 homicides averted annually during this period cost more than $1 billion per year, an expenditure that led the researcher to speculate that society would be safer, with more lives saved, if the same amount of money were instead spent on preschool programs and other prevention efforts.

The research on the 1990s yields several conclusions. First, rising incarceration had a small but still significant effect on the crime rate. Second, the size of this effect was much smaller than that of the other factors cited above, including a thriving economy and changing demographics. Third, incarceration is extremely expensive and thus not a cost-efficient way of lowering the crime rate. The tens of billions of dollars spent on incarceration now and over the last few decades would be much more effective in lowering the crime rate were they spent instead on crime prevention programs such as early childhood intervention programs and on drug treatment and other programs for convicted offenders.

If this is true, then the costs of the surge in imprisonment have not been worth it. The surge has also been creating other problems called collateral consequences that again cast doubt on the wisdom of the get-tough policy. More than 650,000 prison inmates are now being released back to their communities, most of them inner-city neighborhoods, every year. Once back home, the ex-inmates have few job prospects because they have few job skills, and up to 60 percent remain unemployed one year after their release. These problems drive up unemployment rates in their communities and, as a result, crime rates. The surge in incarceration and massive increase in prisoner reentry have led to other problems as well. For example, many inner-city communities have seen large numbers of their young males go to prison or jail. The absence of so many young adults weakens these communities' social cohesion and thus indirectly raises their crime rates. Other problems from the great increase in prisoner reentry include increased child abuse and family violence, the spread of HIV-AIDS and other infectious diseases, and an increase in homelessness.

Many observers especially worry about the collateral effects of the surge in imprisonment on the children of inmates. About 1.5 million children now have a parent in prison, and many others also have a sibling behind bars. The Sentencing Project notes the consequence: "As more young people grow up with parents and siblings incarcerated and a view of time in jail as a normal aspect of one's life experience, the deterrent effect of prison is diminished." Parental imprisonment also leads to less effective socialization of children and a disruption in their family stability and financial well-being.

All these problems mean that the U.S. incarceration policy of the last few decades may well be exacerbating some of our most serious social problems and ironically contributing to the crime rates our nation has now and will have in the future. An additional collateral effect of the surge in incarceration has been its disparate impact on the African American community. One-third of all African American males in the 20–29 age group are now under correctional control: They are in prison or jail, or on probation or parole. Because African Americans appear to have higher rates of street crime (see Chapter 3), the rise in incarceration was bound to have a larger impact on this racial group than on others. But this impact has been much greater than what might have been expected because of the nation's war against drugs that has relied on arrest and imprisonment as the primary means of combating drug trafficking, possession, and use. Much of this war has focused on crack cocaine, with much higher mandatory penalties imposed for this drug than for its close cousin, powder cocaine. Whites tend to use powder cocaine, while blacks tend to use crack cocaine. This difference means that the war on drugs, with its focus on crack, has disproportionately affected African Americans. Citing this problem, one scholar calls the war on drugs a "search and destroy" mission against African American males.

LIFE BEHIND BARS: MYTH VERSUS REALITY

Many Americans probably think prisons are like country clubs, with inmates enjoying large flat-screen televisions, DVD players, other electronic equipment, and gyms and exercise equipment. Like many aspects of public beliefs about crime and justice, the reality is far different from this country-club image, which is little more than a myth. Most prisons are forbidding places, and life behind bars is anything but a country club. Why should we care about life behind bars? No one wants to see prisoners pampered, but neither should anyone want to see them treated less than humanely. On a practical level, what prisoners experience behind bars may affect how they behave when they are released. The worse their experience behind bars, the more likely they are to commit new crimes when they go back to society, and the less safe society will be. For this reason, it is important to understand what life behind bars is really like.

The Pains of Imprisonment

In his 1958 classic book, *The Society of Captives*, Gresham M. Sykes discussed several "deprivations or frustrations of prison life." He referred to these deprivations as the **pains of imprisonment** and said they "pose profound threats to the inmate's personality or sense of self-worth" and for this reason can be just as painful in their own way as the physical punishment used against offenders in the past.

The first deprivation is the loss of liberty. Prison inmates are obviously not free to leave the prison, and they are cut off from family, friends, and relatives. For these reasons, wrote Sykes, the deprivation of liberty impairs inmates' self-esteem.

The second deprivation is the loss of goods and services. Although inmates' basic needs—food, clothing, shelter—are met, and some may actually be better off inside prison because of this, most inmates think they could have more possessions and be leading a better life outside prison. As a result, they consider the loss of goods and services a painful deprivation. They want not just the necessities of life but the amenities that most of us take for granted: their own clothing rather than prison uniforms, a variety of good food rather than monotonous prison fare, privacy rather than a shared, overcrowded cell. In a society that values an individual's worth in terms of the material possessions he or she enjoys, the loss of goods and services is a painful deprivation for inmates.

The third deprivation is the loss of heterosexual relationships. Although home furloughs and conjugal visits have become more common since Sykes wrote his book, the lack of heterosexual outlets for prison inmates remains a serious problem. Their involuntary celibacy is obviously frustrating sexually and emotionally, but it can also further impair their self-worth.

The fourth deprivation is the loss of autonomy. Sykes wrote that inmates are "subjected to a vast body of rules and commands which are designed to control [their] behavior in minute detail." The loss of autonomy again impairs inmates' self-worth by treating them as helpless, dependent children and, in this way, humiliating them.

The final deprivation is the loss of security. Inmates' physical safety is always at risk, most often by other inmates, but also by corrections officers. Recall that many inmates have been previously convicted for violent offenses; others may have no violent convictions but still have violent tendencies. Life behind bars aggravates these tendencies. Inmates are very aware that they will be tested physically and that they must be prepared to fight to defend both themselves and their possessions. Their loss of security makes them very anxious, because they realize they can never be safe.

Overcrowding and Other Conditions

In 2000, a troubling report cited substandard conditions at Washington, DC's Central Detention Facility, more commonly known as the D.C. jail, which holds almost 1700 inmates and employs about 450 corrections officers. The jail's ventilation system was so dirty that summertime room temperatures would sometimes become dangerously high; the lack of adequate ventilation was also implicated in a case of Legionnaires' disease that almost killed a corrections officer. Many sinks and showers simply did not work. Neither did the laundry machines, forcing the inmates to wash their clothes and sheets in their toilets. A lack of jumpsuits meant that some inmates would wear only their underwear for days on end. Mice and rats scurried everywhere, and roaches and other insects bred in puddles of water caused by leaky plumbing.

The conditions in the D.C. jail sound eerily similar to those afflicting its 18th-century counterparts and are found in jails and prisons around the country. In addition to decrepit living conditions, another common problem in our jails and prisons is overcrowding, as many prisons and jails hold many more inmates than the facility was meant to house. Overcrowding can have psychological consequences and is thought to be a prime contributor to prison violence. Overcrowding is so serious that Human Rights Watch, an international human rights organization, charged a few years ago that U.S. prisons are "dangerously overcrowded." A federal judge who toured an Alabama jail found inmates sleeping on the floor next to toilets and on top of shower drains because there was no room for beds. The judge later wrote, "The sardine-can appearance of its cell units more nearly resemble the holding units of slave ships during the Middle Passage of the 18th century than anything in the 21st century."

Other problems contributing to the harsh and painful experience behind bars include prison violence and inadequate health care. We now turn to these problems, which combine with decrepit living conditions and overcrowding to suggest that life behind bars is anything but a country club.

Prison Violence

One of the facts of life behind bars is violence. In saying this, we do not wish to stereotype prisoners as monsters but rather to acknowledge harsh reality. Many prisoners have violent pasts. When they enter prison, their violent tendencies are not likely to stop. They are there against their will, and for many the pains of imprisonment combine with overcrowding and substandard living conditions to foster even more violence and create even more hostility. In fact, if we wanted for some reason to design a correctional facility that would be guaranteed to produce violence, it would look very much like the prisons and jails we now have. Violence in prison, then, results not only from inmates' personal backgrounds and attitudes, but also from the ugly aspects of prison life.

How much prison violence occurs? Surveys of prison inmates find that 10 to 14 percent have been assaulted physically in the previous six months. Prison gangs account for much of the violence that occurs. Much of this violence is meant to help prison gangs secure economic profits illegally, for example, through drug trafficking and extortion. In this regard, prison gangs closely resemble organized crime gangs in the outside world. Like their organized crime counterparts, prison gangs sell protection, drugs, and other contraband. Some have contacts with youth gangs and other illegal enterprises in the outside world, and, when inmates from the gangs are released, they work for these groups.

A specific type of violence in prisons that continues to be a serious problem is prison rape and other sexual assault, either completed, attempted, or threatened. Sexual assault in prisons and jails is a common fact of life behind bars, especially in prisons that are larger and more overcrowded. Surveys of male prison inmates find that about 20 percent to 33 percent are sexually assaulted and that at least 10 percent are raped. Given the large numbers of inmates nationwide, these proportions amount to a significant number, about 600,000 annually, of rapes and sexual assaults behind bars. Most inmates who commit prison rape use rape as a weapon to exert dominance. Inmates at higher risk for sexual victimization include those who are young, small, suffering from mental disorders, not gang-affiliated, thought to be gay or effeminate, or convicted of sexual crimes themselves. Inmates who have squealed on another inmate are also at greater risk for sexual victimization.

One other risk factor is race, as racial tension within prisons contributes to interracial sexual assaults.

Although prison sexual assault stems from inmates' violent tendencies and overcrowding and other conditions, prison administration and staff also bear some blame for the problem. There are too few correctional officers in overcrowded facilities, and the officers receive little or no training on rape and other sexual assault. Mental health counseling for rape victims is rare. Many guards believe that rape victims are gay and consent to the sex that victimizes them. Even when guards do believe a sexual assault has occurred, they often find it easier to look the other way, and many feel that prison rape is to be expected. Some scholars paint a more ominous picture of the complicity of prison staff in prison rape, saying that the staff allows and sometimes even encourages prison rape because it helps maintain order in the prison. This happens in at least two ways, First, prison rape channels prisoners' aggression from guards to other prisoners. Second, inmates know that, if they do not do what guards want, they may end up in cells with known sexual aggressors. Guards may also use the threat of rape to turn inmates into informers as they may threaten to put an inmate into contact with a known aggressor if he refuses to squeal. Some guards have taken bribes from inmates to put young, new prisoners into their cells.

Health and Health Care

Another harsh aspect of life behind bars is poor health and even poorer health care. Many inmates enter jail and prison with a history of mental and physical problems and/or a history of drug and alcohol abuse. These problems make it especially important that inmates receive adequate care while incarcerated. How adequate is this care?

Although health care in jails and prisons has improved during the past two decades because of court rulings, many serious problems remain, and health care overall is inadequate. Prison guards and medical personnel often doubt the claims of prisoners that they are ill and instead think inmates with health complaints are merely malingering. It is also difficult to recruit qualified physicians to work in prisons and jails: most physicians do not view these settings as desirable places to practice, and their compensation is generally lower than in the outside world. As a result, prisons and jails typically attract and employ physicians who are less qualified than their non-corrections counterparts and who are apt to have faced disciplinary charges by medical review boards. In another problem, prisons lack adequate health facilities and diagnostic equipment. Many prisons have no way to isolate prisoners with infectious diseases, and their infirmaries lack 24-hour nursing care. Prisoners with chronic illnesses are not kept separate from other inmates.

Two groups of inmates, women prisoners and older inmates, face special healthcare problems. The medical needs of women prisoners differ from those of male prisoners, yet treatment behind bars for these needs is lacking. We have more to say about this problem in the next section. Older inmates (more than 50,000 prisoners are age 55 or older) have especially high rates of health problems and thus a high need for adequate medical care. The lack of adequate care in prisons thus affects women and older inmates disproportionately.

WOMEN IN PRISON: SEXISM IN CAPTIVITY

Ordinarily women inmates are kept in separate prisons from male inmates. From 1995 to 2006, the number of women prisoners grew about 64 percent, while the number of male prisoners grew about 38 percent. In 2006, women accounted for 8 percent of all state and federal prisoners, up from 5.7 percent in 1990. The media have often depicted the rise in the number of women prisoners as a sign that women are catching up to men in violence and other criminal behavior. This media depiction is yet another myth. Although the number of women prisoners has risen at a greater rate than the number of male prisoners, that is because there were very few women prisoners to begin with. Moreover, many of the new women prisoners have been convicted of drug offenses, as the legal war on drugs has disproportionately affected women. About 29 percent of female prisoners were convicted of drug offenses, compared to only 19 percent of male offenders. It is also quite possible that police are more likely than in the past to arrest women suspects and that prosecutors are more likely to file charges. If so, the rise in women prisoners probably reflects decisions by police and prosecutors more than any real rise in women's criminality.

In any event, the incarceration rate for men is still 13 times greater than that for women, and about 50 to 55 percent of male prisoners were convicted of a violent offense, compared to only about 35 percent of female prisoners (often against a male intimate who had been abusing them). Prison is still very much a man's world, and, partly because of that, women inmates experience certain problems that male inmates either do not experience at all or do not experience as often or as severely.

One of these problems is a history of being abused. Women inmates are much more likely than male inmates to have been physically or sexually abused before their incarceration: in surveys of prisoners, about 60 percent of women inmates and only 16 percent of male inmates report such abuse, and more than one-third of women inmates report being raped before entering prison. Partly for this reason, women inmates are also more likely than male inmates to have histories of drug abuse.

Sexual Victimization

As in men's prisons, sexual victimization in women's prisons is a common problem, but, unlike men's institutions, the offenders are usually prison guards, not other inmates. In male prisons, no more than 20 percent of sexual victimization is committed by prison staff. In women's prisons, though, almost all sexual violence and other types of sexual victimization are committed by prison staff. Although women's prisons developed decades ago in part to reduce sexual abuse by male inmates and guards, most of the guards in women's prisons are still men.

Although the exact prevalence of sexual victimization is difficult to estimate, a report by Amnesty International found more than 1000 cases of sexual abuse in prisons in 49 of the 50 states from 1997 to 1999. Because of fear of retaliation, most of the victims did not report their abuse. The report called such abuse a "major systemic problem" not limited to just a "few bad apples." Amnesty International also found that all but three states permit male guards to do pat-down searches of female inmates, which the report called "inherently abusive." Between 1995 and 1999, the sheriff's department in one Massachusetts county strip-searched every one of the 5500 women arrested by Boston police no matter what the charge; female guards would also perform body cavity searches for drugs and other contraband. The sheriffs did not strip-search male suspects or perform body cavity searches on men.

Medical Needs and Services

As noted above, the medical needs of women prisoners differ from those of male prisoners. This is true for several reasons. First, as also noted above, women inmates are more likely to have prior histories of drug abuse and sexual victimization. Second, women obviously need adequate gynecological care. We have already seen that prisons lack adequate medical and mental health care and services. Given that women have more medical and mental health needs, the lack of adequate medical care in prison affects women especially severely. For example, many state prisons do not offer mammograms and Pap smears. Also, although women inmates have a greater need for drug treatment, they are less likely than male inmates to receive it.

Women Prisoners as Mothers

Women inmates are much more likely than male inmates to have children and also, before incarceration, to be the primary caretakers for their children. As a result, incarceration poses a more serious emotional and practical problem for women inmates than it does for male inmates The children of newly incarcerated women must now be taken care of by other people, usually relatives but sometimes the state, and some women face losing custody of their children permanently. Prisons vary in their visitation rights and facilities for young children, and the majority do not permit extended visits. The majority of incarcerated mothers (and also fathers) report never having been visited by their children since entering prison, in part because their institutions are usually more than 100 miles from their last place of residence.

About 5 percent of women offenders are pregnant when they enter prison. These women face special problems. Like other women inmates, they already come from backgrounds of poverty, drug and alcohol abuse, smoking, poor nutrition, and other problems that render their pregnancies high risk. Once they enter prison, conditions can further increase the chance of pregnancy complications. As with other health services in prison, obstetrical care can be inadequate. When the baby is born, inmates in most prisons have to give up the baby within a few weeks, although a few institutions have nurseries where the children may stay up to age one or so.

A final problem facing some pregnant inmates is shackling during childbirth. The Amnesty International report found that 18 states allow pregnant inmates to be shackled while they give birth even though there is obviously no chance the woman will try to escape. The organization considered such shackling a form of abuse of women in custody.

Treatment and Vocational Programs

For both female and male inmates, treatment and vocational programs in the nation's prisons are missing altogether, or, when present, inadequate and poorly funded. Even so, women's prisons typically have fewer treatment and vocational programs than men's prisons. Because most prisoners are men and because men are thought to be more dangerous to society, the attention and funding of the corrections system traditionally has focused on male prisoners. This, in turn, has led to a neglect of the needs of female prisoners. A more subtle bias also exists in the nature of vocational programming in women's prisons. Although the situation is improving, vocational programming typically has focused on preparing women inmates for traditional female jobs such as secretarial work, sewing, and food service and not for higher-paid jobs such as carpentry, plumbing, and auto repair. Training for the latter jobs remains much less available in women's prisons than in men's prisons. The fact that treatment and vocational programs continue to be less available in women's prisons adds to the pains of imprisonment that women inmates face because of their gender.

THE DEATH PENALTY DEBATE

Capital punishment remains one of the most controversial issues in criminal justice today, as the United States is the only Western nation that still uses it. We highlight here the major questions asked about the death penalty and the evidence concerning these questions.

Does the Death Penalty Reduce Murder?

Many studies have tried to answer this question, and almost all fail to find a deterrent effect. States with the death penalty do not have lower homicide rates than states without it. In the last century, states that eliminated the death penalty did not see their homicide rates increase, and states that adopted the death penalty did not see their homicide rates decline. Studies of well-publicized executions generally fail to find that executions lower the homicide rate. Some even find a brutalization effect, with the homicide rate increasing after publicized executions, perhaps because murderers are imitating the killing that execution involves or because they have a death wish and want to be executed.

Drawing on our discussion of deterrence in Chapter 8, it is not very surprising that the death penalty does not deter homicide. For it to do so, a potential murderer would have to have the death penalty in mind while he is planning his crime and have time to weigh carefully the risk of getting caught and being executed. This, in turn, implies that homicide would have to be the type of crime that is planned carefully by someone who is acting rationally and unemotionally. Yet homicide tends to be a relatively emotional and spontaneous crime in which someone becomes angry or vengeful and lashes out without really thinking through the possible consequences for his or her future. If homicide is indeed such a crime, then capital punishment probably cannot deter it.

Does the Death Penalty Save Money?

Many people believe the death penalty is cheaper than life imprisonment. What do the numbers say? It turns out that the death penalty is actually two to three times more expensive than life imprisonment. Capital cases are extraordinarily time consuming and, therefore, costly. Various procedures at the pretrial and trial stages must be followed closely that are not followed in noncapital cases. Appeals following conviction are mandatory, time consuming, and, again, costly. At all these stages, the state incurs its own costs for prosecution and appeals, and it must almost always also pay these costs for capital defendants, who are typically too poor to afford their own attorneys. As a result, the average capital case costs $2 million to $3 million. In contrast, life imprisonment costs about $1 million in current collars, assuming 40 years in prison and $25,000 in annual incarceration costs per capital defendant. Thus, the average capital case costs $1 million to $2 million more than life imprisonment. If we apply this extra cost to the 3350 people on death row in early 2007, then this extra cost is between about $3.3 billion and $6.7 billion. Capital punishment turns out to be quite costly indeed.

Is the Death Penalty Racially Biased?

Probably. Some but not all studies find that when the defendant is African American, prosecutors are more likely to lodge capital charges (without which there could be no death penalty), and jurors are more likely to find the defendant guilty and to sentence him to death. But other studies do not find these differences, and thus the evidence of discrimination based on the race of the offender is inconsistent.

However, evidence of discrimination is clearer when the race of the victim is considered. Simply put, when whites are victims, death penalty charges and sentences are much more likely than when African Americans are victims. In some jurisdictions, prosecutors are four to five times more likely to seek the death penalty when the victim is white than when the victim is African American. In effect, the judicial system places more value on the lives of white victims than on the lives of black victims.

Are Innocent People Executed?

Innocent people have been sentenced to death, and it is very likely that some have been executed over the years. In January 2000, the governor of Illinois suspended executions in his state after newspaper investigations revealed that 13 innocent men had been sentenced to death in Illinois over the past dozen years. He eventually commuted the sentences of all 167 death row inmates in his state. Reflecting the Illinois experience, by 2007, 124 death row inmates had been released nationwide since the early 1970s after DNA and other evidence cast serious doubt on their guilt.

Death penalty supporters say these cases prove that the system works, because these individuals were not executed, and that it has never been proven that an innocent person has been executed since *Furman*. Opponents reply that some of these men nonetheless came dangerously close to being executed. They concede that no wrongful execution has been proven, but they cite several cases in which the guilt of an executed person was at least highly questionable. One example was the case of Gary Graham, who was executed in Texas in June 2000 for a 1981 murder of a drug dealer. The only evidence against him was one witness's testimony, which was contradicted by two other witnesses whom the defense attorney never called to the stand or even interviewed. The gun used in the murder was never found, and no physical evidence implicated Graham. Another such case was that of Roy Roberts, who was executed in Missouri in 1999 for murder without any physical evidence of his involvement in the crime and despite changing stories by witnesses. Death penalty scholars think that at least 12 innocent people, and probably more, have been wrongly executed since the mid-1970s.

Why are innocent people sentenced to death, and why are some of them apparently executed? Police, prosecutors, and witnesses may lie or conceal evidence, and honest mistakes by police or witnesses can also occur. In addition, capital defendants are often represented by attorneys who are unqualified to handle complex capital cases. Even if they are skilled, they do not have the estimated $250,000 that it takes to defend a capital defendant adequately, as states typically provide them no more than about $2000 to $3000 in legal expenses. Worse yet, some of these attorneys have been notoriously inept. They fail to call important witnesses to the stand or to present important evidence, and some even fall asleep during the trial. In one capital trial in Houston, the defense attorney slept throughout an afternoon's testimony. About one-fourth of death row inmates in Kentucky were defended by attorneys who were later disbarred or resigned to avoid disbarment. An investigation by the *Chicago Tribune* of 131 executions in Texas while George W. Bush was governor found 43 cases where the defense attorney was later disbarred or disciplined. More generally, it concluded that "Texas has executed dozens of Death Row inmates whose cases were compromised by unreliable evidence, disbarred or suspended defense attorneys, meager defense efforts during sentencing and dubious psychiatric testimony."

KEY TERMS

incarceration: Confinement in prison or jail

incapacitation: Depriving offenders of the ability to commit new offenses and thereby improving public safety

jails: Penal institutions run by towns, cities, and counties that houses several kinds of offenders

maximum-security prisons: Large, forbidding prisons that hold inmates who are thought to be the most violent and dangerous to the outside community

medium-security prisons: Prisons that hold property offenders and other inmates who are considered less dangerous than those in maximum-security institutions

minimum-security prisons: Prisons that hold inmates, typically property and drug offenders, who are considered the least dangerous and who are serving the shortest sentences

pains of imprisonment: Gresham Sykes's term for the deprivations of prison life

penitentiaries: Early prisons that were built to hold inmates so they could repent for their offenses

prisons: Penal institutions that house offenders who have been found guilty of felonies and are sentenced to at least one year of incarceration

SUGGESTED READINGS

Beckett, Katherine, and Theodore Sasson. 2004. *The Politics of Injustice: Crime and Punishment in America.* Thousand Oaks, CA: Sage Publications.

Blumstein, Alfred, and Joel Wallman (eds.). 2006. *The Crime Drop in America.* Cambridge: Cambridge University Press.

Gottschalk, Marie. 2006. *The Prison and the Gallows: The Politics of Mass Incarceration in America.* Cambridge: Cambridge University Press.

Petersilia, Joan. 2003. *When Prisoners Come Home: Parole and Prisoner Reentry.* New York: Oxford University Press.

Pollock, Jocelyn M. 2006. *Prisons Today and Tomorrow.* Sudbury, MA: Jones and Bartlett Publishers.

Travis, Jeremy, and Christy Visher (eds.). 2005. *Prisoner Reentry and Crime in America.* New York: Cambridge University Press.

"Speed Bump" © Dave Coverly/Dist. By Creators Syndicate, Inc.

chapter thirteen

Community Corrections and Juvenile Justice

If mass imprisonment costs tens of billions of dollars annually while reducing crime only to a small degree, perhaps other measures might keep society at least as safe while costing far less money. Several such measures exist and together are called community corrections. We first discuss probation and parole before turning to intermediate sanctions.

PROBATION AND PAROLE

Probation refers to the sentence of a convicted offender to supervision in the community by an agent of the court called a probation officer; this sentence is in lieu of incarceration as long as the offender obeys all terms of the probation. These conditions vary from one case to another but typically include reporting to the probation officer regularly, abstaining from alcohol and other drugs, not engaging in criminal behavior, being home by a certain time at night, looking for work, and showing up for work if the offender obtains a job. If an offender violates any of these conditions, he or she risks being incarcerated. **Parole** is often confused with probation, because it also occurs in the community, but it refers to the supervision of an offender who is released from prison after serving a term of incarceration but before his or her full sentence expires. The person who supervises an offender on parole is the parole officer. The conditions of parole are very similar to those for probation, and an offender who violates any condition also risks being returned to prison.

The public hears little about probation or parole unless a probationer or parolee commits a new, serious crime. Yet millions of people are in one of these statuses at any one time. In any one year, more than 4.2 million individuals are on probation, and about 800,000 are on parole. These numbers, when added to the 2.2 million prison and jail inmates, yield a total of more than 7 million individuals under some form of correctional supervision, a figure that amounts to more than 3 percent of all American adults. More than 1 million of all probationers and parolees live in Texas and California.

How Probation Works

Although probation is one type of community corrections, probation officers actually are involved in all stages of the criminal justice system. They first gather information to help judges determine whether defendants should be released on their own recognizance or on bail. If the defendant is convicted, a probation officer gathers more information and files a presentence report with the court. The judge's sentence normally follows the sentence recommendation contained in the probation officer's report. For every 100 defendants convicted of a felony, about 28 percent are sentenced to straight probation (with no jail or prison time to serve), and 70 percent are sentenced to prison or jail.

What determines whether someone is sentenced to probation instead of incarceration? The seriousness of the offense and the defendant's prior record play the greatest role, but personal factors like race play more subtle, complex roles. One of the best probation studies analyzed the cases of 16,500 California males convicted of various felonies. The researchers found that defendants were more likely to receive probation instead of imprisonment if, among other factors, they had fewer than two prior convictions, were not using drugs, did not use a weapon to commit their crime, and did not injure their victim seriously. They were also more likely to receive probation if they had a private attorney instead of a public defender and if they had been out on bail before their trial. All these factors accounted for the sentencing of 75 percent of the cases in the study; this means that no identifiable factors could explain why 25 percent of the defendants got either probation or imprisonment. This finding led the authors to conclude that many offenders who get probation are no different from those who are incarcerated.

When judges sentence defendants to probation, they list the conditions the probationers must follow; these conditions form a contract between the court and the offender. Usually, these conditions have been recommended by the probation officer in the presentence report, but judges also sometimes come up with their own. Three types of conditions exist. Standard conditions, as the name implies, are given to virtually all probationers and involve such requirements as looking for work, reporting changes of address, and not associating with known criminals. Punitive conditions are given to the more serious offenders and include the intermediate sanctions we discuss below. Treatment conditions are sometimes required for offenders with various personal problems such as drug or alcohol abuse. Court rulings have established that all these conditions must be reasonable and constitutional; for example, they cannot take away a probationer's freedom of speech.

If a probationer violates the terms of probation, several outcomes are possible. First, the judge may simply continue the probationary sentence without any changes in the conditions. Second, the judge may extend the length of the sentence and/or impose new, more restrictive conditions. Third, the judge may revoke probation and order the probationer to prison or jail.

Certain personal characteristics predict who is most likely to complete probation successfully. In general, successful completers include those who are married and who have children, who are employed, who are more educated, and who have fewer prior arrests and convictions. Probationers with fewer or no substance abuse problems also are more likely to complete probation successfully.

How Effective is Probation?

Probation obviously allows an offender to be in the community but costs much less than imprisonment would have cost. Whether probation is effective depends on what criteria we have in mind. The two most common criteria in measuring effectiveness are cost and recidivism. When it comes to dollars, probation is very effective, with probation estimated to cost $2000 per year for each probationer, a figure much lower than the $20,000 to $25,000 it costs to imprison one person. If we venture to say that probation saves at least $18,000 per year per person, and then assume for the sake of argument that half of all probationers, or about 2 million people, would need to be added to our nation's prisons if probation did not exist, imprisonment costs would rise by about $40 billion annually. This figure does not even include the extra 2000 prisons that would have to be built (assuming 1000 persons per prison) at a cost of more than $200 billion. Probation obviously saves a lot of money.

But does probation keep society safe? When it comes to recidivism, whether probation is effective depends on how we view the data. On the one hand, about 60 percent of probationers end their terms successfully. But about 15 percent are incarcerated after committing a new offense or violating the terms of their probation. These new offenses amount to thousands of new crimes. Close to 30 percent of prisoners overall were on probation at the time they committed the crime that put them behind bars. This amounts to several hundred thousand crimes. If all the felons who fail probation had been put in prison originally instead of on probation, their many crimes would not have occurred. However, it would have been impossible to predict at the time of sentencing which probationers would fail. Thus the only way to prevent new crimes during probation would be to end probation altogether. As we have seen, that would be prohibitively expensive.

Although this evidence indicates that probation does not always keep society safe, a more relevant question concerns the effectiveness of putting offenders on probation compared to the effectiveness of putting them in prison. Research on this issue finds that the recidivism rates of felons sentenced to probation are very similar to those of matched groups of felons sentenced to prison. Put another way, the recidivism of offenders put on probation is not higher than that of similar offenders put in prison, but neither is it lower. Although probation is no more or less effective than prison, it saves billions of dollars compared to imprisonment. Probation may be more effective if probation officers had smaller caseloads, which are five to eight times larger than optimal, and if many more programs and services were made available for probationers.

This last point suggests that more money should be put into probation. The needed dollars could come from a decreased emphasis on imprisonment, say for the almost half of all inmates now in prison for property and drug offenses not involving violence who could instead be put on probation. This increased funding for probation would have two important benefits. First, it would save money, because the increased cost of probation would still be much less than the cost savings achieved by the reduced use of incarceration. Second, it would make society safer because it would permit hiring many more probation officers, reduce caseloads and thus increase supervision, and pay for more numerous and effective treatment programs and other services. Where it has been tried, the combination of increased supervision and greater treatment and programming achieves significant reductions in recidivism (see the following text).

How Parole Works

As noted earlier, parole involves supervising offenders released to the community after serving a prison sentence. Two types of parole exist. Discretionary parole occurs when parole boards grant early release to inmates serving indeterminate (e.g. five to ten years) prison sentences. Mandatory parole occurs without a parole board decision and occurs because of good-time provisions under determinate sentencing (e.g., exactly five years) statutes. Inmates receiving either type of parole typically leave prison before their full sentences expire. Regardless of the reasons for their release, they report to a parole officer and must heed various restrictions on their behavior. They also take part in one or more intermediate sanctions.

In many respects, parole is very similar to probation. As noted earlier, the key difference is this: probation is used in lieu of imprisonment, while parole is used after imprisonment. Like probationers, parolees can be sent back to prison if they commit either new crimes or technical violations of the conditions of their supervision.

Discretionary parole used to be very common, but it came under attack during the 1970s. Some believed that dangerous criminals returned to the community too soon, while others claimed the process was arbitrary and racially discriminatory. In response, several states, led by Maine in 1976, abolished discretionary parole; 16 states have abolished it as of this writing, including California, Illinois, Virginia, and Washington. These states use determinate sentencing and automatic release from prison.

Ironically, the abolition of discretionary parole for good behavior has made it even more difficult for inmates to succeed in the outside world after release from prison, as they have less incentive to take part in rehabilitation programs while still behind bars. This abolition has also probably increased misbehavior in prison because inmates no longer have incentive to behave well. Another problem concerns the length of prison terms in the states that abolished discretionary parole. One reason for its abolition was the belief that discretionary parole allowed dangerous criminals to be let out too early.

Ironically, however, inmates released in states that have abolished discretionary parole serve terms seven months shorter than those released in states that still have it. The reason for this surprising outcome stems from the discretion of parole boards. Although they have been criticized for letting dangerous inmates out early, they, in fact, can identify these inmates and have them stay behind bars longer. States that have moved away from indeterminate sentences and discretionary parole to determinate sentences and automatic release, in effect, instituted automatic parole for most inmates, including ones who are still dangerous. The abolition of discretionary parole has thus had the opposite effect that its critics intended.

How Effective is Parole?

In July 2007, two convicted burglars who had been released from prison on parole allegedly murdered three members of a wealthy Connecticut family after breaking into their home. Because the suspects were on parole, their crime led Connecticut officials to examine their parole procedures and prompted criticism of the concept of parole.

When parolees do commit new, serious crimes, the public is justifiably outraged. Because many parolees are serious offenders who have been released from prison, their recidivism rates are of special importance in judging the effectiveness of parole. These rates are apt to be high for two reasons. First, as the last chapter noted, ex-inmates have many of the same personal problems, and, in some cases, even worse problems as when they entered prison. These include drug and alcohol abuse, functional illiteracy and a lack of employable skills, and mental and emotional disorders. They also have a prison record that may limit their opportunities for gainful employment and for friendships with law abiding citizens.

Parolees' personal histories and difficulties explain much of their high recidivism rates. But some of the blame also must go to the lack of services for them. As states have spent billions of dollars to build new prisons, they have cut back on services for parolees. A 1999 study found that California had only 200 shelter beds for more than 10,000 homeless parolees, only four mental health clinics for 18,000 parolees with mental disorders, and only 750 spaces in residential treatment programs for 85,000 parolees with drug and alcohol problems.

For all these reasons, it would be surprising if parolees did not have high recidivism rates. Unfortunately, such high rates do exist. About 60 percent of parolees are rearrested for a felony or serious misdemeanor within three years of their release from prison, and just over 40 percent are sent back to prison for these offenses. About one-third of all new prison admissions are these parolees. The factors predicting their recidivism are similar to those for probationers and include a longer prior criminal record, unemployment, criminal friends, and drug use.

Observers continue to criticize parole for multiple reasons. Some criticize it for allowing dangerous offenders to leave prison early and thus endanger public safety, while others criticize it for being an arbitrary process that also lacks adequate rehabilitation services. Our discussion of intermediate sanctions below suggests it is possible to have ex-prisoners in the community on parole without endangering public safety unduly so long as they are in programs combining adequate supervision with adequate treatment. A lack of interest in providing appropriate funding has hampered the establishment and development of such programs. Yet a decreased reliance on imprisonment for many types of offenders would free up the funds needed for the types of community corrections for parolees that, like those for probationers, show strong promise of satisfying public concerns for both punishment and safety. Many scholars think that discretionary parole should be restored in the states that abolished it. It could be made less arbitrary and discriminatory with appropriate standards and due process procedures. The restoration of discretionary parole could also help ensure that dangerous offenders stay longer in prison.

Intermediate Sanctions

A promising crime reduction strategy involves the use of intermediate sanctions. As their name implies, intermediate sanctions occur in between probation and incarceration. They are meant to be more punitive than probation but less punitive than incarceration. Today about 10 percent of adult probationers and parolees participate in them. Common forms of intermediate sanctions include: (a) house arrest and electronic monitoring; (b) boot camps involving young offenders in a paramilitary setting; (c) intensive supervision, involving surveillance of offenders that is more intense than routine probation in frequency and behavioral restrictions; (d) day-reporting centers where offenders attend required treatment and other programs during the daytime and then leave to go home for the night; (e) community service; and (f) fines and restitution.

These sanctions all cost much less than imprisonment, so the key question regarding their effectiveness is whether the recidivism rates of the offenders experiencing them are higher than the rates of imprisoned offenders. Although we do not have space to discuss the research on this issue, intermediate sanctions as a whole do not produce lower recidivism rates than incarceration, but neither do they produce higher recidivism rates. In fact, intensive supervision combined with mandatory drug treatment does appear to produce lower recidivism rates than imprisonment for comparable offenders. This body of evidence is important in assessing the effectiveness of intermediate sanctions, as it suggests that these sanctions keep society as safe (or no more unsafe) as imprisonment while saving billions of dollars. The greater use of probation and intermediate sanctions for the many people now in prison who do not pose a threat of violence to the public would save billions of dollars while not making our society any more unsafe.

JUVENILE JUSTICE: YOUTHS AT RISK?

Long ago our nation recognized the need to treat juvenile offenders differently from adult offenders. Officials thought juveniles were too young to be fully responsible for their actions, and they would respond to proper treatment and services. As youth crime began to rise a few decades ago, a crackdown on juvenile offenders began, with many more sentenced to juvenile detention centers or tried in (adult) criminal court. We discuss juvenile delinquency and the juvenile justice system here and address the contemporary treatment of juvenile offenders.

Juveniles and Delinquency

About 12 percent of the U.S. population are teenagers, and they account for about 17 percent of all arrests, including 16 percent of all violent crime arrests and 32 percent of all property crime arrests. The juveniles who commit the most serious offenses are called serious and violent juvenile (SVJ) offenders. The roots of their behavior are generally identified in the explanations discussed in Chapter 8, but scholarly research has pinpointed more specific factors underlying their behavior. In general, the youths who become SVJ offenders exhibited relatively high levels of aggression and impulsiveness during early childhood and tend to begin committing their serious offenses early in their teenage years rather than later. They are overwhelmingly male and generally grew up in poverty and in dilapidated urban neighborhoods. They are also disproportionately likely to have been abused or neglected by their parents during their childhoods and to have grown up in families beset by marital conflict. They are also more likely to have seriously delinquent friends and to use illegal drugs.

Overall, then, and without trying to defend or excuse their actions, it is fair to say that SVJ offenders typically come from very troubled backgrounds. This fact forces us to ask whether these youths are fully responsible for their crimes. It also suggests that, to reduce SVJ offending, our society must adequately treat the problems these backgrounds cause for SVJ offenders and also alleviate the problems that lead to such backgrounds in the first place. We return to this issue below when we discuss ways of reducing delinquency.

Trends in Juvenile Crime

During the 1990s, highly publicized shootings at Columbine High School and other schools combined with other examples of youth violence to help fuel a myth regarding youth crime. This myth is that youth crime rose dramatically throughout the 1990s and into this decade and that adolescents are becoming more violent all the time. Alarmist declarations by some social scientists in the mid-1990s that a new generation of superpredators was threatening America reinforced this myth. Media coverage also fueled the myth of unprecedented teenage violence. One study found that 40 percent of newspaper stories featuring youths and 48 percent of the stories on television network news shows focus on violence and other illegal behavior. Another study of local television news stories found that 68 percent of the stories about violence concern youth violence, and 55 percent of the stories of youths concern their violence.

Despite the impression these warnings and media coverage create, the reality is quite different. Although youth violence in the United States did rise in the late 1980s and early 1990s, it has actually been declining since 1994. And although up to 55 percent of news stories on youths focus on their violence, teenagers commit only about 16 percent of all violent crime, and less than one-half of 1 percent of all adolescents are arrested each year for a violent crime.

The myth of rising juvenile violence spills over into school shootings. Many of these shootings receive wide publicity, fueling concern over school safety and fear that schools are becoming more dangerous. Once again, however, the reality is quite different. The number of school homicides and other acts of serious violence actually declined after the early 1990s, as did the amount of violence reported in surveys of secondary school students and high school principals. Given that about 50 million children attend school and that about 50 homicides of children occur each year at school, the odds of a child being killed at school are about one in a million. Less than one percent of all murdered children are killed at school. When it comes to serious violence, children are actually much safer at school than elsewhere.

One consequence of the myth of rising youth violence is that it has prompted more and more youth crimes to be removed from the juvenile justice system to the adult criminal justice system. Has this shift in the treatment of youth crime helped or hurt public safety? What does it mean for the future of the juvenile justice system? We try to answer these questions below after looking at the origins of the juvenile justice system to see what lessons the past might hold for the present and the future. Before doing so, we take a brief look at juvenile gangs. Even if it is true that juvenile crime is down from a decade ago, gangs and gang violence remain a serious problem in many of our cities.

Juvenile Gangs

Juvenile gangs (or youth gangs) have existed in the United States for over 100 years. While there is no commonly shared definition of a youth gang, it is typically thought to have the following characteristics:

- three or more individuals, typically between the ages of 12 and 24
- a name and sense of identity usually expressed by clothing and manner of dress, the use of graffiti, and/or the use of hand signs
- some degree of permanence and organizational structure
- an elevated level of involvement in delinquent and criminal activity

According to the 2002–2003 National Youth Crime Gang Survey of police departments nationwide, gang activity is especially prevalent in cities with more than 100,000 residents. Between 1996 and 2003, over 90 percent of large cities reported gang activity in each of those years.

Nearly half of all gang members in the United States are thought to be Hispanics, many of whom were born in Guatemala, Honduras, El Salvador, or Mexico. Approximately 30 percent of the nation's gang population is African American, with the remainder comprised of white and Asian youths. Thousands of gang members born outside the United States are here illegally and have been deported numerous times.

The estimates that law enforcement officials make about the number of gangs and gang members in their jurisdictions may not be accurate. Because the money communities receive from federally funded law enforcement programs may depend on the amount of crime and delinquency in the communities, police have an incentive to exaggerate the number of gangs. On the other hand, fearing the loss of tourist dollars and investment money, local officials may pressure police to publicly underestimate the gang activity in a given community. The fact of the matter is that we do not have an accurate count of the number of gangs and gang members in this country.

Most criminologists and law personnel agree that gang activity increased dramatically during the 1980s and early 1990s. There are at least three explanations for this rise: 1) the interrelated factors of deindustrialization and the loss of jobs in cities combined with poverty and racism that increased the size of the underclass in urban America; 2) the spread of the gang culture across the country, especially via the media; and 3) the explosion of crack cocaine that provided lucrative drug-trade opportunities.

The mechanism by which youths become gang members has been as hotly disputed as the issues of size, growth, number of gangs and overall membership. Although the public and law enforcement officials typically believe that teens are recruited into gangs by existing members, research on this issue yields a different picture: Youths join gangs to be around family members—siblings and cousins in particular—who are already gang members and for the protection that gang membership affords.

Gang members in large metropolitan areas are often responsible for a significant amount of crime committed by adolescents in those locales. According to self-report studies in cities such as Rochester, NY and Seattle, WA, gang members commit between 65 percent and 85 percent of all youth violence. The police chief of Los Angeles has stated, "There is nothing more insidious than these gangs. They are worse than the Mafia. Show me a year in New York where the Mafia indiscriminately killed 300 people. You can't."

Gang scholars have been devoting more attention to girls' involvement in youth gangs. Three types of situations exist: 1) gangs made up exclusively of females that exist independently of males, 2) females as subordinates in male-dominated gangs, and 3) coed gangs comprising males and females with the former outnumbering the latter. According to one national report, 42 percent of all gang-problem jurisdictions stated that the majority of these organizations had female members. As many as one out of three gang members in early adolescence is female. However, the proportion of female to male gang members diminishes as members age, because females leave gangs at an earlier age than males.

Juvenile Justice Then and Now

A separate juvenile justice system is actually a relatively recent invention, dating only to the late 19th century. In colonial America, the task of preventing and handling children's misbehavior was left up to their parents. If parents were thought incapable of controlling their children's behavior, authorities would intervene by administering corporal punishment, by taking children away and having them raised by someone else, or, in extreme cases, by expelling an adolescent from the community.

During the 19th century, the growth of industry and of cities produced increased poverty and more juvenile crime. In many U.S. cities, mobs of teenagers would roam the streets and attack and rob people. Many were arrested and imprisoned with adult criminals, who taught them even more criminal ways and sometimes sexually abused them. Social reformers emphasized that these and other juvenile offenders children were not responsible for their misbehavior and instead were victims of problems such as poverty, neglectful or abusive parents, or substandard living conditions in their neighborhoods. They also said that that juvenile offenders should be kept apart from adult offenders and, more generally, from the adult criminal justice system, because the youths needed help and not just punishment. By the 1870s this understanding of juveniles and juvenile crime led several states to separate juvenile offenders from adult criminals at arraignments, trials, and other legal proceedings. The nation's first juvenile court was established in Chicago in 1899, and almost all states had juvenile courts by the early 1920s.

The goal of these courts was to help and rehabilitate youthful offenders, not punish them. Courts were meant to provide children the equivalent of parental care and treat their problems individually. The procedures they followed were more personalized and informal than those in adult courts in order to project a caring atmosphere and to keep juveniles from being stigmatized as criminals. Written transcripts were not kept, lawyers often were not allowed, and witnesses could not be confronted. Juveniles were also kept out of jail and prison whenever possible, and they were especially kept from contact with adult offenders. Instead of being punished, most juveniles were put on probation and ordered to take part in various kinds of treatment programs.

This philosophy began to change during the 1970s and 1980s, as concern grew over a supposed increase in juvenile crime. The more general get-tough approach to crime control involved calls for the harsher treatment of juvenile offenders. States passed laws requiring the cases of juveniles accused of certain serious offenses to be handled in adult criminal court. Other laws gave prosecutors the discretion to transfer these cases to criminal court without first needing the permission of a juvenile court judge. Still other laws provided mandatory penalties for serious juvenile offenses.

How the Juvenile Justice System Works

The stages of the juvenile justice system generally parallel those in the adult criminal justice system. However, juvenile justice procedures typically are more informal than the corresponding procedures in the adult system, and they include slightly different terminology.

When a juvenile allegedly breaks the law, the police exercise their discretion in deciding whether to arrest the offender or to take some other action, such as talking with his or her parents. If an arrest does occur, the police then decide, often in consultation with juvenile justice officials and the offender's parents, whether to divert the case from juvenile court or to refer it to the court. The police divert about one-fourth of all arrests and refer about 70 percent to juvenile court. Juveniles awaiting court disposition must not be held with adult offenders. Instead, they simply live at home or, occasionally, are sent to juvenile detention facilities.

The intake process begins when a case is referred to juvenile court. Similar to the prosecution stage for adult offenders, the juvenile probation department or the prosecutor's office generally handles the intake process. Its main purpose is to determine whether to dismiss the case, to divert it from the court by handling it informally (perhaps also requiring certain conditions such as restitution, curfews, and treatment programs), or to petition it to juvenile court. About half of all cases making it to the intake stage are handled informally. When this is the outcome, juveniles and their parents typically sign a consent decree in which the juvenile admits to having committed the act. A juvenile probation officer usually monitors the youth's compliance with any conditions of this informal probation. As with adult offenders, juveniles who violate their probation conditions risk having their case go back to juvenile court with possible incarceration in a juvenile detention facility the result.

If, instead, a case is petitioned into juvenile court, the juvenile court judge's task then becomes whether to transfer the case to adult court, if so requested by intake officers, or to adjudicate the youth as delinquent if no transfer request is made. Almost all petitioned cases result in adjudicatory hearings, the equivalent of a trial in the adult criminal justice system. At these hearings, the juvenile court judge hears testimony from various witnesses and decides whether the youth was delinquent.

If a youth is adjudicated as delinquent, the judge's next decision is the disposition, or sentencing, of the case. Before the disposition hearing, juvenile probation officers prepare a set of recommendations for the judge after interviewing the youth, speaking with his or her parents, and, perhaps, having psychological evaluations performed. The judge's disposition can include formal probation, with conditions such as restitution and drug treatment attached, or placement in a juvenile detention facility. About half of all dispositions result in formal probation, and about 30 percent result in residential placement. The juvenile's offense and prior record help determine whether placement is in a secure facility resembling a prison or in a more open facility resembling cottages or homes.

As we have just seen, a series of decisions accompanies every stage of the juvenile justice process: Should a police officer arrest a juvenile? If an arrest does occur, should the case be diverted or referred to juvenile court? If referred, should the case be handled informally or petitioned to juvenile court? If petitioned, should it be transferred to criminal court? If it stays in juvenile court, should the juvenile be adjudicated as a delinquent? If such adjudication occurs, what should the disposition be?

What factors affect all these decisions? The most important factors are the seriousness of the offense and the prior arrest record of the juvenile. Youths accused of serious offenses and those with previous arrests are more likely at all decision points of the juvenile justice process to receive the more punitive outcome—to be arrested, to be petitioned to juvenile court, and so forth. Some evidence also suggests that girls receive more lenient treatment than boys for criminal offenses but harsher treatment for status offenses like running away from home and promiscuity.

Race and ethnicity also seem to make a difference, as African American youths and other minorities are greatly overrepresented in the juvenile justice system. For example, although 15 percent of all youths are African Americans, African Americans are about 26 percent of all the youths who are arrested, 31 percent of those referred to juvenile court, 46 percent of those whose cases are waived to adult criminal court, 40 percent of those placed in juvenile detention centers, and 58 percent of those placed in adult prisons. A 2000 report found that African American youths charged with drug offenses are 48 times more likely than their white counterparts to be sentenced to juvenile prison. Among youths accused of violent crime, blacks spend an average of 254 days in prison after trial and Hispanics 305 days, compared to only 193 days for whites.

Do these disparities reflect racial discrimination, or do they just reflect the greater likelihood of African American youths to be involved in serious crime? Many studies addressed this issue, and they generally find that race, ethnicity, and poverty do affect juvenile case processing even after differences in criminal involvement are considered. A study in California found that minority youths were twice as likely as white youths accused of the same offenses to have their cases transferred to criminal court.

How Should We Treat Juvenile Offenders?

We now treat juvenile offenders more harshly than in the past, and in particular, we transfer many cases of serious juvenile offending to adult criminal court. Underlying this get-tough approach is the belief that harsher punishment will reduce delinquency by deterring potential juvenile offenders and by reducing recidivism by the juveniles tried in adult court.

Has this approach reduced both kinds of delinquency? The research evidence suggests it has not. Studies that examine the impact in several states of new legislation that required or eased the transfer of serious juvenile offenses to criminal court fail to find that delinquency decreased after the legislation was implemented. Studies that compare the recidivism of juvenile offenders tried in adult criminal court with matched samples of offenders processed in juvenile court likewise fail to find that the juveniles transferred to adult court have lower recidivism rates; some studies show that juveniles transferred to adult court actually have higher recidivism rates. Sending juvenile cases to adult court seems to endanger public safety in the long run, not protect it. The harsher treatment of juvenile offenders thus seems to be making our society less safe, not more safe.

If transferring juvenile offenses to adult criminal court does more harm than good, how else should the United States deal with these offenses? Some observers suggest that all offenders, juveniles and adults, be tried in an integrated criminal justice system, with youth discounts given at sentencing to adolescents in recognition of their diminished responsibility. Under this plan, a 14-year-old may receive, say, 25 percent of the adult penalty, and a 16-year-old, 50 percent. Youthful offenders would still be kept separate from adults at all stages of the proceedings and would have the same legal rights that adults have. Other observers object that this plan would essentially abolish the juvenile court. They fear that the more formal legal proceedings that juveniles would face would stigmatize them and produce harsher punishments.

In a different strategy, some scholars urge that new or better paid and designed forms of intermediate sanctions be used for serious juvenile offenders. We saw earlier that intermediate sanctions that combine surveillance with drug counseling, vocational training, and other programs offer some hope in reducing recidivism among adults. The same seems true for juvenile offenders. For these sanctions to work, they must be better funded so that important kinds of treatment, counseling, and training are readily available.

The most promising strategy, however, is one aimed at preventing serious juvenile offending before it begins. This strategy involves early childhood and adolescent intervention programs aimed at the children and youths most at risk for such offending. For example, some programs involve prenatal and postnatal visits by nurses, social workers, and other professionals to the homes of poor, teenaged pregnant women. Their offspring are more likely than those born in more advantaged circumstances to become delinquent. These visits help improve the physical health of the mothers and their infants and the latter's cognitive development. Studies show they more than pay for themselves by helping to prevent childhood cognitive and behavioral problems, child abuse and neglect, and later delinquency. School-based programs focusing on conflict resolution and after-school recreation also have been shown to reduce serious juvenile offending. Again, greater funding of these programs is necessary if they are to succeed in preventing such offending to any great degree, but any dollars spent on these programs save many more dollars in reduced crime, prison costs, and other problems.

KEY TERMS

intermediate sanctions: Social controls that are more punitive than routine probation but less punitive than incarceration

parole: Supervision of an offender released from prison after serving a term of incarceration

probation: A substitute for incarceration that involves placing a convicted offender in the community under the supervision of a probation officer as long as the offender meets certain standards of behavior

SUGGESTED READINGS

Greenwood, Peter W. 2006. *Changing Lives: Delinquency Prevention as Crime-Control Policy.* Chicago: University of Chicago Press.

Lundman, Richard J. 2001. *Prevention and Control of Juvenile Delinquency.* New York: Oxford University Press.

MacKenzie, Doris L. 2006. *What Works in Corrections: Reducing the Criminal Activities of Offenders and Delinquents.* New York: Cambridge University Press.

Petersilia, Joan. 2003. *When Prisoners Come Home: Parole and Prisoner Reentry.* New York: Oxford University Press.

Welsh, Brandon C., and David P. Farrington (eds.). 2007. *Preventing Crime: What Works for Children, Offenders, Victims and Places.* New York: Springer.

"Speed Bump" © Dave Coverly/Dist. By Creators Syndicate, Inc.

chapter fourteen

Conclusion: What Every American Should Know

In this last chapter we take a look back at the major points we have made about crime and justice in the United States in order to draw some conclusions and lessons that "every American should know."

WHAT DO WE KNOW ABOUT CRIME?

We know many things about crime, thanks to the scholarly research on this subject that we have discussed in earlier chapters.

First, we know that crime is a significant problem in the United States and has been for many decades. Our nation has always been concerned about crime and no doubt will continue to be concerned beyond our lifetimes.

Second, we know that crime is rooted in the social environment. Yes, people choose to commit crime, but their social backgrounds heavily influence their choices. People of color commit crime at higher rates than whites not because of anything intrinsically wrong with them but because they are more likely than whites to live in poverty and to live in high-crime areas where the chances of anyone committing crime are relatively higher. Here the familiar saying "there but for the grace of God go I" applies. If any of us would have been born into the social backgrounds and circumstances that help generate higher levels of criminal behavior, we would have been at greater risk for committing crime ourselves. This does not mean we should excuse anyone's criminal behavior and let it go unpunished, but it does mean that our society must recognize that crime is not just an individual problem and instead has social roots that must be addressed.

Third, we know that families matter. Harmonious families with strong bonds between parents and children produce less delinquency.

Fourth, we know that the friends and acquaintances for adolescents also matter tremendously for so many aspects of their behavior and attitudes. Their friends influence their taste in music and clothing, and they also influence their likelihood of breaking or not breaking the law. Parents cannot just lock their teens in their bedrooms, but responsible parents will still try to keep a watchful eye on whom their children associate with and where they go.

Fifth, we know that men commit much more crime than women for two reasons. One is socialization: in the United States as in most modern societies, men are socialized to be more aggressive than females. This socialization contributes to higher rates of violent crimes on the part of men. The second reason is opportunity: males traditionally have many more opportunities than females to commit crime of all types, including street crime, white-collar crime, and political crime. Even with the progress women have made since the 1960s, the upper ranks of both the political structure and corporate boardrooms are dominated by men. Just as men dominate these ranks, so do they dominate the ranks of corporate and political criminals.

Sixth, we know that the poor commit more street crime than the middle classes and the rich, while the rich obviously commit more corporate and other white-collar crime than the poor. Again this is a matter of opportunity. Poor people, especially underclass teenagers, are often motivated by the desire for material goods in a society that measures success largely in terms of possessions. Lacking the opportunity to commit occupational crime (embezzling , for example) they commit street crime. Yet the high amount of white-collar crime indicates that the poor do not have a monopoly on breaking the law for personal gain.

Seventh, we know that crime victims suffer in many ways: financially, psychologically, and behaviorally. Despite these problems, they traditionally have been neglected as the lawmakers and the criminal justice system have placed much more emphasis on capturing and punishing criminals than they have on helping crime victims restore their lives.

Only in the past 25 to 30 years has the plight of crime victims been addressed by society and then in only a token way, as the amount of money allocated each year to crime victims is dwarfed by the amount spent on crime control.

Eighth, we know that criminals will adapt technological innovations in order to victimize the public. For example, identity theft was a less serious problem before the widespread use of computers and access to the Web. Today, it is one of the fastest-growing areas of criminal activity, a particularly insidious crime that can wreak sometimes irrevocable havoc on an individual's life. Law enforcement plays a never-ending game of catch up as the success it realizes in any one area of technological crime prevention is eclipsed by criminals adapting new technological breakthroughs to their benefit.

Ninth, and perhaps most important, we know that the get-tough approach to crime control has cost the United States hundreds of billions of dollars over the past three decades while making little or no difference in the crime rate. More than 650,000 prisoners are now released every year after experiencing the pains of imprisonment with little drug, alcohol, or vocational services. Back in their communities, they face strong prospects of unemployment and the great temptation of drug and alcohol abuse and crime. If prisoner reentry fails for these reasons, as it so often does, these ex-prisoners will commit new crimes and make our society less safe.

Tenth, we know that the war on crime and especially the war on drugs have affected African Americans and Hispanics disproportionately. About one-third of young African American men are now under correctional supervision—in prison or jail or on parole or probation. The racial bias that continues to exist in our criminal justice system should disturb all Americans who believe in the ideals of freedom and equality on which this nation was founded.

Eleventh, we know that an ounce of prevention is worth a pound of cure. Crime prevention efforts like early childhood intervention programs with children at risk for crime, such as those born to teenaged mothers, have been shown to reduce the prospects for crime down the line, and to do so much more cost effectively than incarceration.

Finally, we know that the financial impact of white-collar crime, and especially corporate crime, is significantly greater than the monetary losses of street crime. We also know, contrary to public perceptions, that corporate crime can be deadly. As globalization increases in the 21st century, the opportunities for corporate criminals to victimize the American public as well as people in foreign countries will also increase. There is no reason to believe that corporate criminals will ignore these opportunities.

WHAT SHOULD WE DO ABOUT CRIME?

Here's what we should *not* do: exactly what we have been doing since the 1970s. By this, of course, we mean the get-tough approach. It has cost too much, both financially and socially, with too little payoff in terms of crime reduction. There must be a better way.

What is this better way? The public health approach, outlined at the beginning of the book, points to prevention as the better strategy for achieving true crime reduction at significantly lower cost. Such a prevention strategy should involve the following measures and policies at a minimum:

1. A government-funded employment policy to create jobs with adequate wages for the poor and near-poor;
2. Efforts that address the many problems confronting urban neighborhoods, including overcrowding and dilapidation;
3. An expansion of the early childhood intervention programs for at-risk children to reduce their likelihood of antisocial behavior during childhood and delinquency during adolescence; and
4. A massive effort to improve the nation's schools, many of which are overcrowded and dilapidated and lack adequate books, equipment, and facilities.

HOW CAN WE HELP CRIME VICTIMS?

Our nation now recognizes the plight and problems of crime victims much more than just a few decades ago, but it needs to strengthen its effort to help the many victims of violent and property crime. The following measures and programs should be undertaken at a minimum:

1. Expanded government-funded restitution to victims of violent crime and possibly to victims of property crime;
2. Expanded, free or low-cost psychological counseling and other services for crime victims who need such help, including long-term counseling for victims of rape and other violent crimes who would benefit from this service;

3. Expanded victim services in the criminal courts to help guide victims through their involvement in criminal prosecutions; and

4. Counseling for survivors of murder victims.

HOW CAN WE IMPROVE THE JUSTICE SYSTEM?

We have been critical at several places in this book of our criminal justice system, and rightly so. It is incredibly expensive, it does not reduce crime in a cost-effective manner, it does not afford defendants adequate due process, and it houses inmates in decrepit and dangerous conditions. These considerations suggest that several measures are necessary to improve the criminal justice system. At a minimum they include:

1. A greater use of community corrections in lieu of imprisonment, following the example of many western European nations. These nations have lower rates of serious violent crime than does the United States. We should be more receptive to criminal justice system innovations from these countries and experiment with new crime reduction strategies when and where feasible. Many of these strategies may be culture specific, and not applicable in this country. However, those techniques that do travel well should be adopted in the United States. Community corrections is one such strategy. Studies in the United States and elsewhere show that this strategy keeps society at least as safe as incarceration at significantly lower cost. About half of state prisoners, or some 600,000 individuals, are serving time for nonviolent offenses: property, drug, and public order (e.g., prostitution). Many and perhaps most of these inmates would be reasonable candidates for community corrections, saving the nation billions of dollars at no greater risk to the public safety;

2. Because the prison and jail experience affects recidivism rates, correctional institutions should be improved to reduce and preferably eliminate overcrowding and decrepit living conditions;

3. Well-funded, effective drug and alcohol counseling and vocational training should be established in every prison;

4. Because the death penalty does not deter homicide, costs much more money than life imprisonment, involves the conviction and possibly execution of individuals who are probably innocent, and is racially discriminatory, the death penalty should be abolished;

5. Where necessary, increase funding for training small town and rural-area police officers so that their skill level is commensurate to that of urban police officers;

6. As the nation becomes ever more ethnically and racially diverse, it will be important to recruit police officers, prosecutors, probation and parole officers from these groups;

7. Research indicates that a disturbing number of factually innocent people are convicted of crimes, including serious offenses such as rape and murder. Both state and federal governments should provide additional funding for additional public defenders so that everyone charged with a crime has adequate legal representation. In addition, money should be provided for DNA testing where needed as well as funds to process any backlog of rape kits. While forensic crime scene investigation is no panacea for solving criminal cases, it is unjust and disgraceful that innocent people should be convicted for a lack of crime analysis equipment in the world's wealthiest nation.

KEEPING YOUR FAMILY SAFE

Much of what happens to us in life is a matter of good luck or bad luck, and crime victimization is no different. Just as a car or other motor vehicle may go out of control and strike us at any time no matter how carefully we are driving, riding a bike, or walking, so can criminals strike at any moment. If we are in the wrong place at the wrong time, we will become a crime victim, or perhaps our household will be victimized. We do not have any special wisdom when it comes to keeping your family safe, but there are several measures and precautions:

1. Lock your house/apartment and motor vehicle when not at home or in the vehicle; if living in a high-crime neighborhood, a burglar alarm system or the presence of a dog may be advisable and may deter possible break-ins;

2. Limit the opportunity to become a victim of violent crime by limiting late-night visits to hot places for crime such as bars, taverns, and nightclubs;

3. Because a perhaps surprisingly high number of crime victims are victimized by people they know, caution should be exercised in selecting the people with whom we choose to spend time;

4. Young children should know whom to call in an emergency without a moment's hesitation; teach them to dial "911" if trouble arises;

5. Remain in sight and voice distance of young children in public places. Be wary of strangers who approach them. Teach older children to be watchful of younger siblings;

6. Teach children to be respectful of police officers and not to hesitate to approach law enforcement personnel if they are frightened, lost, or confused;

7. To protect against identity theft, be especially careful with personal information, especially social security numbers, credit card numbers, medical insurance information, bank account numbers, and anything that has to do with personal finances. Buy a paper shredder and use it as you open and discard mail and clean out old files. Make sure old computers are disposed of properly, that is, hard drives are wiped clean. To protect against hackers, take several steps: (a) update your operating system and software regularly as manufacturers attempt to stay one step ahead of hackers; (b) install, update, and run regularly antivirus, antispyware, and firewall software; sometimes these types of software can be purchased as one package; (c) be especially vigilant with your passwords and personal identification numbers, and never store them in your browser; (d) do not be a victim of a pharming attack, which occurs when you type a Web address in your browser and the address is hijacked and sent to the scam site.

KEEPING YOUR FAMILY OFF THE POLICE BLOTTER

This last section is meant for parents or people who expect to be parents. The mother of one of the authors used to say that she accepted the credit for anything good her son did but refused to accept the blame for anything bad he did. We can raise our children in the best way possible, and they will still fail us by committing crime and getting into other kinds of trouble; we can raise our children in the worst way possible, and they will still come through by not committing crime, doing well in school, and so forth. That said, parents can increase the odds that their children will turn out to be law-abiding citizens with the following practices and measures:

1. Mothers should have adequate diets during pregnancy and avoid smoking, drinking, or taking other drugs;

2. Parents should raise their children in a firm but fair way, with little or no corporal punishment;

3. Teenage mothers and other parents of at-risk children should seek whatever social services are available to ensure that they and their children enjoy optimal nutrition and parenting practices;

4. As noted above, parents of adolescents should keep a watchful eye on their children's friends and acquaintances and on their children's activities; reasonable curfews will help limit their teens' opportunities to get into various kinds of trouble;

5. Parents should raise their sons in a way that limits their sons' affinity for using violence and other coercive behavior to resolve conflicts and achieve other goals. In particular, they should raise their sons to be respectful of females. No physical violence of any type should ever be used against a girl or woman under any circumstances, and no man should ever force or coerce someone into sex;

6. Do your best to maintain a trusting relation with your children, especially during the teenage years. Remember that respect is a two-way street. Maintaining a mutually trusting relationship with your children can prevent a lot of problems, including those involving crime and the criminal justice system;

7. Remember we all make mistakes (think of some of the things you did as a teenager!). If necessary, consult a marriage and family therapist experienced in dealing with the problems your child is facing. A firm but fair response is often required to address rule- or law-violating behavior by a child. Understanding and forgiveness can also go a long way toward correcting behavioral problems.

Taken together, the measures, programs, policies, and recommendations outlined in the preceding sections certainly will not eliminate the crime problem altogether, but they still offer great promise for reducing crime and victimization and for keeping our families safe. Our nation has been following a sadly misguided get-tough effort to reduce crime since the 1970s. It is now time for a better strategy that promises to be much more effective in reducing crime at far lower cost. The various recommendations in this chapter point the way to such a strategy, one that follows the public health model that has helped so significantly to reduce the risk from so much disease and illness. Our nation would do well to follow this strategy to reduce crime in the years and decades ahead.

References

Chapter 1. What No One is Telling You About Crime and Justice

p. 3   *only about 10 percent of all street crimes . . . are serious violent crimes:* Rand, Michael, and Shannan Catalano. 2007. *Criminal Victimization, 2006.* Washington, DC: Bureau of Justice Statistics, U.S. Department of Justice.

p. 4   *only about 10 percent of street crimes result in an arrest:* Federal Bureau of Investigation. 2007. *Crime in the United States, 2006.* Washington, DC: Federal Bureau of Investigation.

p. 4   *only about 40 percent of all serious violent crimes result in an arrest:* Federal Bureau of Investigation, 2007.

p. 4   *study of several hundred college students:* Vandiver, Margaret, and David Giacopassi. 1997. "One Million and Counting: Students' Estimates of the Annual Number of Homicides Occurring in the U.S." *Journal of Criminal Justice Education* 8:135–143.

p. 4   *more than 90 percent of criminal cases are resolved by means of a plea bargain:* Regoli, Robert M., and John D. Hewitt. 2008. *Exploring Criminal Justice.* Sudbury, MA: Jones and Bartlett Publishers.

p. 5   *the expenditure of tens of billions of dollars:* Maguire, Kathleen, and Ann L. Pastore (eds.). 2007. *Sourcebook of Criminal Justice Statistics [Online].* Available at: http://www.albany.edu/sourcebook.

p. 5   *More than 2.2 million individuals are in jail or prison:* Sabol, William J., Todd D. Minton, and Paige M. Harrison. 2007. *Prison and Jail Inmates at Midyear 2006.* Washington, DC: Bureau of Justice Statistics, U.S. Department of Justice.

p. 5   *highest incarceration rate in the Western world:* The Sentencing Project. 2007. *Facts About Prison and Prisoners.* Washington, DC: The Sentencing Project.

p. 5   *have had to reduce higher education budgets:* Schiraldi, Vincent, and Rose Braz. 2003. "Crisis in California Education: The Choice Between Prisons and Schools." *San Francisco Chronicle* January 23: http://www.justicepolicy.org/article.php?id=62; Steptoe, Sonja. 2007; "California's Growing Prison Crisis." *Time* June 21: http://www.time.com/time/nation/article/0,8599,1635592,00.html; Sterngold, James, and Mark Martin. 2005. "Hard Time: California's Prisons in Crisis." *San Francisco Chronicle* July 3:A1.

Chapter 2. The Crime Problem

p. 9   *Americans suffer more than 20 million violent and property crimes:* Rand, Michael, and Shannan Catalano. 2007. *Criminal Victimization, 2006.* Washington, DC: Bureau of Justice Statistics, U.S. Department of Justice.

p. 9   Discussion of lifetime chances of criminal victimization: Koppel, Herbert. 1987. *Lifetime Likelihood of Victimization.* Washington, DC: U.S. Department of Justice, Bureau of Justice Statistics.

p. 9   Discussion of women becoming victims of domestic violence and of rape and sexual assault: Tjaden, Patricia, and Nancy Thoennes. 2000. *Full Report of the Prevalence, Incidence, and Consequences of Violence Against Women.* Washington, DC: National Institute of Justice and the Centers for Disease Control and Prevention.

p. 9   *The United States has the highest homicide rate:* Zimring, Franklin E., and Gordon Hawkins. 1997. *Crime is Not the Problem: Lethal Violence in America.* New York: Oxford University Press.

p. 9   *our nation spends about $200 billion:* Maguire, Kathleen, and Ann L. Pastore (eds.). 2007. *Sourcebook of Criminal Justice Statistics [Online].* Available at: http://www.albany.edu/sourcebook.

p. 10   *Americans have always considered crime a serious problem:* Barkan, Steven E. 2009. *Criminology: A Sociological Understanding.* Upper Saddle River, NJ: Prentice Hall; quote in presidential commission report: Pepinsky, Harold E., and Paul Jesilow. 1984. *Myths That Cause Crime.* Cabin John, MD: Seven Locks Press, p. 21.

p. 10   *According to the FBI:* Federal Bureau of Investigation. 2007. *Crime in the United States, 2006.* Washington, DC: Federal Bureau of Investigation.

p. 10   *top ten causes of death:* Mokdad, Ali H., James S. Marks, Donna F. Stroup, and Julie L. Gerberding. 2004. "Actual Causes of Death in the United States, 2000." *Journal of the American Medical Association* 291:1238–1245.

p. 10–11 Discussion of estimated financial cost and death toll of white-collar crime: Barkan 2009.

p. 11   Discussion of how the media overdramatize crime: Kappeler, Victor E., and Gary W. Potter. 2005. *The Mythology of Crime and Criminal Justice.* Prospect Heights,

IL: Waveland Press; Surette, Ray. 2007. *Media, Crime, and Criminal Justice: Images, Realities, and Policies.* Belmont, CA: Wadsworth Publishing Co.

p. 11 *homicides are more than 25 percent of the crime stories on the evening news:* Dorfman, Lori, and Vincent Schiraldi. 2001. *Off Balance: Youth, Race and Crime in the News.* Washington, DC: Building Blocks for Youth; Feld, Barry C. 2003. "The Politics of Race and Juvenile Justice: The 'Due Process Revolution' and the Conservative Reaction." *Justice Quarterly* 20:765–800.

p. 11 Discussion of TV news coverage during 1990s: Dorfman and Schiraldi, 2001; studies of Baltimore and Philadelphia residents: Bunch, William. 1999. "Survey: Crime Fear is Linked to TV News." *The Philadelphia Daily News* March 16:A1; Farkas, S., and A. Duffett. 1998. *Crime, Fears and Videotape: A Public Opinion Study of Baltimore-Area Residents.* Washington, DC: Public Agenda.

p. 12 Discussion of news stories depicting members of races as offenders or victims: Dorfman and Schiraldi, 2001; Sorenson, S.B., J.G. Manz, and R. Berk. 1998. "News Media Coverage and the Epidemiology of Homicide." *American Journal of Public Health* 88:1510–1514.

p. 12 *whites think they are more likely to be victimized by minorities than by whites:* Dorfman and Schiraldi, 2001.

p. 12 *Another provocative study focused on subjects who watched news stories:* Gilliam, F.D., and S. Iyengar. 2000. "Prime Suspects: The Influence of Local Television News on the Viewing Public." *American Journal of Political Science* 44:560–573.

p. 12 Discussion of coverage of youth crime: Dorfman and Schiraldi, 2001.

p. 13 Discussion of then-Vice President George Bush: Mendelberg, Tali. 1997. "Executing Hortons: Racial Crime in the 1988 Presidential Campaign." *Public Opinion Quarterly* 61:34–57; Isikoff, Michael. 1989. "Drug Buy Set Up for Bush Speech." p. A1 in *The Washington Post*; President Bill Clinton engaged in harsh crime rhetoric: Chambliss, William J. 1999. *Power, Politics, and Crime.* Boulder: Westview Press.

p. 14 Discussion of Americans' views of punishment and how to prevent crime and deal with criminals: Maguire and Pastore, 2007.

p. 14 Discussion of racial prejudice affecting whites' views: Barkan, Steven E., and Steven F. Cohn. 2005. "Why Whites Favor Spending More Money to Fight Crime: The Role of Racial Prejudice." *Social Problems* 52:300–314; Soss, Joe, Laura Langbein, and Alan R. Metelko. 2003. "Why Do White Americans Support the Death Penalty?" *The Journal of Politics* 65:397–421; Unnever, James D., and Francis T. Cullen. 2007. "Reassessing the Racial Divide in Support for Capital Punishment." *Journal of Research in Crime and Delinquency* 44:124–158.

p. 15 *"policymakers must be careful not . . . ":* Barkan and Cohn 2005, p. 312.

p. 15 Discussion of African Americans' and Hispanics' views about the police: Maguire and Pastore 2007; Weitzer, Ronald, and Steven A. Tuch. 2006. *Race and Policing in America: Conflict and Reform.* New York: Cambridge University Press.

p. 16 Discussion of incarceration data: The Sentencing Project. 2007. *Facts About Prison and Prisoners.* Washington, DC: The Sentencing Project; criminal justice expenditure data: Maguire and Pastore 2007.

p. 16 Discussion of small impact of get-tough approach on crime: Blumstein, Alfred, and Joel Wallman (eds.). 2006. *The Crime Drop in America.* Cambridge: Cambridge

University Press; discussion of greater crime reductions: Greenwood, Peter W. 2006. *Changing Lives: Delinquency Prevention as Crime-Control Policy.* Chicago: University of Chicago Press.

p. 17   Discussion of impact of get-tough approach on racial and ethnic minorities: The Sentencing Project 2007; Mauer, Marc. 2006. *Race to Incarcerate.* New York: New Press.

p. 17   Discussion of additional problems caused by get-tough approach: Mauer, Marc, and Meda Chesney-Lind (eds.). 2003. *Invisible Punishment: The Collateral Consequences of Mass Imprisonment.* New York: The New Press; Petersilia, Joan. 2003. *When Prisoners Come Home: Parole and Prisoner Reentry.* New York: Oxford University Press.

p. 18   Discussion of the funnel effect: Barkan 2009.

p. 18   Discussion of the wedding-cake model: Walker, Samuel. 2006. *Sense and Nonsense About Crime and Drugs: A Policy Guide.* Belmont, CA: Wadsworth Publishing Company.

p. 20   Discussion of the discretionary model: Hawkins, Keith (ed.). 1993. *The Uses of Discretion.* New York: Oxford University Press; Regoli, Robert M., and John D. Hewitt. 2008. *Exploring Criminal Justice.* Sudbury, MA: Jones and Bartlett Publishers.

p. 20–21   Discussion of the consensual model: Eisenstein, James, and Hebert Jacob. 1977. *Felony Justice: An Organizational Analysis of Criminal Courts.* Boston: Little, Brown and Company; Regoli and Hewitt 2008.

p. 21–22   Discussion of the crime control and due process models: Packer, Herbert L. 1993. "Two Models of the Criminal Process." pp. 14–27 in *Criminal Justice: Law and Politics,* George F. Cole (ed.). Belmont, CA: Wadsworth Publishing Company.

p. 22   Discussion of U.S. government actions after September 11: Cole, David, and Jules Lobel. 2007. *Less Safe, Less Free: Why America is Losing the War on Terror.* New York: New Press.

Chapter 3. How Much Crime Is There and Who Commits It?

p. 25   *The police hear about only 40 percent:* Rand, Michael, and Shannan Catalano. 2007. *Criminal Victimization, 2006.* Washington, DC: Bureau of Justice Statistics, U.S. Department of Justice.

p. 26   Washington Post article: Eggen, Dan. 2007. "Violent Crime, a Sticky Issue for White House, Shows Steeper Rise." *Washington Post* September 25.

p. 26   Discussion of 2006 crime statistics according to the FBI and BJS: Federal Bureau of Investigation. 2007. *Crime in the United States, 2006.* Washington, DC: Federal Bureau of Investigation.

p. 26   Discussion of problems with Uniform Crime Report (UCR) measurement of crime: Catalano, Shannan M. 2006. *The Measurement of Crime: Victim Reporting and Police Recording.* New York: LFB Scholarly Publishing; O'Brien, Robert M. 2000. "Crime Facts: Victim and Offender Data." pp. 59–83 in *Criminology: A Contemporary Handbook,* Joseph F. Sheley (ed.). Belmont, CA: Wadsworth.

p. 28   *the police record only about two-thirds of all citizen complaints:* Warner, Barbara D., and Glenn L. Pierce. 1993. "Reexamining Social Disorganization Theory Using Calls to the Police as an Increase of Crime." *Criminology* 31:493–517.

p. 29   Discussion of crime-reporting scandals since 1990 by police: Butterfield, Fox. 1998. "As Crime Falls, Pressure Rises to Alter Data." p. A1 in *The New York Times*; Fazlollah, Mark, Michael Matza, Craig R. McCoy, and Clea Benson. 1999. "Women Victimized Twice in Police Game of Numbers." p. A1 in *The Philadelphia Inquirer*; Matza, Michael, Crag R. McCoy, and Mark Fazlollah. 1998. "Panel to Overhaul Crime Reporting." p. A1 in *The Philadelphia Inquirer*.

p. 29   *as critics say that college administrators and campus security:* Matza, Michael. 1997. "Auditors to Eye Penn's Campus-Crime Data." p. A1 in *The Philadelphia Inquirer*.

p. 29–30   *In a well-known study from the 1960s:* Black, Donald. 1970. "Production of Crime Rates." *American Sociological Review* 35:733–748.

p. 30   Discussion of why reported rapes rose in the 1970s and 1980s: Baumer, Eric P., Richard B. Felson, and Steven F. Messner. 2003. "Changes in Police Notification for Rape, 1973–2000." *Criminology* 41:841–872.

p. 30–31   Discussion of inaccuracy of NCVS: Catalano 2006.

p. 32   *People are more or less likely . . . :* Federal Bureau of Investigation 2007.

p. 32   Discussion of why the South has high violent crime rate: Bailey, Frankie Y. 2004. "Honor, Class, and White Southern Violence: A Historical Perspective." pp. 331–353 in *Violent Crime: Assessing Race and Ethnic Differences*, Darnell F. Hawkins (ed.). Cambridge: Cambridge University Press; Lee, Matthew R., William B. Bankston, Timothy C. Hayes, and Shaun A. Thomas. 2007. "Revisiting the Southern Subculture of Violence." *The Sociological Quarterly* 48:253–275.

p. 32   Discussion of why urban areas have higher crime rates: Roncek, Dennis W., and Pamela A. Maier. 1991. "Bars, Blocks, and Crimes Revisited: Linking the Theory of Routine Activities to the Empiricism of 'Hot Spots'." *Criminology* 29:725–753; Stark, Rodney. 1987. "Deviant Places: A Theory of the Ecology of Crime." *Criminology* 25:893–911.

p. 33   Discussion of climates and crime rates: Hipp, John R., Daniel J. Bauer, Patrick J. Curran, and Kenneth A. Bollen. 2004. "Crimes of Opportunity or Crimes of Emotion? Testing Two Explanations of Seasonal Change in Crime." *Social Forces* 82:1333–1372.

p. 33–34   Discussion of age and higher crime rates: Steffensmeier, Darrell, and Emilie Allan. 2000. "Looking for Patterns: Gender, Age, and Crime." pp. 85–127 in *Criminology: A Contemporary Handbook*, Joseph F. Sheley (ed.). Belmont, CA: Wadsworth.

p. 34–35   Discussion of gender and crime rates: Belknap, Joanne. 2007. *The Invisible Woman: Gender, Crime, and Justice.* Belmont, CA: Wadsworth Publishing Company; Steffensmeier and Allan 2000.

p. 34–35   Discussion of early answers regarding why men commit more crime than women: Klein, Dorie. 1973. "The Etiology of Female Crime." *Issues in Criminology* 8:3–30.

p. 35   Discussion of socialization differences between the sexes: Stockard, Jean. 2006. "Gender Socialization." in *Handbook of the Sociology of Gender*, Janet Saltzman Chafetz (ed.). New York: Springer. pp. 215–227.

p. 35   Discussion of why female arrests rose in the 1990s: Stanley, Kameel. 2007. "Juvenile Crime by Girls is On the Rise." *The Grand Rapids Press* May 20:B1; Steffensmeier, Darrell, Jennifer Schwartz, Hua Zhong, and Jeff Ackerman. 2005. "An Assessment of Recent Trends in Girls' Violence Using Diverse Longitudinal Sources: Is the Gender Gap Closing?" *Criminology* 43:355–405.

p. 36   Discussion of social class differences in offending: Braithwaite, John. 1981. "The Myth of Social Class and Crime Reconsidered." *American Sociological Review* 46:36–47; Dunaway, R. Gregory, Francis T. Cullen, Velmer S. Burton, Jr., and T. David Evans. 2000. "The Myth of Social Class and Crime Revisited: An Examination of Class and Adult Criminality." *Criminology* 38:589–632; Tittle, Charles R., Wayne J. Villemez, and Douglas A. Smith. 1978. "The Myth of Social Class and Criminality: An Empirical Assessment of the Empirical Evidence." *American Sociological Review* 43:643–656.

p. 36   *"Yet, through it all, . . .":* Stark 1987, p. 894.

p. 37   Statistics on race and arrest/imprisonment: Maguire and Pastore 2007; The Sentencing Project. 2007. *Facts About Prison and Prisoners.* Washington, DC: The Sentencing Project.

p. 37   Discussion of disagreement by researchers over race and crime statistics: Mann, Coramae Richey. 1993. *Unequal Justice: A Question of Color.* Bloomington: Indiana University Press; Miller, Jerome G. 1996. *Search and Destroy: African American Males in the Criminal Justice System.* New York: Cambridge University Press; Wilbanks, William. 1987. *The Myth of a Racist Criminal Justice System.* Monterey: Brooks/Cole Publishing Company.

p. 37–39   Discussion of evidence on race and crime and of explanations for relationship between race and crime: Bellair, Paul, and Thomas L. McNulty. 2005. "Beyond the Bell Curve: Community Disadvantage and the Explanation of Black-White Differences in Adolescent Violence." *Criminology* 43:1135–1168; Harris, Anthony R., and James A. W. Shaw. 2000. "Looking for Patterns: Race, Class, and Crime." pp. 129–163 in *Criminology: A Contemporary Handbook*, Joseph F. Sheley (ed.). Belmont, CA: Wadsworth; Sampson, Robert J., and William Julius Wilson. 1995. "Toward a Theory of Race, Crime, and Urban Inequality." pp. 37–54 in *Crime and Inequality*, John Hagan and Ruth D. Peterson (eds.). Stanford: Stanford University Press; Walker, Samuel, Cassia Spohn, and Miriam DeLone. 2007. *The Color of Justice: Race, Ethnicity, and Crime in America.* Belmont, CA: Wadsworth Publishing Company.

p. 39   *"where the probabilities . . .":* Stark 1987, p. 906.

p. 39–40   Findings from a 2000 national survey: authors' analysis of data from the 2000 General Social Survey, National Opinion Research Center.

p. 39–40   Discussion of evidence on immigration and crime rates: Martinez, Ramiro, Jr., and Abel Valenzuela (eds.). 2006. *Immigration and Crime: Race, Ethnicity, and Violence.* New York: New York University Press; Rumbaut, Rubén G., and Walter A. Ewing. 2007. *The Myth of Immigrant Criminality and the Paradox of Assimilation: Incarceration Rates Among Native and Foreign-born Men.* Washington, DC: American Immigration Law Foundation; Vélez, María B. 2006. "Toward an Understanding of the Lower Rates of Homicide in Latino versus Black Neighborhoods: A Look at Chicago." pp. 91–107 in *The Many Colors of Crime: Inequalities of Race,*

Ethnicity, and Crime in America, Ruth D. Peterson, Lauren J. Krivo, and John Hagan (eds.). New York: New York University Press.

Chapter 4. Robbers, Rapists, and Serial Killers: Violent Crime in America

p. 45   *"The evil mind or malice . . ."*: Gifis, Steven. 1996. *Barron's Law Dictionary* 4th ed. Woodbury, NY: Barron's Educational Series, p. 82.

p. 45–47   Discussion of homicide and statistics: Various Uniform Crime Reports from 1980 to 2006, Federal Bureau of Investigation.

p. 47   *"What is there about residing . . . ?"*: Luckenbill, David, and Daniel P. Doyle, 1989. "Structural Position and Violence—Developing a Cultural Position." *Criminology* 27:419–436.

p. 47–48   Discussion of a number of serial killers: Jenkins, Philip, 1988. "Myth and Murder: The Serial Killer Panic of 1983–85" *Criminal Justice Research Bulletin* 3(11):1–7.

p. 48   Discussion of international serial killers. "World's Worst Killers" *BBC News*, October 30, 1999, http://news.bbc.co.uk; "Pedro Alonso Lopez—The Monster of the Andes" 2007. *About.Com*, http://www.crime.about.com; Matthews, Owen, and Anna Nemtsova, "From Russia, with Love" *Newsweek*, October 25, 2007, http://newsweek.com.

p. 49–53   Typology of serial killers: Fox, James A., and Jack Levin. 1994. *Overkill: Mass Murder and Serial Killing Exposed.* New York: Plenum Press.

p. 49   *"My sexual fantasy is . . ."*: Rader, Dennis. "Secret Confessions of BTK." Dateline NBC interview, August 12, 2005.

p. 50   Discussion of Washington D.C. area sniper killers: Bryjak, George J. 2002. "The Horror of Serial Killers." *San Diego Union-Tribune*, p. B4.

p. 51   Discussion of serial killers as cruel or crazy: Fox and Levin 1994.

p. 51   *"the stalking stage is when you . . ."*: Rader, Dennis. Dateline NBC interview, August 12, 2005.

p. 51–52   *A study of 15 inmates on Florida's death row:* Fox and Levin 1994.

p. 52   *"if head trauma were as strong . . ."*: Fox and Levin 1994, p. 90.

p. 52   Discussion of XYY "super male": Hickey, Eric C. 1991. *Serial Murderers and Their Victims.* Pacific Grove, CA: Brooks/Cole Publishing.

p. 52   *"who drink heavily and indulge . . ."*: Hickey 1991, p. 65.

p. 52   *"risk factors . . . they are neither . . ."*: Fox, James A., and Jack Levin. 2005. *Extreme Killing: Understanding Serial and Mass Murder.* Thousand Oaks, CA: Sage Publications, p. 111.

p. 52   *"neuropsychiatric problem alone don't make you violent . . ."*: Fox and Levin 2005, p. 111.

p. 53   Discussion of FBI data: Fox and Levin 2005.

p. 53   Discussion of mass murderers: Fox and Levin 2005.

p. 53–57   Discussion of the typology of mass murderers: Holmes, Ronald M., and Stephen T. Holmes. 2001. *Mass Murder in the United States*, Upper Saddle River, NJ: Prentice Hall.

p. 54   *"were highly selective . . ."*: Fox and Levin 2005, p. 87.

p. 54   Discussion of suicide by proxy: Fox and Levin 2005, p. 166.

p. 54   *"Socially isolated, they regard work as . . ."*: Fox and Levin 1994, p. 179.

p. 54–55   Discussion of Jonestown killing: Locke, Michelle. 2000. "Jonestown Survivors Remember," The Rick A. Ross Institute for the Study of Destructive Cults, Controversial Groups and Movements. April 5, 2000; Nelson, Stanely. 2003. Jonestown: *The Life and Death of Peoples Temple*, aired PBS, The American Experience, April 2006.

p. 56   *"'spillover effect' took hold: His anger . . ."*: Fox and Levin 1994, p. 223.

p. 56–57   *"can sometimes breed feelings . . ."*: Fox and Levin 2005, p. 216.

p. 57   *"both logistically and symbolically, . . ."*: Fox and Levin 2005, p. 216.

p. 57   *"They're angry and they want . . ."*: Begley, Sharon, 2007, *"The Anatomy of Violence"* Newsweek, April 30, pp. 40–46.

p. 57   *"They may think . . ."*: Begley 2007, p. 44.

p. 57   *The National Crime Victimization Survey defines:* U.S. Department of Justice, Office of Justice Statistics, Bureau of Justice Statistics, Washington, DC, available at http://www.ojp.gov.bjs.

p. 58   Discussion of serial rapists and FBI study: Hazlewood, Robert, R. and Janet Warren, 1990, "The Criminal Behavior of the Serial Rapist." FBI Law Enforcement Bulletin, February, pp. 11–16.

p. 59   Discussion of rape myths: Groth, A. Nicholas. 1979. *Men Who Rape: The Psychology of the Offender.* New York: Plenum Books.

p. 59   *"for by their mutual consent and contract . . ."*: Van Hasselt, Vincent B. 1988. *The Handbook of Family Violence.* New York: Plenum Press.

p. 60   *"may be the most predominant . . ."*: Groth 1979, p. 5.

p. 60   Discussion of the three categories of rape: Bergen, Raquel Kennedy. 2006. "Marital Rape: New Research Directions," *National Online Resource Center on Violence Against Women,* February 2006, http://www.vawnet.org.

p. 60   *"When a woman is raped by a stranger . . ."*: Abuse Counseling and Treatment, Inc. "Marital Rape." p. 2, http://www.actabuse.com.

p. 60   *In one study of marital rape, 17 percent of victims reported an unwanted pregnancy:* Campbell, J.C. 1989. "Women's Responses to Sexual Abuse in Intimate Relationships." *Health Care for Women International* 10(4): 335–346.

p. 61   Discussion of one study of two- and four-year schools . . . Sampson, Rana, 2002, "Acquaintance Rape of College Students" *U.S. Department of Justice,* Office of Community Oriented Policing Services, www.cops.usdoj.gov.

p. 61–62   Discussion of three drugs related to sexual assault and acquaintance rape: "Date Rape Drugs," 2004, March, *National Women's Health Information Center,* Washington, DC: U.S. Department of Health and Human Services, Office of Women's Health.

p. 62   *"When my victim screamed, . . ."*: Groth 1979, p. 59.

p. 62   *"The cause of this backlog . . ."*: Cantlupe, J. 2003, "Rape Evidence Sits Unprocessed" *San Diego Union Tribune,* January 23, p. A3.

p. 63   *"It strikes me that . . ."*: Arnold, Ben. 2006. "A Vexing Leap in U.S. Robbery Rate," *Christian Science Monitor*, December 21, p. 3.

p. 63   Discussion of street robbers: Katz, Jack. 1988. *Seductions in Crime.* New York: Basic Books.

p. 63   Discussion of the study of street robbers: Miller, Jody. 1998. "Up it Up: Gender and the Accomplishment of Street Robbery." *Criminology* 36:37–65.

p. 64–66   Discussion of four parts of a street robbery: Wright, Richard, and Scott Decker. 1997. "Creating the Illusion of Impending Death: Armed Robbers in Action." *The HFG Review of Research* 2(1):10–18.

p. 67   Discussion of anecdotal evidence: Roman, J. and A. Chalfin. 2007. "Is There an iCrime Wave?" *Urban Institute, JusticePolicy Center.* Available at: http://www.urban.org/url.cfm?ID=411552, accessed March 18, 2008.

p. 67   *"unlike a jacket of a sneaker . . .":* Roman and Chalfin 2007, p. 1.

p. 67   *"was predictable and could have . . .":* Roman and Chalfin 2007, p. i.

p. 68   Discussion of findings from the Roman Catholic Church and John Jay School of Criminal Justice in New York: Bono, A. 2006. "John Jay Study Reveals Extent of Abuse Problem" *Catholic News Service, The American Catholic.* Available at: http://www.americancatholic.org/news/clergysexabuse/johnjaycns.asp, accessed February 19, 2008.

p. 69   Discussion of Catholic Church and Philadelphia grand jury: Cipriano, R., 2005, "Philadelphia Cardinals 'Excused and Enabled Abuse', Covered Up Crimes" *National Catholic Reporter,* October 7. Available at: http://natcath.org/NCR_Online/archives2/2005d/100705/100705a.php, accessed February 19, 2008, p. 1.

p. 69   *"We mean rape. Boys who were raped . . .":* Catholic Church and Philadelphia grand jury in Cipriano, R., 2005, "Philadelphia Cardinals 'Excused and Enabled Abuse', Covered Up Crimes" *National Catholic Reporter,* October 7. Available at: http://natcath.org/NCR_Online/archives2/2005d/100705/100705a.php, accessed February 19, 2008, p. 1.

p. 69   Discussion of the Archdiocese of Los Angeles: Bona, A., and Jerry Filteau. 2007. "Catholic Clergy Sex Abuse Report Finds Drop in Reported Victims" April 11, *Catholic News Services.* Available at: http://www.catholic.org/national/national_story.php?id=23713, accessed February 19, 2008.

p. 70   Discussion of San Diego diocese and bankruptcy: Sauer, Mark. 2007. "San Diego Diocese Filing for Bankruptcy Over Sex-Abuse Scandal." *San Diego Union-Tribune,* February 7.

p. 70   Sexual abuse in Protestant churches: *Reformation.com.* "Sexual Abuse of Minors by Protestant Clergy." Available at www.reformation.com, accessed February 19, 2008.

p. 70   Discussion of Christian Ministry Resources: Clayton, Mark. 2002. "Sex Abuse Spans Spectrum of Churches" *The Christian Science Monitor,* April 5.

p. 70   *"The incidence of sexual abuse by clergy . . .":* Mattingly, Terry. 2002 "Baptists' Traditions Make It Hard to Oust Sex-Abusing Clergy." *Knoxville News-Sentinel,* June 22, p. C2.

p. 70   *"30 to 35 percent of ministers from all . . .":* Trull, Joseph E., and James E. Carter. 1993. *Ministerial Ethics.* Knoxville, TN: Broadman & Holman.

p. 71   *"You name me a denomination . . .":* Johnson, Michelle. 2002. "Catholic Church not only religion facing abuse problems, expert says" *Catholic News Service,* April 2, http://catholic.org, 2007.

p. 71   Discussion of sexual abuse and Jehovah's Witnesses in Kentucky: Goldstein, Laurie. 2002. "Ousted Members Contend Jehovah's Witnesses' Abuse Policy Hides Offenses" *New York Times*, August 11.

p. 71   Discussion of rate of sexual abuse among Jewish clergy: Gross-Schaefer, Arthur. 1999. "Rabbi Sexual Misconduct: Crying Out for a Communal Response" *The Reconstructionist* 63(2):58.

p. 71–72   Discussion of clergy abuse in the Jewish community: The Jewish Coalition Against Sexual Abuse/Assault. "Clergy Abuse: Rabbis, Cantors & Other Trusted Officials" *The Awareness Center*. Available at: http://www.theawarenesscenter.org/clergyabuse.html, accessed February 19, 2008.

p. 72   Discussion of Minneapolis psychotherapist Gary Schoener: Watanabe, Teresa. 2002. "While Sexual Misconduct Has Rocked Many Religions, Leaders of Some Have Acted Far More Quickly Than Others" *Los Angeles Times*, March 25.

p. 72   *"has varied dramatically . . .":* Watanabe 2002, p. 1.

p. 72   *"what we are doing in creating safe environments . . .":* Bono and Filteau 2007, p. 1.

p. 72   *"What drove leaders to respond . . .":* Clayton 2002, p. 3.

Chapter 5. Hookers, Dopers, and Corporate Crooks

p. 76   Discussion of burglary statistics: FBI Uniform Crime Report, 2006, Federal Bureau of Investigation. Available at: www.fbi.gov, accessed March 7, 2009.

p. 76–77   Study of burglars in St. Louis: Wright, Richard, and Scott Decker. 1995. *Burglars on the Job: Streetlife and Residential Break-ins.* Boston: Northeastern University.

p. 77   *"novices are not firmly entrenched . . .":* Miethe, T.D., and R. McCorkle, 1998. *Crime Profiles: The Anatomy of Dangerous Persons, Places, and Situations.* Los Angeles: Roxbury Publishing Company, p. 140.

p. 77–78   *"The military taught me what . . .":* Cromwell, Paul F., James N. Olson, and D'aunn Wester Avary. 1991. *Breaking and Entering: An Ethnographic Analysis of Burglary.* Newbury Park, CA: Sage Publications, p. 108.

p. 78   Discussion of false alarms: Sampson, Rana, 2007. "The Causes, Effectiveness, & Costs of False Burglar Alarms"*Center for Problem-Oriented Policing*. Available at: http://www.popcenter.org.

p. 78   Discussion on motor vehicle theft: FBI Uniform Crime Report, 2007. Federal Bureau of Investigation. Available at: www.fbi.gov, accessed March 7, 2009.

p. 78–80   Discussion of typology of auto theft: New York State Division of Criminal Justice Services, 2006. Division of Criminal Justice Services, Office of Strategic Planning, Bureau of Justice Funding. Albany, NY.

p. 79   *This is especially true of automobiles such as . . .:* Insurance Information Institute, "The Topic," *Auto Theft.* Available at: http://www.iii.org/media/hottopics/insurance/test4/, accessed March 7, 2008.

p. 79   *"Would you rather pay $500 . . .":* Insurance Information Institute. *Auto Theft.* Available at: www.iii.org.

p. 80   *"in order to disguise . . .":* New York State Division of Criminal Justice Services, 2006. Division of Criminal Justice Services, Office of Strategic Planning, Bureau of

Justice Funding. Albany, NY: p. 3.

p. 80   *"not only help police find . . ."*: Insurance Information Institute. *Auto Theft*, May 2007.

p. 81   *According to one estimate. . .*: Gaebler Ventures, 2008. "Retailers and Product Recalls." Available at: http://www.gaebler.com.

p. 81   Discussion of Cost of corporate crime: Reiman, Jeffery. 2001. *The Rich Get Richer and the Poor Get Prison*. Boston: Allyn & Bacon.

p. 81   Discussion of the National Commission on Product Safety data: Simon, David R. 2002. *Elite Deviance*. Boston: Allyn & Bacon.

p. 81   *"compulsory consumption of violence"*: Green, G. 1990. *Occupational Crime*, Chicago: Nelson-Hall, p. 135.

p. 81   Discussion of Rocky Flats nuclear weapons site: Abramson, Rudy, and Dean Takahasi, "Rockwell Agrees to $18.5 Million Fine: The firm will plead guilty to illegal disposal of radioactive waste that shut down Rocky Flats nuclear weapons site, sources say." *Los Angeles Times*, March 26, 1992, p. A3.

p. 81–82   *At any one time the FBI is investigating . . .*: Federal Bureau of Investigation, "Environmental Crime" *2003 Facts and Figures*. Available at: http://www.fbi.gov/libref/factsfigure/enviro.htm, accessed March 7, 2008.

p. 82   *The Los Angeles County Environmental Crimes Strike Force . . .*: Hammet, T. M., and J. Epstein. 2007. "Prosecuting Environmental Crime: L.A. County", Lectric Law Library Stacks. Available at: http://www.lectlaw.com.

p. 82   Discussion on crime in oil, auto, and pharmaceutical industries: Clinard, Marshall, and Peter Yeage. 1980. *Corporate Crime*, New York: Free Press.

p. 82   *Australian criminologist John Braithwaite . . .*: Braithwaite, John. 1986. "The Corrupt Industry" *The New Internationalist*, No. 165, November, Available at: www.newint.org.

p. 82   *"Rats die in clinical trials . . ."*: Braithwaite, John. "The Corrupt Industry" *The New Internationalist*, No. 165, p. 1.

p. 82   *"Corrupt executives are no better . . ."*: Radelat, A. 2002. "Achcroft Likens WorldCom to Common Thieves" *Gannet New Service*, August 11.

p. 83   *"With a non-prosecution agreement, . . ."*: "Corporate Crime and Punishment" 2005, *Multinational Monitor*, 26 (11, 12): p. 1.

p. 83   *"There are too many ways . . ."*: Pontell, Henry. 2002. "Corporate Crime and Punishment" *Today@UCI*. Available at: http://today.uci.edu.

p. 83–84   Discussion of identity theft: Cantwell, Maria. 2004. "Identity Theft Statistics" *Fighting Identity Theft*. Available at: http://cantwell.senate.gov,.; "Understanding Identity Theft Part 1—Identity Theft Explained" 2003, Identity Theft 911, November. Available at: http://www.identitytheft911.com; "Identity Theft Statistics—Cases Per Year" 2004; "My ID Fix—Identity Theft Prevention & Victim Assistance Center." Available at: http://www.myfix.com.

p. 83   *"the single greatest type . . ."*: Bock, Wally. 2007. "Don't be 'Phish' Bait—Easy Steps to Protect Yourself." *FrugalFun.com*. Available at: http:www.frugalfun.com, accessed March 7, 2008, p. 2.

p. 84   *"ID theft usually occurs not because of . . ."* Consumer Reports. 2003. "Stop Thieves from Stealing You". Available at: http://www-erights.prod.consumerreports.org/cro/money/credit-loan/identity-theft/identity-theft-

1003/overview/?resultIndex=4&resultPageIndex=1&searchTerm=stolen, accessed March 7, 2008.

p. 84–86   Discussion of most common ways identity thieves steal information: Consumer Reports. "Stop Thieves from Stealing You"; Consumer Reports. 2007. "Net Threats." September, pp. 32–33.

p. 85   *"is a better way to make money . . ."*: Paller, Allan. 2007. "Net Threats," *Consumer Reports*, September p. 29.

p. 85   *"The majority of victims . . .."*: Finch, James. 2007. "Operation Pot Roast", Federal Bureau of Investigation, Press Release, June 13, p. 2. Available at: www.fbi.gov.

p. 85–86   *"A botnet could shut down. . ."*: CNN.com. 2007. "FBI: Millions of Computers Roped Into Criminal 'Robot Networks'" November 29. Available at: www.cnn.com.

p. 86   *According to the Identity Theft Resource Center:* Cantwell 2004.

p. 86   "Anyone who stores information . . .": Consumer Reports 2007, p. 17.

p. 86–87   Discussion of medical identity theft: Dixon, Pam. 2006. "Medical Identity Theft: The Information Crime that Can Kill You" *The World Privacy Forum*. Available at: www.worldprivacyforum.org; CBS News, "Protect Against Medical ID Theft" October 9. Available at: www.cbsnews.com,; Koemer, Brian. 2007. "The Consequences of Medical Identity Theft" About.com. Available at: www.idtheft. about.com, "ID Theft Infects Medical Records." *Los Angeles Times*, October 3.

p. 86   *"deeply entrenched in the healthcare . . ."*: Dixon 2006, p. 88.

p. 87–89   "I have come to the view that humans . . .": Beaty, J. 1989. "Do Humans Need to Get High?" *Time Magazine*, August 21, p. 58.

p. 88   Discussion of characteristics of drug dependence: Goode, E. 1991. *Deviant Behavior*, Englewood Cliffs, NJ: Prentice Hall.

p. 88–89   Discussion of the addictive personality: Nelson, B. 1983. "The Addictive Personality: Common Traits are Found" *New York Times*, January 18, 1983; "Addictive Personality" *Encyclopedia of Drugs, Alcohol, and Addictive Behavior*, 2008. http://www.enotes.com; Helm, J. 2008. "Myth of the 'Addict Gene.'" *Addiction-Info*. Available at: http://www.addictioninfo.org/articles/961/1/Myth-of-an-AddictGene/Page1.html, accessed March 17, 2008.

p. 89   *"There are almost no examples . . ."*: Helm 2008.

p. 89   *"are important not just . . ."*: Helm 2008.

p. 89   Discussion of drug use in the country: Substance Abuse and Mental Health Services Administration (SAMHSA) statistics. 2006. "Youth Drug Use Continues Downward Slide: Older Adults Rates of Use Increase" United States Department of Health and Human Services. Available at: http://www.samhsa.gov/news/newsreleases/060907_nsduh.aspx, accessed March 7, 2008.

p. 90   Discussion of drug deaths in this country: Drug War Facts. 2007. "Annual Causes of Death in the United States" Available at: http://www.drugwarfacts.org/causes.htm, accessed March 7, 2008.

p. 90   Discussion of drugs and crime: Wilson, James Q., and R.J. Herrnstein 1985, *Crime and Human Nature*, New York: Simon & Schuster.

p. 90   *"All of these studies . . ."*: Inciardi, James. 1999. *The Drug Legalization Debate*, Thousand Oaks, CA: Sage, pp. 45–79.

p. 91   Discussion about prisoners and drug related crime: Bureau of Justice Statistics. 2006. "Drug Use and Crime" U.S. Department of Justice, Washington, DC. Available at: www.ojp.usdoj.gov.

p. 91–92   Discussion of the international drug distribution system: Lyman, Michael. 1987. *Narcotics and Crime Control*, Springfield, IL: Charles C. Thomas, pp. 62–63; Albini, Joseph. 1992. "The Distribution of Drugs: Models of Criminal Organization and Their Integration." *Drugs, Crime and Social Policy: Research Issues and Concerns*, Thomas Mieczkowski (ed.), Boston: Allyn & Bacon, pp. 62–63.

p. 92   United Nations 2007 World Drug Report. 2007. *United Nations Office on Drugs and Crime*. Available at: www.unodc.org.

p. 93   *Almost 90 percent of all coca. . .*: United Nations Office of Drugs and Crime 2007. "World Drug Report." Available at: www.unodc.org.

p. 93   *"One can safely conjecture . . ."*: Albini 1992, p. 98.

p. 94   Discussion of dealing drugs: Levitt, S.D., and S. Venkatesh "An Economic Analysis of a Drug-Selling Gang's Finances, *Quarterly Journal of Economics* 115(3):755–789; Venkatesh, S. 2008. *Gang Leader for A Day: A Rogue Sociologist Takes to the Streets*, New York: Penguin Press; Kingsbury, A. 2008. "Dispelling the Myths About Gangs: Q &A with Sudhir Venkatesh," *U.S. News & World Report*, January 8, 2008. Available at: http://www.usnews.com/articles/news/national/2008/01/08/qa-sudhir-venkatesh.html, accessed March 17, 2008; King, R. 2003. "The Economics of Drug Selling: A Review of the Research", *The Sentencing Project*. Available at: www.sentencingproject.org; Hagedorn, J. M. 1994. "Homeboys, Dope Fiends, Legits, and New Jacks" *Criminology* 32(2):197–219; Mohamed, A.R. and E. Fritsvold. 2006. "Damn, It Feels Good to be a Gangsta: The Social Organization of the Illicit Drug Trade Servicing a Private College Campus" *Deviant Behavior* 27(1):97–125.

p. 94   Discussion of the Sentencing Project . . .: King, Ryan S. 2006. "The Next Big Thing: Methamphetamine in the United States," *The Sentencing Project*, Washington DC. Available at: http://www.google.com/search?q=The+Next+Big+Thing%3A+Methamphetamine+in+the+United+States&ie=utf-8&oe=utf-8&aq=t&rls=org.mozilla:en-US:official&client=firefox-a, accessed March 7, 2008.

p. 94   *"drug selling was not a specialized . . ."*: King 2003, p. 3.

p. 95   *"drug dealing can be seen . . ."*: King 2003, p. 5.

p. 95   *"I don't know, I really just do it . . ."*: Mohamed and Fritsvold 2006, pp. 114–115.

p. 95   *"when I was a freshman . . ."*: Mohamed and Fritsvold 2006, p. 116.

p. 95   *"It was an easy way to make money."*: Mohamed and Fritsvold 2006, p. 116.

p. 96   *"If you said, where'd you get . . ."*: Mohamed and Fritsvold 2006, p. 118.

p. 96   *"rich white college drug dealers . . ."*: Mohamed and Fritsvold 2006, p. 119.

p. 96   *"can kiss my ass . . ."*: Mohamed and Fritsvold 2006, p. 119.

p. 96   *"He must have some very influential parents."*: Mohamed and Fritsvold 2006, p. 120.

p. 96   Discussion of methamphetamine use: Substance Abuse and Mental Health Service Administration (SAMSHA). 2004. "Overview of Drug Use in the United States" *National Survey on Drug Use and Health, 2003 and 2004*. Available at: www.oas.samsha.gov.

p. 97   The Arrest Drug Abuse Monitoring (ADAM) data: King 2006. pp. 10–11.

p. 97–98 Discussion of Angela Valdez's research: Valadez, Angela. 2006. "Meth Madness: How the *Oregonian* Manufactured an Epidemic, Politicians Bought It and You're Paying" *Willamette Weekly Newspaper*, March 22, 2006.

p. 98 *"If meth causes property offenses . . .":* Valdez, *Willamette Weekly Newspaper*, March 22, 2006.

p. 98 *Frank Owen notes that Mexican . . .:* Owen, Frank. 2007. *No Speed Limit: The Highs and Lows of Meth*, St. Martin's: New York.

p. 98 *". . . it is our national responsibility to use . . .":* King 2006, p. 28.

p. 98–100 Discussion on drug decriminalization: Edlemann, Nathan. 2001. "An Unwinnable War on Drugs" *New York Times*, April 26; "Drug Policy Issues," January 2007, *News Batch*. Available at: www.newsbatch.com; Centers for Disease Control and Prevention. 2002. "Drug Associated HIV Transmission Continues in the United States."Available at: www.cdc.gov; National Institute on Drug Abuse. 2007. "NIH Survey Shows Most People with Drug Use Disorders Never Get Treatment." Available at: www.nida.nih.gov.

p. 98–100 Discussion of decriminalization: Petty, L. 2001. "Marijuana's Globalization" Medill School of Journalism, Evanston, IL: Northwestern University. Available at: http://docket.medill.northwestern.edu/archives/000045.php, accessed March 17, 2008; Australian Institute of Criminology, 2008. "Illicit and Alcohol: International and Overseas responses." Available at: http://www.aic.gov.au/research/drugs/international/, accessed March 17, 2008; Keizer, R. 2001. "The Netherlands' Drug Policy" Parliament of Canada. Available at: http://www.parl.gc.ca/37/1/parlbus/commbus/senate/com-e/ille-e/presentation-e/keizer-e.htm, accessed March 17, 2008; U.S. Drug Enforcement Administration 2007. "Portugal: Decriminalization of All Illicit Drugs" Available at: http://www.justice.gov/dea/ongoing/portugal.html, accessed March 17, 2008.

p. 99 *"Decriminalization ignores the fact . . .":* Petty 2001, p. 2.

p. 100 *"not many young people . . .":* Keizer 2001.

p. 101 *"lives in a world of pimps, . . .":* Little, Craig B. 1995. *Deviance and Control: Theory Research and Social Policy*, 3rd edition. Itasca, IL: Peacock Publishers, p. 187.

p. 102 Potterat, John J. 2004. "Morality in a Long-Term Open Cohort of Prostitute Women." *American Journal of Epidemiology* 159(8):778–785.

p. 102 *"The high homicide and overall mortality . . .":* Potterat, John J. 2004, p. 783.

p. 101–102 Discussion of the study of San Francisco street walkers: Farley, Melissa. 1998. "Prostitution, Violence, and Post Traumatic Stress Disorder" *Women and Health*, 27(3): 37–49.

p. 101 *"acute anxiety, depression, insomnia . . .":* Farley 1998.

p. 102 Discussion of the categories of prostitution: Flowers, R. Barri 1998, *The Prostitution Women and Girls*. Jefferson, NC: McFarland & Company, Inc.

p. 102–103 Discussion of pimps: Siegel, Larry. 1999. *Criminology*, Belmont, CA: West/Wadsworth.

p. 103 Discussion of the pimp-prostitute relation: Bracey, D.H. 1983. "The Juvenile Prostitute: Offender and Victim" *Victimology* 8 (3): 151–159; Maddan, Sean A. 2002. "Occupational Risks of Teenage Prostitution," M.D. McShane and F.P.

Williams (eds.), *Encyclopedia of Juvenile Justice*, Thousand Oaks, CA: Sage, pp. 305–306.

p. 103–105   Discussion of call girls: Simon, David R. 2002. *Elite Deviance*, Boston: Allyn & Bacon.

p. 103   *"frequently retained to help close . . .":* Simon 2002.

p. 105–106   Discussion of child prostitution: Estes, Richard J., and Neil Alan Weiner. 2001. "The Commercial Sexual Exploitation of Children in the U.S., Canada, and Mexico" Philadelphia: University of Pennsylvania. Available at: http://caster.ssu. upenn.edu.

p. 105   *"Child sexual exploitation is the most . . .":* Estes, Richard. 2001. "Commercial Child Exploitation: 'The Most Hidden From of Child Abuse. National Association of Social Workers (NSAW). Available at: http://www.socialworkers.org/ pressroom/2001/091001.asp, accessed March 7, 2008.

p. 105   Discussion of Statistics on child prostitutes in the United States: Federal Bureau of Investigation. 2005. "Innocence Lost Arrests: National Crackdown on Child Prostitution", December 16. Available at: http://www.fbi.gov/page2/dec05/inno- cence_lost_arrest3.htm, accesses March 7, 2008; Sugg, John F. 2006. "A Horri- ble Achievement: Atlanta's Child Prostitution Business is Booming" *Atlanta Creative Loafing,* Available at: http://atlanta.creativeloafing.com/gyrobase/ Content?oid=oid%3A66047, accessed March 7, 2008; Miko, Francis T. 2006. "Trafficking in Persons: The U.S. and International Response" CRS Report for Congress, *Congressional Research Service*, The Library of Congress; "Study: Child Sex Abuse 'Epidemic' in the U.S." 2001, *CNN.com.* Available at: http://archives.cnn.com/2001/LAW/09/10/child.exploitation/, accessed March 7, 2008.

p. 106   Discussion about sex tourism: Child Exploitation and Obscenity Section. 2007. "Child Sex Tourism" U.S. Department of Justice. Available at: http://www. usdoj.gov/criminal/ceos/sextour.html, accessed March 7, 2008.

p. 106   *"On this trip I've had sex with a 14 year-old girl . . .":* Child Exploitation and Obscenity Section. 2007.

p. 106   *Percentage of males who frequent prostitutes.* Scott, M.S., and K. Dedel, 2006. "Street Prostitution, Second Edition" Community Oriented Policing Services, U.S. Department of Justice. Available at: http://www.cops.usdoj.gov.

p. 106   *". . . the commercial sexual exploitation of children . . .":* Dr. Peter Piot, Execu- tive Director, UNAIDS. 2001. "Stockholm Congress: Keynote Address" *World Congress Against Commercial Sexual Exploitation of Children.* Available at: www.csecworldcongress.org.

p. 106–108   Discussion of men who frequent prostitutes: Hughes, Donna M. 2004. "Best Practices to Address the Demand Side of Sex Trafficking." University of Rhode Island. Available at: http://www.uri.edu/.

p. 107   *Hughes cites an unpublished . . .:* Hughes 2004.

Chapter 6. Victims and Victimization: Will You Be Next?

p. 111   *the primary purpose of the justice system. . .:* McDonald, William. 1977. "The Role of the Victim in America." *Environmental Criminology*, R.E. Barnett and J. Hagel III (eds.), Thousand Oaks, CA: Sage Publications, pp. 37–52.

p. 112   *"It is difficult to understand . . .":* Fattah, E. A. 1997. "Toward a Victim Policy Aimed at Healing, Not Suffering." *Victims of Crime.* R.C. Davis, A. J. Lurigio, and W.B Skogan (eds.), Thousand Oaks, CA: Sage Publications, pp. 257–272.

p. 112   *"criminology could be characterized as 'offenderology'":* Karmen, Andrew. 1990. *Crime Victims: An Introduction to Victimology.* Pacific Grove, CA: Brooks Cole, p. 8.

p. 112–113 Discussion of crime victimization statistics: National Crime Victimization Survey. September 2006. *Criminal Victimization, 2005.* Bureau of Justice Statistics, U.S. Department of Justice, Washington DC; *Homicide Trends in the U.S., 2005.* Bureau of Justice Statistics, U.S. Department of Justice, Washington DC.

p. 116–117 Discussion by economist David Anderson: Anderson, D. 2007. "Cost of Crime: It Just Doesn't Pay! New Crime Study Pegs Cost at $1.7 Trillion Annually." *Davidson News & Events,* Davidson University. Available at: http://www2.davidson.edu; Anderson, D. 1999. "The Aggregate Cost of Crime." *Journal of Law and Economics* 42(2):611–642.

p. 116   Discussion of economic cost of crime: The National Center for Victims of Crime. 2005. "Cost of Crime." Available at: www.ncvc.org; Office of Justice Publications, U.S. Department of Justice. 2005. "Costs of Crime and Victimization." Available at: www.ojp.usdoj.gov; Anderson, D. 1999. "The Aggregate Cost of Crime." *Journal of Law and Economics,* 42(2): 611–642; "The Cost of Crime: It Just Doesn't Pay! New Crime Study Pegs Cost at $1.7 Trillion Annually," 2007. *Davidson News Events,* Davidson College. Available at: www2.davidson.edu.

p. 116   *Based on the nationally. . .:* Cohen, Mark, and Ted Miller. 1998, "The Cost of Mental Health Care for the Victims of Crime." *Journal of Interpersonal Violence* 13(1): 93–110.

p. 117   *"Submission in our culture is viewed . . .":* Bard, Morton, and Dawn Sangrey. 1986. *The Crime Victim's book,* 2nd edition. Seacaucus, NJ: Citadel, p. 82.

p. 117   Discussion of routine activities of Americans: Cohen, Lawrence, and Marcus Felson. 1979. "Social Change and Crime Rate Trends: A Routine Activity Approach." *American Sociological Review,* 44: 588–607.

p. 117   Discussion of absolute and probabilistic exposure to risk: Gottfredson, Michael. 1981. "On the Etiology of Crime Victimization." *Journal of Criminal Law and Criminology,* 72(2): 714–734.

p. 118   Discussion of Minneapolis crime study: Sherman, Lawrence, D.C. Gottfredson, D.L. McKenzie, J.Eck, P.Reuter, and S.D. Bushaway. 1998. *Preventing Crime: What Works, What Doesn't, What's Promising.* Washington DC: Office of Justice Programs, National Institute of Justice.

p. 118   *"the amount of crime of every type . . .":* Roneck, Dennis, and Pamela Maier 1991. "Bars, Blocks, and Crimes Revisited: Linking the Theory of Routine Activities to the Empiricism of 'Hot Spots.'" *Criminology* 29(4): 725–753.

p. 119   *"It takes money and effort to overcome distance.":* Brantingham, P.L. and P.J. Brantingham. 1981. "Notes on the Geometry of Crime." *Environmental Criminology,* P.J. Brantingham and P.L Brantingham (eds.), Thousand Oaks, CA: Sage Publications, pp. 27–57.

p. 119–120 Discussion of victim precipitation: Wolfgang, Marvin. 1958. *Patterns of Criminal Homicide.* Philadelphia: University of Pennsylvania Press.

p. 120   *In one of the most controversial pieces. . .:* Amir, Menachem. 1967. "Victim Precipitated Rape." *The Journal of Criminal Law, Criminology, and Police Science,* 58(4): 493–502.

p. 120   *"Although studies of victim precipitated rape . . .":* Miethe, Terance, and R. McCorkle. 2005. *Crime Profiles: The Anatomy of Dangerous Persons, Places, and Situations.* Los Angeles, CA: Roxbury.

p. 120   Discussion of physical characteristics of victims: Finkelhor, David, and N. Asdigian. 1996; "Risk Factors for Youth Victimization: Beyond a Theory of Lifestyles/Routine Activity Theory Approach." *Violence and Victims* 11(1):3–19.

p. 121   *"because these characteristics . . .":* Finkelhor, David, and N. Asdigian. 1996. p. 6.

p. 121   Discussion of female robbers: Sommers Ira, and Deborah Baskins. 1993. "The Situational Context of Violent Female Offending." *Journal of Research in Crime and Delinquency,* 30(2): 136–162.

p. 121   Further discussion on repeat victimization: Weisel, D.M., 2005. "Analyzing Repeat Victimization." Problem-Oriented Guides for Police, Community Oriented Police Services, U.S. Department of Justice. Available at: www.cops.usdoj.gov.

p. 121   Discussion of repeat victimization: Farrell, Graham and Ken Pease. 2000. "Why Repeat Victimization Matters," *Repeat Victimization: Crime Prevention Studies, Vol 12,* G. Farrell and K. Pease (eds), St. Louis, MO: Willow Tree Press, pp. 1–4.

p. 122   *"that some repeat offenders . . .":* Weisel, D. L. 2004. *Analyzing Repeat Victimization.* Problem-Oriented Guides for Police Problem-Solving Tools Series, U.S. Department of Justice, Washington DC. Available at: www.cops.usdoj.gov, p. 17.

p. 123   *"most offenses are highly concentrated . . .":* Weisel, D. L. 2004, p. 17.

p. 124   *"the roles of 'victim' and 'victimizers' . . .":* Fattah, E.A. 1993. "The Rational Choice/Opportunity Perspective as a Vehicle for Integrating Criminological and Victimological Theories." In *Rational Activity and Rational Choice,* R.V. Clarke and M. Felson (eds.), New Brunswick, NJ: Transaction Books, p. 239.

p. 124   *"interchangeable and may be assumed . . .":* Fattah, E.A. 1993, p. 239.

p. 124   Discussion of study of active armed robbers: Wright, Richard and Scott Decker. 1998. *Armed Robbers in Action: Stickups and Street Culture,* Boston: Northeastern University Press.

p. 124   Discussion of victimization and delinquent lifestyle: Lauristen, Janet, R.L. Sampson, and J.H. Laub. 1991. "The Link Between Offending and Victimization Among Adolescents." *Criminology* 29(2):265–292.

p. 125   Discussion of "van Dijk chains": Felson, Marcus. 1998. *Crime and Everyday Life.* Thousand Oaks, CA: Pine Forge Press.

p. 125–126 Discussion of crime displacement and victimization: Trasler, G. 1986. "Situational Crime Control and Rational Choice: A Critique." *Situational Crime Prevention: From Theory to Practice,* K. Heal and G Laycock (eds), London: Home Office Research and Planning Unit, Her Majesty's Stationary Office, pp. 17–24.

p. 126   Discussion of Mary Byron: *Women in Criminal Justice: A Twenty Year Update* 1998. Chapter 3, "Women and Girls as Victims" U.S. Department of Justice. Available at: www.ojp.usdoj.gov, see also The Mary Byron Foundation, available at: www.marybyronfoundation.org.

p. 127   Discussion of Kitty Genovese: Dorman, M. 2007. "The Killing of Kitty Genovese" Newsday.com. Available at: http://www.newsday.com/community/guide/lihistory/ny-history-hs818a,0,7944135.story, accessed March 7, 2008.

p. 128   *"It is simply unfair that victims should . . .": President's Task Force on Victims of Crime, 1982.* Office for Victims of Crime, Office of Justice Programs. Available at: www.ojp.usdoj.gov, p. 79.

p. 128   *"growing body of evidence . . .":* Umbreit, Mark. 1998. "Restorative Justice through Victim-Offender Mediation: A Multi-Site Assessment." *Western Criminology Review* 1(1). Available at: http://wcr.sonoma.edu/v1n1/umbreit.html, accessed March 7, 2008.

p. 129–130 Discussion of evaluation of VORPs in four cities: Umbreit, Mark. 1994. *Victim Meets Offender: The Impact of Restorative Justice and Mediation,* Monsey, NY: Willow Tree Press; Umbreit 1998.

p. 130   Discussion of the effectiveness of VORPs: Zehr, Howard, 1991. "Restitution Reduces Recidivism" *Crime and Justice Network Newsletter,* October, University of Colorado. Available at: www.colorado.edu; Niemeyer, M. and D. Shichor. 1996. "A Preliminary Study of a Large Victim/Offender Reconciliation Program." *Federal Probation* 60(3): 30–34; "Victim-Offender Reconciliation Programs," 1998. John Howard Society of Alberta. Available at: www.johnhoward.ab.ca; Evje, A. and R.C. Cushman "A Summary of the Evaluations of Six California Victim Offender Reconciliation Programs." California Courts. Available at: www.courtsinfo.ca.gov.

p. 130–132 Discussion of victims of Crime Act: "Victims of Crime Act Victims Funds," 1999. Washington DC, Office of the Victims of Crime, U.S. Department of Justice, Office of Justice Programs; "Making Good Use of the Crime Victims Fund," 2007, *National Association of Crime Victim Compensation Boards.* Available at: www.nacvcb.org; "Crime Victim Compensation: Resources for Recovery," 2007, *National Association of Crime Victim Compensation Boards.* Available at: www.nacvcb.org.

p. 132–134 Discussion of Crime Victims' Rights Act: U.S. Department of Justice, 2008. "Crime Victims' Rights Act" Office for Crime Victims. Available at: http://www.ojp.usdoj.gov/ovc; Fontana, D. 2004. "The New Crime Victims' Rights Act" *FindLaw Legal News and Commentary.* Available at: http://writ.news.findlaw.com.

p. 133–134 Discussion of victims and sentencing: Kelly, D. 1990. "Victim Participation in the Criminal Justice System," *Victims of Crime: Problems, Policies, and Programs,* A. J. Lurigo, W.G. Skogan, and R.C. Davis (eds.), Newbury Park, CA: Sage Publications, pp. 172–177; Wells, R. 1990, "Considering Victim Impact: The Role of Probation." *Federal Probation,* 54(3) pp. 26–29; Alexander, E.K. and J.H. Lord. 1994. U.S. Department of Justice. "A Victim's Right to Speak: A Nation's Responsibility to Listen." Impact Statements. Available at: www.ojp.usdoj.gov; Doerner, W.G. and S. P. Lab. 1995. *Victimology,* Cincinnati: Anderson; The National Center for Victims of Crime. 1999. "Victim Impact Statements." Available at: www.ncvc.org.

p. 133   Discussion of victim impact statements: Schuster, M.L., and A. Popen, 2006. "2006 WATCH Victim Impact Statement Study," Minnesota Center Against Violence & Abuse. Available at: http://www.mincava.umn.edu.

Chapter 7. Crime and Criminal Law: Order, Liberty, and Justice for All?

p. 138   Provisions of Patriot Act and other post-9/11 actions: Chang, Nancy. 2002. *Silencing Political Dissent.* New York: Seven Stories Press; Cole, David, and James X. Dempsey. 2006. *Terrorism and the Constitution: Sacrificing Civil Liberties in the Name of National Security.* New York: W.W. Norton.

p. 139   Discussion of elements of criminal law: Davenport, Anniken U. 2006. *Basic Criminal Law: The U.S. Constitution, Procedure, and Crimes.* Upper Saddle River, NJ: Prentice Hall; Samaha, Joel. 2008. *Criminal Law.* Belmont, CA: Wadsworth.

p. 144–145   Discussion of legal defenses to criminal charges: Davenport 2006; Samaha 2008.

p. 145–153   Discussion of rights of criminal suspects and defendants: Davenport 2006; Samaha 2008.

p. 149   Discussion of whether rights of suspects and defendants has hampered crime control: Walker, Samuel. 2006. *Sense and Nonsense About Crime and Drugs: A Policy Guide.* Belmont, CA: Wadsworth Publishing Company.

p. 150   Quotation by police chief: Brewster, T. 1991. "Law and Order." *Life* 14:85.

p. 150   Study of 118 interrogations: Journal, Yale Law. 1967. "Note, Interrogations in New Haven: The Impact of Miranda." *Yale Law Journal* 76:1519–1648; Salt Lake City study: Cassell, P.G., and B.S. Hayman. 1996. "Police Interrogations in the 1990s: An Empirical Study of the Effects of Miranda." *UCLA Law Review* 43:839–898.

p. 150   Discussion of reasons why suspects waive Miranda rights: Hoffman, Jan. 1998. "As Miranda Rights Erode, Police Get Confessions from Innocent People." p. A1 in *The New York Times*; Leo, Richard A. 1996. "Miranda's Revenge: Police Interrogation as a Confidence Game." *Law & Society Review*, 30:259–288.

p. 150   Quotation by sheriff: Brewster 1991:86.

p. 151   Study of and quotations by Chicago narcotics officers: Orfield, Myron W. 1987. "The Exclusionary Rule and Deterrence: An Empirical Study of Chicago Narcotics Officers." *University of Chicago Law Review*, 54:1016–1055.

p. 151   Discussion of studies of exclusionary rule: Walker 2006; study of California cases: Fyfe, James J. 1983. "The NIJ Study of the Exclusionary Rule." *Criminal Law Bulletin* 19:253–260.

Chapter 8. Why They Break the Law

p. 158   Surveys of drug and alcohol use by convicted offenders: Mumola, Christopher J. 1999. *Substance Abuse and Treatment, State and Federal Prisoners, 1997.* Washington, DC: Bureau of Justice Statistics, U.S. Department of Justice; quotation by Samuel Walker: Walker, Samuel. 2001. *Sense and Nonsense About Crime and*

Drugs: A Policy Guide, 5th ed. Belmont, CA: Wadsworth Publishing Company, p. 117.

p. 159   Studies that find offenders do not think about getting caught: Tunnell, Kenneth D. 1996. "Let's Do It: Deciding to Commit a Crime." pp. 246–258 in *New Perspectives in Criminology*, edited by John E. Conklin. Boston: Allyn and Bacon; Wright, Richard T., and Scott Decker. 1994. *Burglars on the Job: Streetlife and Residential Break-ins*. Boston: Northeastern University Press.

p. 159   Marvell and Moody study: Marvell, Thomas B., and Carlisle E. Moody. 1995. "The Impact of Enhanced Prison Terms for Felonies Committed with Guns." *Criminology* 33:247–281.

p. 159   Some scholars think three-strikes laws have increased the homicide rate: Marvell, Thomas B., and Jr. Carlisle E. Moody. 2001. "The Lethal Effects of Three Strikes Laws." *Journal of Legal Studies* 30:89–106.

p. 160–161   Discussion of research on punishment and ex-inmates' criminal behavior: Bishop, Donna. 2000. "Juvenile Offenders in the Adult Criminal Justice System." *Crime and Justice: A Review of Research* 27:81–167; Spohn, Cassia, and David Holleran. 2002. "The Effect of Imprisonment on Recidivism Rates of Felony Offenders: A Focus on Drug Offenders." *Criminology* 40:329–357.

p.161   Overall conclusion of research on deterrence: an extensive discussion may be found in Nagin, Daniel S. 1998. "Deterrence and Incapacitation." pp. 345-368 in *The Handbook of Crime and Punishment*, edited by Michael Tonry. New York: Oxford University Press.

p. 161–165   Discussion of biological explanations: Akers and Sellers 2007; Lilly et al. 2007; Barkan, Steven E. 2009. *Criminology: A Sociological Understanding*. Upper Saddle River, NJ: Prentice Hall.

p. 163–165   Discussion of psychological explanations: Akers and Sellers 2007; Barkan 2009; Lilly et al. 2007.

p. 165–173   Discussion of sociological explanations: Akers and Sellers 2007; Barkan 2009; Lilly et al. 2007.

p. 165   Discussion of Rodney Stark's views: Stark, Rodney. 1987. "Deviant Places: A Theory of the Ecology of Crime." *Criminology* 25:893–911.

p. 166   Quotation by Elliott Currie: Currie, Elliott. 1985. *Confronting Crime: An American Challenge*. New York: Pantheon Books, p. 160.

p. 169   Discussion of study by Chambliss: Chambliss, William J. 1973. "The Saints and the Roughnecks." *Society* 11:24–31.

p. 173   Discussion of crime-reduction programs centering on helping at-risk children: Welsh, Brandon C., and David P. Farrington (Eds.). 2007. *Preventing Crime: What Works for Children, Offenders, Victims and Places*. New York: Springer.

Chapter 9. Taking It to the Streets: Cops on the Job

p. 179   *"Many preferred the relative . . . .":* Lundman, R. L. 1978. *Police and Policing: An Introduction*. New York: Holt, Rinehart and Winston.

p. 181   Newspaper headline: The Mongrel Regime, Dulaney, W.M. 1996 *Black Police in America*, Bloomington: Indiana University Press, p. 595.

p. 181   Discussion of African American police: "Georgia's First Black Police Officers May Take Pension Battle to Court," *CNN.com*, March 1, 2008. Available at: http://www.cnn.com/2008/US/03/01/black.officers.pension.ap/, accessed March 27, 2008.

p. 181   *"I do solemnly swear . . .":* Dulaney, W.M. 1996, *Black Police in America*, p. 52.

p. 182   *"There's a sense of limited . . .":* Walsh, E. "Inside the Blue Line, A Racial Gulf of Their Own" *The Washington Post National Weekly Review*, December 18–24, p. 8.

p. 182   St. Louis police officer quote: Walsh, M.S. 1994. "Inside the Blue Line: A Racial Gulf of Their Own." *The Washington Post National Weekly Edition*, December 18, p. 8.

p. 182   Discussion of women police since 1845: Wells, Sandra K. and Betty L. Alt. 2005. *Police Women: Life with the Badge.* Westport, CN: Praeger.

p. 182   *"Just as a mother . . .":* Hamilton, Mary. 1971. *The Policewoman: Her Service and Ideals.* New York: Arno Press (original work published in 1924).

p. 182   *"Officer, I don't mean . . .":* Martin, Susan E. 1980, *Breaking and Entering: Police Women on Patrol.* Berkeley: University of California Press, p. 145.

p. 182   *"My first day on . . .":* Martin. 1980, p. 390.

p. 182   Discussion of how male officers harass female colleagues: Martin 1980.

p. 183   Discussion of gay and lesbian officers: Meers, E. 1998. "Good Cop, Gay Cop." *Advocate*, March 3: 26–35.

p. 183   Study of San Diego Police Department: Frank, N. 2001. "Pink and Blue: Outcomes Associated With Open Gay and Lesbian Personnel in the San Diego Police Department." The Michael D. Palm Center, http://www.palmcenter.org.

p. 183   Discussion of New York City police academy graduation class: Newman, A. 2007. "Two Men Who Are Faces of a New Force That Has Broken the Old Stereotypes" *New York Times*, July 10.

p. 183   *"The police do not prevent crime.":* Bayley, David H. 1994. *Police for the Future.* New York: Oxford University Press, p. 4.

p. 183–184   Discussion of the research done on the number of officers and crime rates: Bayley 1994.

p. 184   *"Police shouldn't be expected to prevent crime":* Bayley 1994, p. 10.

p. 185   Discussion of police as peacekeepers: Bayley 1994; Walker, Samuel. 1999. *The Police in America*, McGraw-Hill: Boston.

p. 185   *Patrol officers "become cynical and hard to convince":* Bayley 1994, p. 20.

p. 186   Discussion of the police occupational world: Van Maanen, John. 1999. "The Asshole" p. 346–367, *The Police and Society*, (ed. V.E. Kappeler), Prospect Heights, IL: Waveland Press.

p. 186   *"I guess what our job really boils down to . . .":* Van Maanen, John. "The Asshole."

p. 186   *"I'll tell ya, as long as . . .":* Van Maanen, John. 1999. "Kinsmen in Repose" p. 220–237.

p. 187   Discussion of citizen request for dusting a crime scene: Bayley 1994.

p. 187   *"As a police chief I know . . .":* "Quote of the Day" 2002, *New York Times*, October 28.

p. 187   *"begin with an identification . . .":* Bayley 1996, p. 27.

p. 187   Discussion of interrogation: Inbau, F.E, J.E. Reid, J.P. Buckley, and B.C Jayne. 2001. *Criminal Interrogation and Confessions.* Gaithersberg, MD: Aspen.

p. 188   Interrogation excerpts, Wilkens, J. and M. Sauer. 1999. "The Arrest." *San Diego Union-Tribune,* May 12.

p. 189   *"Yes, find the murderer . . .":* Dolbee, S. 1999. "The Lying Game—Coaxing Confessions with Lies May be Legal, But Is It Ethical?" *The San Diego Union-Tribune,* May 21.

p. 189   Discussion of traffic officer assignments: Slolnick, J. 1994. *Justice Without Trial: Law Enforcement in a Democratic Society.* New York: Macmillan.

p. 189   *"quotas have been . . .":* Reid, A. 1998. "Metro Police Have Quotas, Are Writing More Tickets." *The Washington Post,* November 24.

p. 189   *"These officers are hustled and harangued . . .":* "Police Picket Traffic as Pact Protests Go On." *New York Times,* January 29, 1997, p. B3.

p. 190   Discussion of traffic tickets quotas in Falls Church: VA: Jackson, Tom. 2004. "Falls Church Police Must Meet Quota for Tickets." *The Washington Post,* August 8.

p. 190   Discussion of New Jersey police officers: Riley, B. 1999. "2 Cops Lose in Appeal of Ticket Quota." *Newark Star Ledger,* March 4.

p. 190   Discussion of police officers as gatekeepers: Walker 1999.

p. 190–191 Discussion of police decisions and perception: Wilson, James Q. 1968. *Varieties of Police Behavior.* Cambridge, MA: Harvard University Press.

p. 191   Discussion of Dolly Keaton: "Woman 97, Jailed for Expired Registration Appears on 'Today'" 2004, NBC5.com Dallas/Fort Worth, http://www.nbc5i.com, accessed April 27.

p. 191   *"You're going to need this,"* and discussion of profiling: Harris, D. 1999. "Driving While Black—Racial Profiling on Our Nation's Highways." *An American Civil Liberties Special Report.* June, http://www.aclu.org.

p. 192   Discussion of New Jersey trooper engaged in profiling: Marks, A. 1999. "Black and White View of Police." *Christian Science Monitor,* June 9.

p. 192   Testimony of Alabama state trooper: Webb, G. 1999. "DWB" *Esquire,* April 1, pp. 118–129.

p. 192   ACLU rolling survey: Harris 1999.

p. 193–195 Discussion of militarizing the police: Kraska, P.E. and V.E. Keppeler. 1997. "Militarizing American Police: The Rise and Normalization of Paramilitary Units." *Social Problems,* 44, pp. 1–17.

p. 193–194 *"We've had special forces folks . . .":* Kraska, Keppeler 1997.

p. 194   Discussion of war on crime: Klockars, C. B. 1999. "The Dirty Harry Problem." *Police and Society* (ed. V.E. Keppeler), Prospect Heights, IL: Waveland Press, pp. 388–394.

p. 194   *"We did a crack-raid . . .":* Kraska and Keppler, "Militarizing the Police," p. 9.

p. 195   *"while courts have been extremely deferential . . .":* Balko, R. 2006. "Overkill: The Rise of Paramilitary Police Raids in America." Washington, DC: Cato Institute.

p. 194–195 Discussion of SWAT teams: Balko, R. 2006. *Overkill: The Rise of Paramilitary Police Raids in America.* Cato Institute, www.cato.org.

p. 195   Discussion of PPUs disaster: McNamara, Joseph. "Policing: The Sultans of Swat"
            1999. *The Economist,* October 28, p. 28.

p. 195   Discussion of National Black Police Association. Hampton, Ron. 1999. "Police
            Culture." *Nation,* May 31, p. 20.

p. 195   *"thrill of chasing criminals . . .":* Lindord, D. 1999. "Police Culture, " *Nation,*
            May 31, p. 20.

p. 195   *"Police officers confront . . .":* "The Sultans of Swat," *The Economist,* October 2,
            1999, p. 29.

p. 195   Discussion of the PPUs increase: McNamara 1999.

p. 195–196 *"To study the history of police . . .":* and the discussion of Lexow Commission:
            Kappeler, V.E., R.D. Sluder, and G.P. Alpert. 1998. *Forces of Deviance—Under-*
            standing the Dark Side of Policing. Prospect Heights, IL: Waveland Press.

p. 196   Report by the Government Accounting Office: Chevigny P. 1995. *Edge of the*
            Knife: Police Violence in the Americas. New York: New Press.

p. 196   Discussion of drug-related corruption in the 1980s: McAlary, M. 1987. *Buddy*
            Boys: When Good Cops Turn Bad. New York: G.P. Putnam's Sons.

p. 196   *"the bloodiest, most violent betrayal . . .":* Wenner, D. "NYPD 'Mafia Cops'
            Convicted of Murders" April 7, 2006, p. 1. *FoxNews.com.* Available at:
            http://www.foxnews.com/story/0,2933,190875,00.html, accessed March 27,
            2008.

p. 197   *"Even though I was a bad guy . . .":* McAlary M. 1987, pp. 170–71.

p. 197   *"The sheer hopelessness of the task . . .":* Fraser, R. 2007. "Drug War Corrupt-
            ing Cops in Hawaii Elsewhere." *Hawaii Reporter,* March 1, p. 1.

p. 197–198 Discussion of police corruption typology: Sherman, L.W. 1974. "Introduction:
            Toward A Typology of Police Corruption." *Police Corruption* (ed. L.W. Sher-
            man), New York: Anchor Books, pp. 1–39.

p. 198   Discussion of police corruption impact on law enforcement: Walker, Samuel and
            Charles M. Katz. 2007. *The Police in America.* New York: McGraw-Hill.

p. 198–199 Discussion of sexual predators typology: Sapp A. 1994. "Sexual Misconduct by
            Police Officers." *Police Deviance* (ed. Thomas Barker and David Carter), pp.
            187–200, Cincinnati, OH.

p. 198–199 Anderson, "When the Good Guys are Bad Guys," Bryjak, George J. 2006 *Adiron-*
            dack Daily Enterprise, August 23.

p. 199   *"I would see it all . . .":* Stamper, N. 2006. *Breaking Rank: A Top Cop's Expose*
            of the Dark Side of American Policing, p. 121.

p. 199   *"Sure I see a good looking fox . . .":* Sapp 1994, p. 189; *"When you drop in a*
            few times . . .": p. 191; *"You bet I get . . .":* p.192; *"I know several dozen guys*
            . . .": p. 194.

p. 200   *"the use of excessive physical force . . .":* Walker, S.C., Spohn, and M. DeLone,
            1996. *The Color of Justice: Race, Ethnicity, and Crime in America.* Belmont, CA:
            Wadsworth, p. 96.

p. 200   Discussion of female drivers and sexual favors: Walker, Samuel and Dawn Irl-
            beck, 2002. "Driving While Female: A National Problem in Police Misconduct."
            Police Accountability, http://www.policeaccountability.org.

p. 200   *"about 5 percent of America's cops are on the prowl for women . . ."*: Stamper, Norm. 2005. *Breaking Ranks: A Top Cop's Expose of the Dark Side of Policing*, p. 174 Nation Books: New York.

p. 200   *"but what's reasonable force . . ."*: Marks, A. 1999. "NYPD as Lab for Reducing Police Brutality." *Christian Science Monitor*, May 13.

p. 201   Report: *Amnesty International*: http://www.amnesty.org. "Amnesty International's Continuing Concerns about Taser Use" 2006, *Amnesty International*, http://www.amnestyusa.org, "Amnesty International Releases Its Briefing on Tasers. Submitted to the U.S. Justice Department," October 2007. *Amnesty International*, http://www.amnestyusa.org.

p. 201   *"to incapacitate dangerous . . ."*: in "USA: Excessive and Lethal Force . . ." pp. 2–3.

p. 201   *"While medical examiners . . ."*: "Amnesty International Releases . . ." p. 1.

p. 201   *"coroners have cited . . ."*: "Amnesty International Releases . . ." p. 1.

p. 202   Wake Forest University Baptist Medical Center Study: "Wake Forest Study Gives Green Light to Tasers" *The Chronicle of Higher Education*, October 2007. http://chronicle.com. "Wake Forest Study Ignores the Elephant in the Room, 294 Dead and Counting" Jayadev, R. *Educate-Yourself*, October 2007, http://educate-yourself.com.

p. 202   *"did not appear to present . . ."*: "USA: Excessive and Lethal . . ." p. 2.

p. 202   *"These results support . . ."*: "Wake Forest Study . . ." Quote *"are far more likely to suffer . . ."*: "Wake Forest Taser Study Ignores . . ."

p. 201–202 Discussion of tasers: "USA: Excessive and Lethal Force? Amnesty International's Concerns about Deaths and Ill-Treatment Involving Police Use of Tasers" November 2004.

p. 203   *"Cops develop the sense . . ."*: Bouza, A. 1991. "Police Officers' Attitude Toward Civilians Causes Police Brutality." *Police Brutality* (ed. W. Dudley), San Diego, CA: Greenhaven, pp. 77–79.

p. 203   *"Brutality like other forms . . ."*: "Corruption in the New York City Police Department," 1995. *Policing the Police* (ed. W. Dudley) San Diego, CA: Greenhaven Press, pp. 28–44.

p. 203–204 Discussion of police work on assumption of suspect's guilt: Fyfe, J.J. 1997. "The Split Second Syndrome and Other Determinants of Police Violence." *Critical Issues of Policing* (eds. R.G. Dunham and G.P. Alpert), Prospects Heights, IL: Waveland Press, pp. 531–547.

p. 204   *"white officers don't understand . . ."*: Dudley W., *Police Brutality*

p. 204   *"typical black gang member* and *typical white ugly cop . . ."*: Chevigny, P. 1995. *Edge of the Knife . . .* p. 122.

p. 204   Discussion of anxiety and fear: "Corruption in the New York City Police Department."

p. 205   *"This is the worst aspect of police culture . . ."*: Chevigny 1995, p. 43.

p. 205   *"What do you do if you see . . ."*: Dudley 1991, pp. 12–13.

p. 205   *"Why does the blue wall exist . . ."*: Marks, A. 1999. "NYPD as a Lab for Reducing Police Brutality." *Christian Science Monitor*, June 9.

p. 205   *"Good officers are . . ."*: Pasco, in Marks, A. 1997. "More Officers are Willing to Break the Code of Silence." Christian Science Monitor, August 20.

p. 205–207   Discussion of policing the police: Bayley, David. 1985. *Patterns of Policing.* New Brunswick, NJ: Rutgers University Press.

p. 206   Discussion of John Crew: "Why the Police are Hard to Police" 1999, *Christian Science Monitor,* September 27.

p. 206   *"Without watchdog organizations . . ."*: "Why the Police are Hard . . ." p. 1.

p. 206   *"If you are going on a . . ."*: "Why the Police are Hard . . ." p. 1.

p. 206   Compstat: Alpert, G.P., R.G. Dunham, and M.S Stroshine. 2005. *Policing Continuity and Change.* Long Grove, IL: Waveland Press.

p. 206   *"even those they know are brutal; . . ."*: "Political Considerations and Aggressive Policing" 1998, p. 1, *Human Rights Watch,* http://www.hrw.org.

p. 206   *This organization "was developed for the purpose . . ."*: Black Cops Against Police Brutality. 2007. *Black Cops Against Police Brutality,* http:www.b-cop.org, p. 1.

p. 207   Research by Ronald Hunter: Hunter, Ronald. 1999. "Officer Opinions on Police Misconduct." *Journal of Contemporary Criminal Justice,* Vol 5 2:155–170.

p. 207   *"According to the Court's interpretation . . ."*: "Fighting for Public Accountability" 2007, p. 1, Bay Area Police Watch, *The Ella Baker Center,* http://ellabaker-center.org.

p. 207   *"they would be branded as anti-law enforcement . . ."*: "Police Accountability Bill Stalled—It's Time for Speaker Nunez to Speak Up." 2007. *California Progressive Report,* http://www.californiaprogressivereport.com.

Chapter 10. Pretrial Procedures and Plea Bargaining: From Arrest to "Let's Make a Deal"

p. 212   Discussion of arrest: Carp R. A. and R. Stidham. 1996. *Judicial Process in America.* Washington, DC: CQ Press, A Division of Congressional Quarterly.

p. 212   Discussion of preliminary hearings: Steury, E. H. and N. Frank, 1995. *Criminal Court Process,* Minneapolis/St. Paul: Wadsworth Publishing; *"prevent hasty, malicious, improvident . . ."*: Steury and Frank 1995, p. 278.

p. 212   Discussion of initial appearance: Neubauer D. 2007. *America's Courts and the Criminal Justice System.* Belmont, CA: Wadsworth Publishing Company; Emanuel, S.L. and S. Knowles. 1999–2000. *Criminal Procedure.* Larchmont, NY.

p. 212–213   Discussion of charging: Neubauer 2007.

p. 213   Discussion of preliminary hearing: Saltzburg, S. A. and D. J. Capra 1992. *American Criminal Procedure—Cases and Procedure.* St. Paul, MN: West Group; Carp and Stidham, *Judicial Process in America.*

p. 213–215   Discussion of grand juries: Saltzburg and Capra 1992; Carp and Stidham 1996; Brenner, S. and L. Shaw. 2002. "Using a Grand Jury to Investigate the September 11, 2001 Terrorists Attacks." *Federal Grand Jury,* University of Dayton. Available at: http://campus.udayton.edu.

p. 215   Discussion of arraignment: Neubauer 2007; Marcus, P. and C.H. Whitebread. 1996. *Gilbert Law Summaries: Criminal Procedures.* Chicago: Harcourt Brace Legal & Professional Publications.

p. 215–216   Discussion of pretrial release and detention: Saltzburg and Capra 1992; Cohen, T.H. and R. A. Reaves. 2007. "Pretrial Release of Felony Defendants in State Courts," *Bureau of Justice Statistics*, U.S. Department of Justice, www.ojp.usdoj.gov.

p. 216   Discussion of bail: Saltzburg and Capra 1992.

p. 216–218   Discussion of bail bondsmen: Saltzburg and Capra 1992; Neubauer 2007; "Bounty Hunters: Loose Cannons on Prowl" 1998, Fulton County Daily Report, January 13; Herbert, B. 1997. "A Wild West Tragedy" *New York Times*, September 4.

p. 218–200   Discussion of discovery and pretrial motions: Gifis, S.H. 2003. *Law Dictionary.* Woodbury, NY: Barron's Educational Series, Inc.; Steury and Frank 1996; Emanuel and Knowles 1999–2000.

p. 218   *"refers to evidence and/or statements . . ."*: Gifis, S. H., 2003. *Law Dictionary, Fifth Edition.* Woodbury, NY: Barron's Educational Series.

p. 218   Discussion of disclosure requirements the prosecution must satisfy: Emanuel and Knowles 1999–2000.

p. 220–221   Discussion of the courtroom as a clubhouse: Blumberg, A. S. 1967. "The Practice of Law as a Confidence Game—Organizational Cooptation as a Profession." *Law and Society Review*, June, pp. 15–39; Eisenstein, J. and H. Jacob. 1991. *Felony Justice: An Organizational Analysis of American Courts.* Lanham, MD: University Press of America.

p. 220   *"You might have 10 or 15 preliminary . . ."*: Clynch, E. J. and D. W. Neubauer. 1981. "Trial Courts as Organizations." *Law and Policy Review*, 3 1:61–94; "Improving State and Local Criminal Justice Systems" 1998, *Office of Justice Programs*, U. S. Government Printing Office, Washington, D.C.

p. 221–223   *"plea bargaining is an essential . . ."*: and discussion of plea bargaining: Acker, J. R. and D. C. Brody. 2004. *Criminal Procedure: A Contemporary Perspective.* Sudbury, MA: Jones & Bartlett Publishers, p. 494; Jost, K. 1999. "Plea Bargaining." *CQ Researcher*, Vol 9 6:115–133; Langbein, J.H. 1992. "On the Myth of Written Constitutions: The Disappearance of the Jury Trial." *Harvard Journal of Law and Public Policy* Vol 15 1:119–128.

p. 223   *"A legal system that comes to depend . . ."*: Palermo, G. B., M. A. White, L. A. Wasserman, and W. Hanrahan. 1996. "Plea Bargaining: Injustice for All?" *International Journal of Offender Therapy and Comparative Criminology*, Vol 42 2:111–123.

p. 223–224   Discussion of prosecutors and plea bargaining: Miller, H. S., W. F. McDonald, and J. A. Cramer. 1978. *Plea Bargaining in the United States.* Washington, DC, National Institute of Law Enforcement and Criminal Justice, U.S. Department of Justice.

p. 223–224   Discussion of the number of plea bargaining: Blankenship, G. 2003. "Debating the Pros and Cons of Plea Bargaining" July 13, 2003, *Golialth Business Knowledge on Demand.* Available at: http://www.goliath.ecnext.com.

p. 224   Discussion of judges and plea bargaining: Heumann, M. 1981. *Plea Bargaining—The Experience of Prosecutors, Judges, and Defense Attorneys.* Chicago: The University of Chicago Press; McConville, M. and C. Mirsky. 1995. "Guilty Plea Courts: A Social Disciplinary Model of Criminal Justice." *Social Problems*, Vol 42 2:216–234. *The circumstances are these: The Experience of Prosecutors*, pp. 142–43.

p. 226   Discussion of public defenders and plea bargaining: Hannah, J. 2007. "Caseloads Crushing, Lawyers Say; Public Defenders Use Boone as Example of Underfunding." *National Association of Criminal Defense Lawyers*, September 22, www.nacdl.org.

p. 226   *"How could you be trying this stupid case . . ."*: Lynch, D. 1999, "Perceived Judicial Hostility to Criminal Trials" Criminal Justice Behavior, Vol 26 2:217–233, Weinstein, H. 2002, "Georgia Fails Its Poor Defendants, Report Says" *Los Angeles Times*, December 13.

p. 226–227 Study of full- and part-time criminal defense lawyers: Liptak, A. 2007. "Public Defenders Get Better Marks on Salary" *The New York Times*, July 14, www.nytimes.com.

p. 227–228 Let's Make a Deal discussion: Emmelmann, D. 1996. "Trial by Plea Bargain: Case Settlement as a Byproduct of Recursive Decision Making." *Law & Society Review*, Vol 30 2:335–361.

p. 228–229 Is plea bargaining here to stay discussion: *The form of plea bargaining becomes:* Fisher, G. 2003. *Plea Bargaining's Triumph: A History of Plea Bargaining in America.* Palo Alto, CA: Stanford University Press; Huemann, M. *Plea Bargaining, As the actors spend more time:* Heumann, p. 157.

p. 228   *"I have a case now . . ."*: Emmleman, D. S. 1996. "Trial by Plea Bagain: Case Settlement as a Product of Recursive Decisionmaking", *Law & Society Review*, 30(2):342.

p. 229   *"guilty pleas characterize high and low volume courts"*: Heumann, M. 1981. p. 287. *Plea Bargaining: The Experience of Prosecutors, Judges, and Defense Attorneys*, Chicago: The University of Chicago Press.

p. 229   *"First and foremost, plea bargaining is not . . ."*: Blankenship 2003.

p. 229   *"as the actors spend more time . . ."*: Heumann 1981, p. 157.

Chapter 11. Criminal Trials and Courtroom Issues: Convicting the Innocent, Exonerating the Guilty

p. 223   *"exciting experiment in the conduct . . ."*: Kalven, H. and H. Zeisel. 1996. *The American Jury*, Chicago: The University of Chicago Press, p. 4.

p. 234   Criminal trial statistics: "The State-of-the-States Survey of Jury Improvement Efforts: Executive Summary" 2007, *National Center for State Courts*, www.ncsconline.org.

p. 234–236 Discussion of jury selection and composition: Saks, M. J. 1996. "The Greater the Jury the More the Unpredictability." *Judicature*, Vol 79 5:263–265. *"If we draw juries at random . . ."*: State Court Organization, 1998, Washington D.C., Bureau of Justice Statistics, U. S. Department of Justice.

p. 235   Discussion of Supreme Court Decision *Williams v. Florida*: Sperlich, P.W. 2007. "U.S. Supreme Court, *Williams v. Florida*." Anwsers.Com. Available at: http://www.answers.com.

p. 236   *"hunch, insight, whim, prejudice, or pseudoscience.":* Neubauer, D. W. 2004. *America's Courts and the Criminal Justice System.* Belmont, CA: West/Wadsworth, p. 341.

p. 236   Discussion of Supreme Court Decision *Batson v. Kentucky:* U.S. Supreme Court, 2007. *Batson v. Kentucky, Justia.com U.S Supreme Court Center.* Available at: http://www.supreme.justia.com.

p. 235–238 Discussion of voir dire: *State Court Organization, 1998,* Gollner, P. M. 1994. Consulting by Peering Into the Minds of Jurors," *New York Times,* January 7; Neubauer, 2007. *American Courts and the Criminal Justice System,* Belmont, CA: Wadsworth; Dominic, C. 2008. "A Letter From the President" *American Society of Trial Consultants,* www.astcweb.org

p. 237–238 Discussion of prospective jurors: American Society of Trial Consultants, 2008, http://www.astcweb.org.

p. 237   Discussion of jury consultants: Hutson, M. "Unnatural Selection" *Psychology Today,* March/April 2007. Available at: http://psychologytoday.com/articles/pto-20070302-000001.xml, accessed March 27, 2008; *Scientific Jury Selection,* Lieberman, J. D. and B. D. Sales, Washington DC: American Psychological Association.

p. 238   *"How can you possibly make sure . . .":* Hutson 2007, p. 6.

p. 238   *"repeated studies and experience . . .":* Small, D. I. 1999. *Going to Trial: A Step-by-Step Guide to Trial Practice and Procedure, Second Edition.* Chicago: American Bar Association, p. 181.

p. 238   Discussion of opening statement: Small, D. I. 1999 *Going to Trial: A Step by Step Guide to Practice and Procedure.* Chicago: American Bar Association.

p. 239   *"must be so conclusive and complete . . .", "It does not require . . .":* Gifis, S. H. *Law Dictionary,* NY: Barron's Educational Series, p. 170.

p. 238–241 Discussion of presenting the evidence: Small 1999; Neubauer 2004; Black, R. 2000. *Black's Law,* New York: Simon & Schuster, *The State-of-the States Executive Summary; cross-examination is the greatest:* Wigmore, pp. 10–11, *Black's Law.*

p. 241–242 Closing arguments: Small 1999; Neubauer 2004.

p. 242–244 Discussion on instructing the jury and jury deliberations: Straun, D. U. and R. W. Buchanan. 1976. "Jury Confusion: A Threat to Justice." *Judicature,* Vol 59 10:478–483; Kalvan, H. and H. Zeisel. 1966. *The American Jury.* Boston: Little-Brown; Reskin, B. F. and C. A. Visher. 1986. "The Impacts of Evidence and Extralegal Factors in Jurors' Decisions." *Law & Society Review,* Vol 20 2:423–428.

p. 242   *"telling the jurors to watch a baseball game . . .":* Adler, S. J. 1994. *The Jury: Trial and Error in the American Courtroom.* New York: Random House, p. 14.

p. 242   *"The outer boundaries of a product . . .":* Adler 1994, p. 129

p. 244   *"the influence of social and demographic factors . . .":* Ford, M. C. 1986. "The Role of Extralegal Factors in Jury Verdicts." *The Justice System Journal,* Vol 11 1:16–39, p. 16.

p. 244–245   Discussion about the verdict: Steury, E. H. and N. Frank. 1996. *Criminal Court Process*. Minneapolis, MN: West; Vidmar, N., S. Beale, M. Rose, and L. Donnelly. 1997. "Should We Rush to Reform the Criminal Jury?" *Judicature*, Vol 80:286–290.

p. 245–246   Discussion of appeals: Cecil, J. S. and D. Stienstra. 1987. "Deciding Cases Without Appeals: An Examination of Four Courts of Appeals." Washington DC: Federal Judicial Center; Neubauer, *American Courts and the Criminal Justice System*.

p. 246–249   Discussion of jury nullification: Barkan S, 1983, "Jury Nullification in Political Trials" *Social Problems*, Vol 31 1:28–45; Simon, R. J. 1992, "Jury Nullification or Prejudice and Ignorance in the Marion Barry Trial?" *Journal of Criminal Law*, Vol 20:261–266; Brown, D. K. 1997, "Jury Nullification Within the Rule of Law" *Minnesota Law Review*, Vol 81 5:1149–1200; Holden, B. A., L. P. Cohen and E. de Lisser 1995, "Race Seems to Play an Increasing Role in Many Jury Verdicts," *The Wall Street Journal*, October 15; Bikupic, J. 1999, "Veto by Jury" *The Washington Post National Weekly Edition*, March 29; Linder, D. 2001, "Jury Nullification" *University of Missouri-Kansas City School of Law*, www.law.umkc.edu; Scheflin, A. W. and J. M. Van Dyke 1995, "Merciful Juries: The Resilience of Jury Nullification" in *Courts and Justice: A Reader*, G. L. Mays and P. R. Gregware editors, pp. 256–285, Prospect Heights, IL: Waveland Press; Balko, R. 2005. "Justice Often Served by Jury Nullification" August 1, *Fox News*, www.foxnews.com.

p. 246   *"Attention Jurors and Future Jurors . . ."*: Scheflin and Van Dyke 1995, p. 257.
p. 247   *"African American jurors are doing . . ."*: Scheflin and Van Dyke 1995, p. 258.
p. 248   *"there is real cynicism . . ."*: Bikupic 1999.
p. 248   *"a society without laws . . ."*: *"Jury Nullification in Jury Trials."*
p. 246–249   Discussion of jury nullification: Biskupic, J. 1999. "Jury Rooms, Form of Civil Protest Grows" *Washington Post*, February 8, 1999; *We the Jury*, Abramson, J. Cambridge: Harvard University Press; *The System in Black and White: Exploring The Connectedness Between Race, Crime, and Justice*, Markowitz, M. and D. D. Jones-Brown, Westport, CT: Praeger; "Justices Say Jurors May Not Vote Conscience," Dolan, M. May 8, 2001, *Los Angeles Times*.

p. 247   *"Our data suggest . . ."*: Abrahamson, p.160.
p. 248   *"There is no reason to believe . . ."*: Abrahamson, p. 160.
p. 248   *"it is not only his right . . ."*: Adams from "Quotes on the Powers and Duties of Jurors" Family Guardian, 2007, p. 1. http://famguardian.org.
p. 248   *"if exercising their judgement . . ."*: Hamilton "Quotes and the Power and Duties of Jurors," p. 1; *"violation of a jurors oath . . ."*: "Justices Say Jurors May Not Vote Conscience" p. 1.
p. 249   *"jury nullification is not explicit . . ."*: "Justices Say Jurors May Not Vote Conscience," *"Members of the jury . . .,"* "Jury Nullification in Political Trial": Barkan, S. 1983. *Social Problems* 31(2):37.
p. 249   Discussion of wrongful convictions: Huff, C. R.; A. Rattner, and E. Sagarin. 1997. "Convicted but Innocent: Wrongful Convictions and Public Policy." *Order*

under Law: Readings in Criminal Justice (ed. R. G. Culbertson and R. A. Weisheit), pp. 253–276. Prospect Heights, IL: Waveland Press, Inc.

p. 249–250 Discussion of eyewitness errors: Ng, W. and R. C. Lindsay. 1994. "Cross-Race Facial Recognition." *Journal of Cross-Cultural Psychology*, Vol 25 4:217–232; Gibeaut, J. 1999. "Yes, I'm Sure That's Him." *ABA Journal.* Vol 87 10:90; Liptak, A. 2004. "Study Suspects Thousands of False Confessions." *New York Times*, April 19; Liptak, A. 2007. "New Study Examines False Confessions." *International Herald Tribune*, July 22, www.iht.com.

p. 250 *"In thousands of cases every year . . .":* Feige, D. 2006. "Witnessing Guilty, Ignoring Innocence?" *New York Times*, June 6, www.nytimes.com.

p. 250–251 Discussion of prosecutorial and police misconduct and error: Lait, M. and S. Glover. 2000. "LAPD Officer Corroborates Perez on Beating." *Los Angeles Times*, March 14; Lait, M. and S. Glover. 2000. "LAPD Chief Calls For Mass Dismissal of Tainted Cases." *Los Angeles Times*, January 27; Gordon, N. 2004. "Misconduct and Punishment: State Disciplinary Authorities Investigate Prosecutors Accused of Misconduct." *The Center for Public Integrity*, www.publicintegrity.org; Elliot, A. and B. Weiser. 2004. "When Prosecutors Err, Others Pay the Price." *New York Times*, March 21, www.nytimes.com.

p. 250 *"in which serious misconduct . . .": New York Times,* March 21.

p. 251 *"The FBI's misconduct was clearly . . .":* "U.S. Must Pay $101.7 Million to Men Framed by FBI" CNN.com July 26, www.cnn.com.

p. 251 *"My recently completed study . . .":* Moran, R. 2007. "The Presence of Malice." *New York Times,* August 2, www.nytimes.com.

p. 251 *"Well, they told us that.":* Herbert, B. 2002. "An Imaginary Homicide" *New York Times*, August 15, 2002, p. D10; "planted the idea . . .": "An Imaginary Homicide."

p. 251–253 Discussion of false confessions: Herber, B. 2002. "An Imaginary Homicide." *New York Times*, October 2, www.nytime.com; Liptak, A. 2004. "Study Suspects Thousands of False Confessions." *New York Times*, April 19, www.nytimes.com.

p. 251–253 Discussion of false confessions: Gudjonsson, G. 2003. *The Psychology of Interrogations and Confessions*, West Sussex, England: John Wiley & Sons, Ltd, pp.194–197.

p. 252 *"Mentally retarded people get . . .":* Robinson, B. A. 2004. "False Confessions by Adults." *Justice Denied: The Magazine of the Wrongly Convicted*, April 19, www.justicedenied.org; "False Confessions & Mandatory Recording of Interrogations." 2007. *The Innocence Project*, www.innocenceproject.org; Moore, S. 2007. "Exoneration Using DNA Brings Change in Legal System." *New York Times*, October 1, www.nytimes.com.

p. 254 Discussion of inadequacy of counsel: David, J. 2004. "The Innocence Project." *Z Magazine*, April, pp. 19–21; "Jimmy Ray Bromgard," 2006, *The Innocence Project*, www.innocneceproject.org; "Bad Lawyering" 2006, *The Innocence Project*, www.innocenceproject.org.

p. 254 *"nothing guarantees the conviction . . .":* Longley, S. B. and B. Sheck. 2000. "Legal Genes," *People*, pp. 11–13, May 15; Lane, C. 2002. "Supreme Court

Turns Down Appeal by Killer Who Cited Bad Lawyering." *The Buffalo News*, May 29.

p. 254   *"yet to see a death . . ."*: Lane, *"Supreme Court Turns Down . . ."*

p. 253   *"Asking a jury to judge the credibility . . ."*: Sealy, G. 2002. "Confused Confessions," September 25 *ABC News*, www.abcnews.com.

p. 255   *"appeared to be the epitome . . ."*: Huff *"Convicted but Innocent . . ."*

p. 256–257   Discussion of race as a factor in convicting the innocent: Liptak. A. 2007. "Study of Wrongful Convictions Raises Concerns Beyond DNA" July 23, 2007. *The New York Times*, http://select.nytimes.com.

p. 255   *"black men accused of raping white women. . ."*: Liptak, A. 2007. "Study of Wrongful Convictions Raises Questions Beyond DNA." July 23, *New York Times*, nytimes.com.

p. 256–257   Discussion of DNA testing: "Robert Clark" 2006, The Innocence Project" www.innocenceproject.org; "DNA Wins Freedom for Man After 26 Years" 2008, CNN.com January 3, www.cnn.com; Taylor, S. 2007, "Guest Shot: Innocents in Prison." August 7, *Truth in Justice Files*, www.truthinjusticefiles.blogspot.com; *only handles cases where postconviction*: "How do you choose your cases?" 2007, *The Innocence Project*, www.innocenceproject.org.

p. 257–258   Discussion of fingerprints: Newman, A. 2001. "Fingerprinting's Reliability Draws Growing Court Challenges." *New York Times*, April 1, www.nytimes.com; "Fingerprints: Infallible Evidence?" 2003, *CBS News*, July 18, www.cbsnews.com; Feige, D. 2004. "Printing Problems" *Slate*, May 27, www.slate.com.

p. 257   *"There's complete disagreement . . ."*: *"Fingerprints: Infallible Evidence?"*

p. 258   *"There isn't a single experiment . . ."*: *"Fingerprints: Infallible Evidence?"*

p. 258   *"There are very few employers who will terminate . . ."*: *"Fingerprinting's Reliability . . ."*

p. 258   *"very good possibility . . ."*: *"Fingerprinting's Reliability . . ."*

p. 258–260   Discussion of the CSI Effect: Willing, R. 2004 "'CSI Effect' Has Juries Wanting More Evidence" USA Today, August 5, http://usatoday.com, Roane, K. 2005, "The CSI Effect" U.S. *News & World Report*, April 25, www.usnews.com, Thomas, A. P. 2006, "The CSI Effect: Fact or Fiction?" 115 YALE L. J. POCKET PART 70, www.thepocketpart.org.

p. 259   *"resulted in either acquittal or hung jury . . ."*: *"The CSI Effect: Fact of Fiction?"*

Chapter 12. Prisons and Jails: Punishment at Any Cost?

p. 263   Discussion of and quotations by Ernie Preate, Jr.: Bleyer, Jennifer. 2001. "Prison Converts." *The Progressive* June:28–30.

p. 264–267   Discussion of history of prisons: Foucault, Michel. 1977. *Discipline and Punish: The Birth of the Prison*. New York: Pantheon Books; Walker, Samuel. 1998. *Popular Justice: A History of American Criminal Justice*. New York: Oxford University Press.

p. 267–268   Discussion of jails and prisons: May, David, Kevin Minor, Rick Ruddell, and Betsy Matthews. 2008. *Corrections and the Criminal Justice System*. Sudbury,

MA: Jones and Bartlett Publishers; Seiter, Richard P. 2008. *Corrections: An Introduction.* Upper Saddle River: Prentice Hall.

p. 269   Discussion of evidence casting doubt on conclusion that increased incarceration reduced crime: Blumstein, Alfred, and Joel Wallman (eds.). 2006. *The Crime Drop in America.* Cambridge: Cambridge University Press; Gainsborough, Jenni, and Marc Mauer. 2000. *Diminishing Returns: Crime and Incarceration in the 1990s.* Washington, DC: The Sentencing Project.

p. 269   *"little support for the . . .":* Gainsborough and Mauer 2000:3, The Sentencing Project.

p. 269–270 Study and conclusion that imprisonment is "incredibly inefficient means of reducing crime": Spelman, William. 2000. "The Limited Importance of Prison Expansion." pp. 97–129 in *The Crime Drop in America* (eds Alfred Blumstein and Joel Wallman). Cambridge: Cambridge University Press (quotation on p. 124).

p. 269   Study of reduction in homicides during 1990s: Rosenfeld, Richard. 2000. "Patterns in Adult Homicide: 1980-1995." pp. 130–163 in *The Crime Drop in America*, edited by Alfred Blumstein and Joel Wallman. Cambridge: Cambridge University Press.

p. 271   *"As more people grow up . . .":* Gainsborough and Mauer 2000:25, The Sentencing Project.

p. 271   Discussion of surge in incarceration: Miller, Jerome G. 1996. *Search and Destroy: African American Males in the Criminal Justice System.* New York: Cambridge University Press.

p. 272   Discussion and quotations by Sykes: Sykes, Giesham M. 1958. *The Society of Captives.* Princeton: Princeton University Press, p. 64.

p. 273   Discussion of substandard conditions: Kovaleski, Serge F. 2000. "D.C. Finds Dangers in Ailing Jail." *The Washington Post* September 17:A1.

p. 273   Discussion by Human Rights Watch: Human Rights Watch. 2001. *Prisons in the United States of America.* Available at: http://www.hrw.org/hrw/advocacy/prisons/u-s.htm. New York: Human Rights Watch.

p. 273   *"The sardine-can appearance . . .":* Firestone, David. 2001. "Crowded Jails Create Crisis for Prisons in Alabama." *The New York Times*, May 1:A1.

p. 274–275 Discussion of prison violence: Cullen, Francis T., and Jody L. Sundt. 2000. "Imprisonment in the United States." *Criminology: A Contemporary Handbook* (ed. Joseph F. Sheley). Belmont, CA: Wadsworth Publishing Company. pp. 473–515; Wooldredge, John D. 1998. "Inmate Lifestyles and Opportunities for Victimization." *Journal of Research in Crime and Delinquency* 35:480–502.

p. 274   Discussion of prison rape and sexual assault: Dumond, Robert W. 2000. "Inmate Sexual Assault: The Plague That Persists." *The Prison Journal* 80:407–414; Lewin, Tamar. 2001. "Little Sympathy or Remedy for Inmates Who are Raped." *The New York Times*, April 15:A1.

p. 275–276 Discussion of health and health care: Dabney, Dean A. and Michael S. Vaughn. 2000. "Incompetent Jail and Prison Doctors." *The Prison Journal* 80:184–209.

p. 276   Discussion of reasons for increase in women inmates: Steffensmeier, Darrell, Jennifer Schwartz, Hua Zhong, and Jeff Ackerman. 2005. "An Assessment of

Recent Trends in Girls' Violence Using Diverse Longitudinal Sources: Is the Gender Gap Closing?" *Criminology* 43:355–405.

p. 277   Discussion of rape and sexual assault of women inmates: Dumond 2000; Fleming, Sue. 2001. "Abuse of Women Inmates Seen Rampant." *The Boston Globe*, March 7:A7.

p. 277   Report by Amnesty International: Fleming 2001.

p. 278   Discussion of women prisoners as mothers: Cullen and Sundt 2000; Enos, Sandra. 2001. *Mothering from the Inside: Parenting in a Women's Prison*. Albany, NY: State University of New York Press.

p. 279–282   Discussion of death penalty: Bohm, Robert M. 2007. *Deathquest: An Introduction to the Theory and Practice of Capital Punishment in the United States*. Cincinnati: Anderson Publishing Company; Kurtis, Bill. 2004. *The Death Penalty on Trial: Crisis in American Justice*. New York: Public Affairs.

p. 281–282   Discussion of executions of capital defendants who may have been innocent: Death Penalty Information Center. 2006. *Innocence and the Death Penalty*. Washington, DC: Death Penalty Information Center.

p. 282   Chicago Tribune investigation and quotation: Mills, Steve, Ken Armstrong, and Douglas Holt. 2000. "Flawed Trials lead to Death Chamber." *Chicago Tribune*, June 11:A1.

Chapter 13. Community Corrections and Juvenile Justice

p. 285   Discussion of probation and parole: Petersilia, Joan. 1998. "Probation and Parole." *The Handbook of Crime and Punishment* (ed Michael Tonry). New York: Oxford University Press. pp. 563–588; Petersilia, Joan. 2003. *When Prisoners Come Home: Parole and Prisoner Reentry*. New York: Oxford University Press.

p. 286   Probation study of California inmates: Petersilia, Joan, and Susan Turner. 1986. *Prison Versus Probation in California: Implications for Crime and Offender Recidivism*. Santa Monica, CA: RAND Corporation.

p. 287   Discussion of the effectiveness of probation: Petersilia 1998.

p. 288   Discussion of recidivism by people on probation compared to recidivism of ex-inmates: Clear, Todd, and Karen Terry. 2000. "Correction Beyond Prison Walls." pp. 517–537 in *Criminology: A Contemporary Handbook* (ed Joseph F. Sheley). Belmont, CA: Wadsworth Publishing Company.

p. 289–291   Discussion of parole: Petersilia, Joan. 2003. *When Prisoners Come Home*. New York: Oxford University Press.

p. 290   Discussion of crime by two convicted burglars: Fernandez, Manny, and Alison Leigh Cowan. 2007. "For 2 Suspects in Killings, Records of Repeat Offenses." *The New York Times*, August 3:B5.

p. 290–292   Discussion of 1999 California study: Petersilia, Joan. 1999. "Parole and Prisoner Reentry in the United States." *Crime and Justice: A Review of Research* 26:479–529.

p. 291–292   Discussion of the effectiveness of intermediate sanctions: Clear and Terry 2000; Tonry, Michael. 1998. "Intermediate Sanctions." pp. 683–711 in *The Handbook of Crime & Punishment* (ed Michael Tonry). New York: Oxford University Press.

p. 292–293   Discussion of serious and violent juvenile offenders: Loeber, Rolf, and David P. Farrington (eds.). 1998. *Serious and Violent Juvenile Offenders: Risk Factors and Successful Interventions.* Thousand Oaks, CA: Sage Publications. Wasserman, Gal A., Laurie S. Miller, and Lynn Cothern. 2000. *Prevention of Serious and Violent Juvenile Offending.* Washington, DC: Office of Juvenile Justice and Delinquency Prevention, U.S. Department of Justice.

p. 293   Studies of news media coverage of youth and crime: Dorfman, Lori, and Vincent Schiraldi. 2001. *Off Balance: Youth, Race and Crime in the News.* Washington, DC: Building Blocks for Youth.

p. 294   Discussion of 2002–2003 National Youth Gang Surveys: Egley, Arlen, Jr. 2005. *Highlights of the 2002–2003 National Youth Gang Surveys.* Washington, DC: Office of Juvenile Justice and Delinquency Prevention, U.S. Department of Justice.

p. 295   Discussion of gang activity increase during 1980s and 1990s: Blumstein, Alfred. 1995. *Youth Violence, Guns, and Illicit Drug Markets.* Washington, DC: National Institute of Justice, U.S. Department of Justice.

p. 295   *"There is nothing more insidious . . . ":* http://feinstein.senate.gov/05releases/r-gangs7.htm.

p. 295–296   Discussion of gang-problem jurisdictions: Institute for Intergovernmental Research, "Frequently Asked Questions Regarding Gangs," http://www.iir.com/nygc/faq.htm.

p. 296–297   History of juvenile justice system: Walker, Samuel. 1998. *Popular Justice: A History of American Criminal Justice.* New York: Oxford University Press.

p. 299   Discussion of 2000 report on race and juvenile justice: Butterfield, Fox. 2000. "Racial Disparities Seen as Pervasive in Juvenile Justice." *The New York Times,* April 26:A1; "a study in California": Lewin, Tamar. 2000. "Racial Discrepancy Found in Trying of Youths." *The New York Times,* February 3:A14.

p. 299   Discussion of research evidence on effects of transferring juveniles: Feld, Barry C. 1998. "Juvenile and Criminal Justice Systems' Responses to Youth Violence." *Crime and Justice: A Review of Research* 24:189-261.

p. 299   Discussion of youth discounts: Feld 1998.

p. 300   Discussion of prevention of delinquency: Welsh, Brandon C., and David P. Farrington (eds.). 2007. *Preventing Crime: What Works for Children, Offenders, Victims and Places.* New York: Springer.

Chapter 14. Conclusion: What Every American Should Know

p. 306   Discussion of prevention strategy, for strategies and approaches with strong promise for reducing crime: Currie, Elliott. 1998. *Crime and Punishment in America.* New York: Henry Holt; Sherman, Lawrence W., Denise C. Gottfredson, Doris L. MacKenzie, John Eck, Peter Reuter, and Shawn D. Bushway. 1998. *Preventing Crime: What Works, What Doesn't, What's Promising.* Washington, DC: Office of Justice Programs, National Institute of Justice; Welsh, Brandon C., and David P. Farrington (eds.). 2007. *Preventing Crime: What Works for Children, Offenders, Victims and Places.* New York: Springer.

p. 306–307 Discussion for improving the criminal justice system, for strategies and approaches to improve the criminal justice system: Jacobson, Michael. 2005. *Downsizing Prisons: How to Reduce Crime and End Mass Incarceration.* New York: New York University Press; Tonry, Michael. 2004. *Thinking About Crime: Sense and Sensibility in American Penal Culture.* New York: Oxford University Press; Travis, Jeremy, and Christy Visher (eds.). 2005. *Prisoner Reentry and Crime in America.* New York: Cambridge University Press.

p. 307–308 Discussion of helping crime victims: for strategies and approaches to help crime victims: Web site of National Center for Victims of Crime, http://www.ncvc.org/ncvc/Main.aspx.

p. 309 For additional advice regarding protection against identity theft may be found in Gralla Preston. 2007. "Safety Net" USAA Magazine, Winter:20–21.

Index